Occurrence, Toxicity & Analysis of Toxic Compounds in Oceanic Biota

Occurrence, Toxicity & Analysis of Toxic Compounds in Oceanic Biota

Thomas Roy Crompton

CRC Press
Taylor & Francis Group
Boca Raton London New York

CRC Press is an imprint of the
Taylor & Francis Group, an **informa** business

CRC Press
Taylor & Francis Group
6000 Broken Sound Parkway NW, Suite 300
Boca Raton, FL 33487-2742

First issued in paperback 2022

© 2017 by Taylor & Francis Group, LLC
CRC Press is an imprint of Taylor & Francis Group, an Informa business

No claim to original U.S. Government works

ISBN-13: 978-1-498-70154-9 (hbk)
ISBN-13: 978-1-03-233986-3 (pbk)
DOI: 10.1201/9781315371887

Library of Congress Cataloging-in-Publication Data

Names: Crompton, T. R. (Thomas Roy)
Title: Occurrence, toxicity & analysis of toxic compounds in oceanic biota / Thomas Roy Crompton.
Other titles: Occurrence, toxicity, and analysis of toxic compounds in oceanic biota
Description: Boca Raton : CRC Press, Taylor & Francis Group, 2017. | Includes bibliographical references and index.
Identifiers: LCCN 2016018841 | ISBN 9781498701549
Subjects: LCSH: Marine organisms--Effect of water pollution on. | Marine ecology. | Water quality bioassay. | Toxicity testing. | Bioaccumulation.
Classification: LCC QH545.W3 C76 2017 | DDC 578.77--dc23
LC record available at https://lccn.loc.gov/2016018841

Visit the Taylor & Francis Web site at
http://www.taylorandfrancis.com

and the CRC Press Web site at
http://www.crcpress.com

Contents

Preface

It is becoming increasingly apparent that the oceans are not an unlimited reservoir into which waste can be dumped and that control of the level of these emissions is necessary if complete destruction of the environment, including fish and invertebrate communities and plant life, is to be avoided.

Many types of pollutants, including metals, organometallic compounds and organic compounds, are particular offenders in this respect. Many types of organic compounds, often very sophisticated molecules, are now being applied to farmland and crops, the majority of which finish up in the marine environment. The analysis of these chemicals in fish, invertebrates and plant life often requires complicated equipment and methods which have only become available recently.

Inputs of pollutants into the marine environment, water and sediments include discharges, either direct, via river inputs or via rainfall, of domestic, industrial and shipboard waste; underwater fumaroles and volcanoes; and discharge into the ocean by rivers or by rainfall of factory emissions, power station smoke emissions, incinerators, plants and fires – all obvious sources of pollution. A further source of pollution is drainage of chemicals added to land or crops, including fertilisers and insecticides. Such chemicals are progressively leached from the land by rain deposition and reach the oceans via streams and rivers or discharge directly from coastal land.

For the purpose of this book, pollution is defined as a change in water quality that causes deterioration effects on fish and invertebrate populations and plant life in the water. The book deals with two aspects.

The first aspect is the determination of the concentration of the pollutants or toxic chemicals in the creatures. Incidentally, the determination of such chemicals in the water in which these creatures live is the subject of my previously published book, *The Analysis of Marine Waters and Sediments* (Taylor and Francis, 2014).

The second aspect is the toxicity of toxic substances that enter creatures or the organs of such creatures in the marine environment and can cause deterioration in the health of or the death of marine animal life or damage to plant life in the aquatic community.

Incidentally, this can affect the health of humans who eat these creatures or, in extreme cases, the mortality of humans.

The levels of pollutants that occur in creatures depend on the levels of these substances present in the water in which they live and, in the case of bottom-feeding fish or invertebrates, the level of pollutants in sediments.

The exposure of creatures to known concentrations of toxicants for stipulated periods of time enables the toxicity of the pollutant to be established, as measured by the relationship between the concentration and the time taken for 50% of the creatures to die (i.e. LC_{50}) or experience adverse effects (i.e. LE_{50}), such as distorted limbs or loss of weight. A further method of assessing the toxicity of a pollutant is based on relating the composition of the water in which the creatures live to the concentration of the toxicant found in the animal tissue or, better still, the concentration in a particular organ of the animal in which the toxicant concentrates. Preferentially, such data can be related to the water composition and the condition of the animal, in terms of ill health or mortality.

In Chapters 8 through 10, we discuss, respectively, the toxic effects of metals, organometallic compounds and organic compounds in fish and invertebrates.

Detailed information from world literature is given on LC_{50} and LE_{50} values for a wide range of toxicants in various types of fish and invertebrates, and the effects on health or mortality of a wide range of creatures upon exposure to toxicants. A wide range of invertebrates are discussed, including lobsters, oysters, shrimps, molluscs and mussels.

Similarly, Chapter 11 discusses the effect of various toxicants on plant life occurring in the marine environment, including phytoplankton, plankton, algae, weeds and diatoms.

The rate of bioaccumulation is effected by several factors, including the body weight of the creatures, the rate of growth, salinity and temperature.

Finally, bioaccumulation is discussed in Chapter 12. Bioaccumulation is a process whereby the concentration of a toxicant in the tissues or organs of fish or crustacea increases steadily over a period to time.

At some point in this process, the toxicant concentration in the creature will exceed the safe dose; that is, the LC_{50} and LE_{50} values will exceed the safe limits for the health or mortality of the creatures.

This book is essential reading for analytical chemists, environmentalists, toxicologists, food technicians and fishery experts.

Author

Thomas Roy Crompton is a consultant and technical author. He was awarded a PhD in polymer technology from the University of Salford, UK. Before retirement, he served for 30 years as head of the Analytical Research Laboratory of the Polymer Research Laboratory, Shell Chemicals UK, and for 15 years as head of Water Analysis Laboratories in the UK. He has published more than 50 books on the analysis of environmental samples, including waters, soils and sediments, as well as on polymer technology, organometallic compounds and power sources. His areas of expertise include analysis of natural and sea waters; soils, sediments and sludges; preconcentration techniques in water analysis; and application of chromatography and mass spectrometry to water analysis.

1 Toxicity Testing of Water

1.1 WATER ANALYSIS–BASED TEST DETAILS

Toxicity testing on animals, as opposed to humans, is centred on exposing a known number of animals to known concentrations of the suspect toxic substance in their diet and ascertaining the number of mortalities or the number of adverse effects (e.g. reduced growth) occurring in a specified time span. The experiments are carried out at a number of different concentrations of the suspect toxicant using a fixed span of T days in all experiments.

From the data obtained, it is possible to derive statistically the concentration of the test substances that will, in T days' exposure, kill 50% of the test animals, the so-called LC_{50}. Also, it is possible to derive the concentration of test substance that will cause a specific adverse effect on 50% of the test animals, the so-called LE_{50}.

The whole set of experiments can then be repeated for a different test duration of T^1 days to establish the effect of exposure time on LC_{50} or LE_{50}. Comparison of LC_{50} or LE_{50} values for a particular animal for different test substances enables the relative toxicities of different substances to be evaluated, while comparison of LC_{50} or LE_{50} values for different test animals using the same test substance enables the relative effect of a given test substance on different species to be evaluated.

The test substance can be fed to the animal in a liquid form or as a solid diet, or in the case of fish and crustacea, it can be added in controlled amounts to the water in which the tests are conducted. By using seawater and river water, it is possible to compare relative effects of particular concentrations of test substances in saline and non-saline media.

An additional test that can be carried out is the measurement of the concentration of test substances in the tissues and organs of the test animal, in which many test substances concentrate to a much higher concentration than is present in the surrounding water. Correlations can then be obtained between LC_{50}, LE_{50} and actual concentrations in animal tissues and organs.

Such tests are, of course, only applicable to sacrificial animals and not to humans. In the latter case, all that can be done is to test the suspect toxicant on a range of laboratory animals (e.g. mice and monkeys) that are believed to react to the test substance similarly to man and apply the LC_{50} and LE_{50} values so obtained as a safety factor (usually 1/10 to 1/100) such that the toxic effect, if any, to man will be absent or minimal. This practice is used extensively in drug evaluation.

1.1.1 EFFECT OF TEST SUBSTANCE CONCENTRATION ON LC$_{50}$ AND LE$_{50}$ DETERMINATIONS IN FISH AND CRUSTACEA

Short-term LC$_{50}$ or LE$_{50}$ tests are run for 4 days, that is, 4-day LC$_{50}$ or LE$_{50}$. Commonly, to obtain more complete information, tests might also be run for 1, 10, 100 and 1000 days. Figure 1.1 illustrates the effect of zinc concentration on LC$_{50}$ for 4-day and 10-day tests, that is, 4-day LC$_{50}$ and 10-day LC$_{50}$. From this plot, it is seen that about 390 µg L^{-1} of the test substance would cause 50% mortality of fish in 4 days, reducing to 115 µg L^{-1} during a 10-day exposure.

1.1.2 EFFECT OF EXPOSURE TIME ON LC$_{50}$ AND LE$_{50}$ DETERMINATIONS IN FISH AND CRUSTACEA

Figure 1.2 shows the effect of exposure time in days, up to 1000 days, in the LC$_{50}$ test on percent mortalities of salmon and non-salmonid fish. This shows that in the case of cadmium:

1. Salmonid fish are more sensitive to the test substance than non-salmonids.
2. The lower the concentration of test substance in the water, the longer the fish survive, as would be expected.

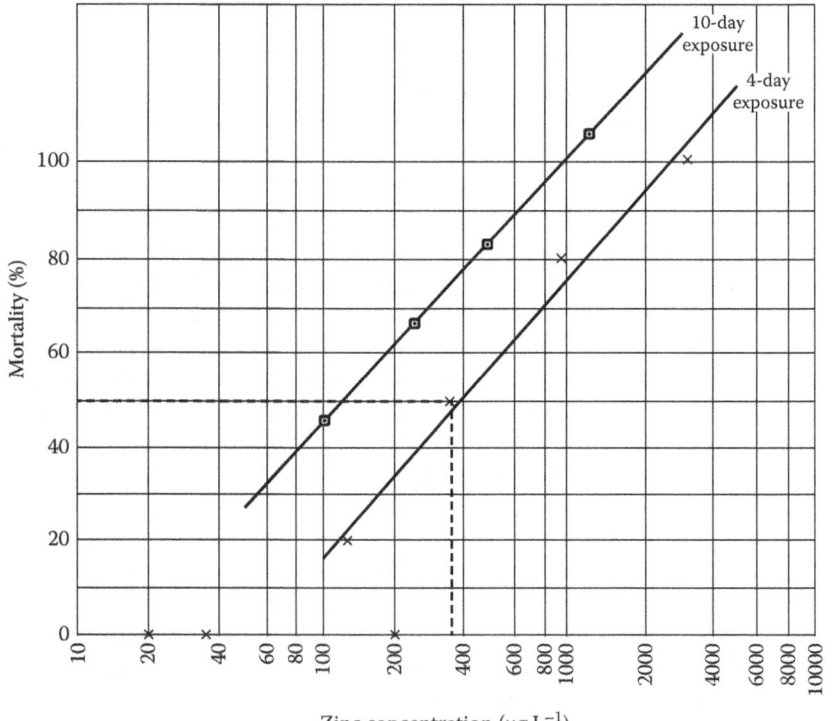

FIGURE 1.1 Method of obtaining LC$_{50}$ by interpolation for the toxicant zinc.

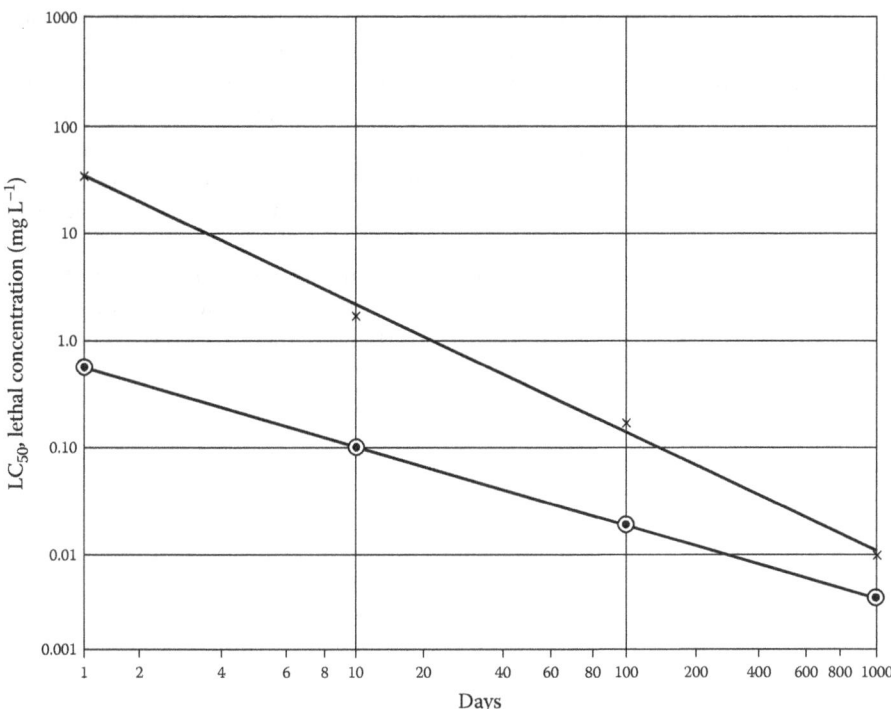

FIGURE 1.2 Effect of exposure time of fish to cadmium on LC$_{50}$. ×, non-salmonid fish; ○, salmonid fish.

It is seen above that the toxic effect of the substance on fish is a consequence of not only the concentration of the substance, but also the duration of exposure.

The lethal (LC$_{50}$) adverse effect (LE$_{50}$) concentration becomes lower as the duration of exposure increases.

1.1.3 EFFECT OF OTHER EXPERIMENTAL PARAMETERS ON LC$_{50}$ AND LE$_{50}$ DETERMINATIONS

LC$_{50}$ or LE$_{50}$ values reported in the literature for a particular toxicant and particular exposure time, for example, 4-day LC$_{50}$, vary considerably. An example of the ranges of values obtained for a particular toxicant is given below:

Duration of Toxicity Test (days)	Range of LC$_{50}$ Values Reported in the Literature
1	6.6–124
10	1.2–23
100	0.23–4.3
1000	0.05–0.8

Such variability in reported LC_{50} values is a consequence of the lack of control of experimental parameters when LC_{50} measurements are being made, as well as the fact that some types of fish are more sensitive to a particular test substance than others. Experimental parameters that affect the results obtained in LC_{50} or LE_{50} measurements include temperature (Figure 1.3), pH (Figure 1.4), water hardness (Figure 1.5), salinity, dissolved oxygen content, the presence or absence of light, water flow rate through the test chamber, the chemical form of toxicant and acclimatisation of fish to test conditions.

To carry out LC_{50} or LE_{50} measurements, as many as practicable of these parameters should be controlled in order to make intermeasurement reproducibility as high as possible, and all test parameters should be quoted with the experimental result.

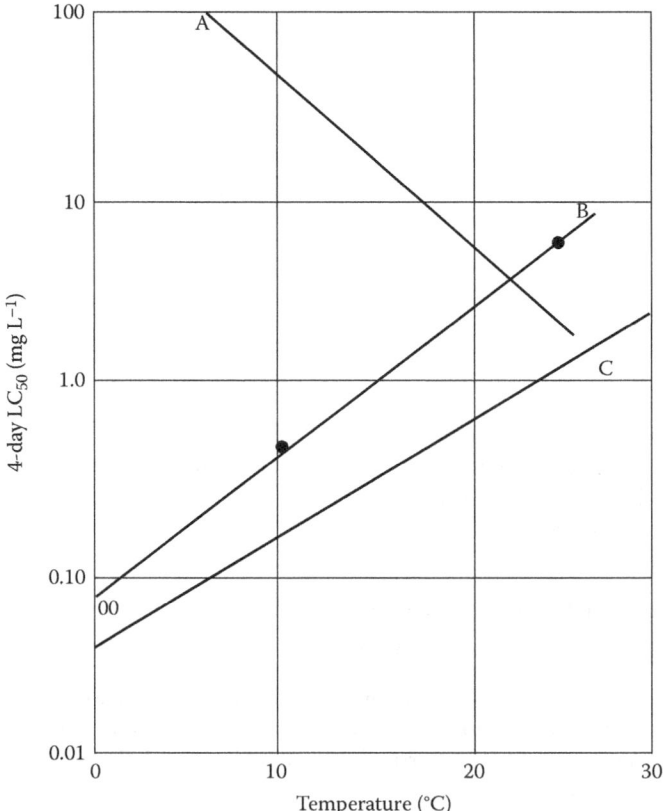

FIGURE 1.3   Effect of test temperature on 4-day LC_{50}. Line A shows silver (as silver nitrate) with salmonid fish; that is, temperature increases as toxicity increases. Line B shows copper with salmonid fish. Line C shows cadmium with non-salmonid fish. For both lines B and C, an increase in temperature reduces the toxicity. For chromium, zinc and silver with non-salmonid fish, toxicity is independent of temperature. For mercury, cadmium, copper and zinc with freshwater invertebrates, an increase in temperature increases toxicity.

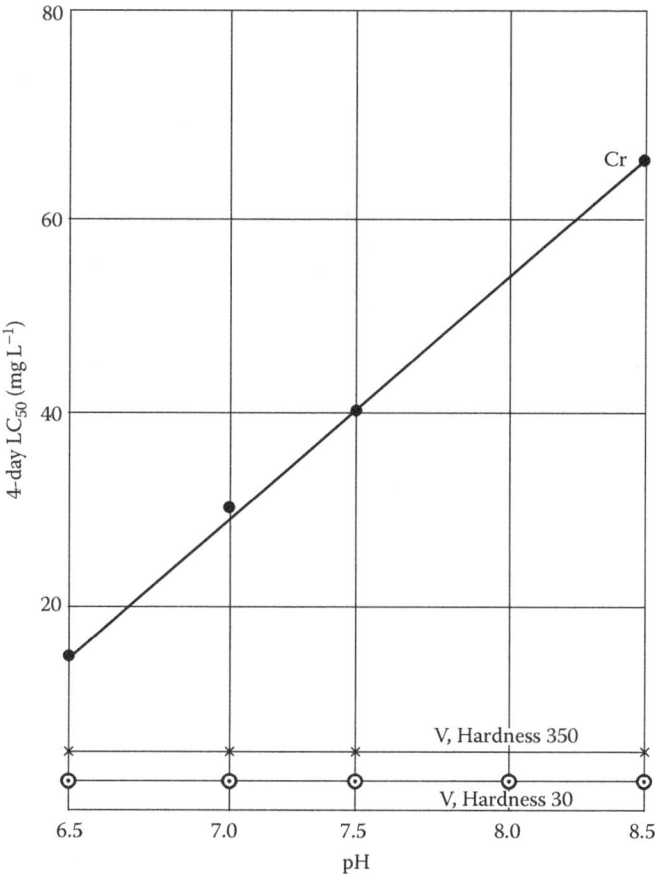

FIGURE 1.4 Effect of pH on LC$_{50}$. Freshwater fish and invertebrates for the toxicant chromium.

1.1.4 DERIVATION OF MAXIMUM SAFE CONCENTRATION STANDARD (S) OF A TEST SUBSTANCE FOR CONTINUOUS EXPOSURE TO FISH

A plot of the range of available LC$_{50}$ data versus the exposure time data for a particular test substance and type of fish is shown in Figure 1.6. In selecting potential values of the standard, a boundary line (dotted) drawn to enclose the lower limits of the reported lethal effect (LC$_{50}$) or adverse effect (LE$_{50}$) concentration (i.e. a conservative estimate) would describe a continuous standard in the form of an equation predicting the maximum acceptable concentration of the test substance (with no safety margin) permissible for a specified duration of time. Thus, for continuous exposure of fish, the 365-day asymptote.

$S_x = 220 \mu g\, L^{-1}$ would represent a potential standard (S) for the survival of fish for 1 year in the test water. However, as in practice the concentration of the test substance in the water is not measured continuously, it is desirable to apply a safety factor to the

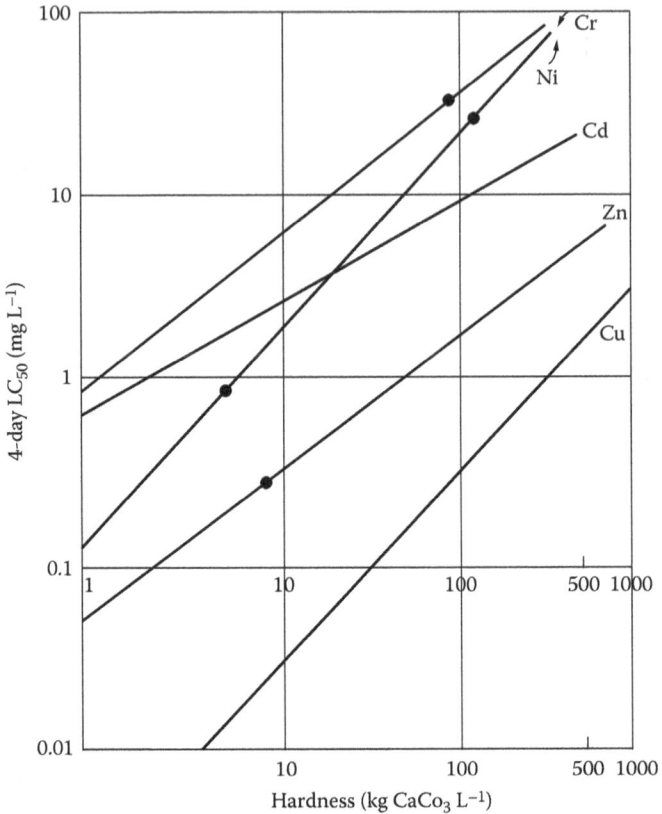

FIGURE 1.5   Effect of water hardness on LC_{50} for the toxicants chromium, nickel, cadmium, zinc and copper.

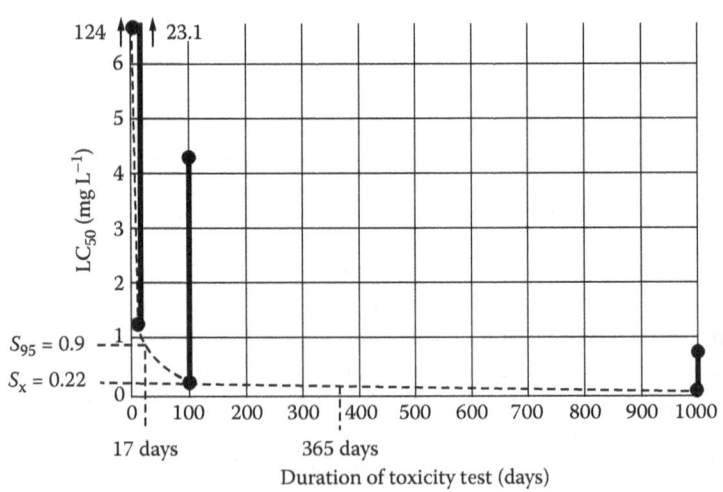

FIGURE 1.6   Test duration vs. LC_{50} plot for nickel.

$S_x = 220\,\mu g\,L^{-1}$ figure to ensure that any hour-to-hour variations in concentration of the test substance in the test water do not exceed $220\,\mu g\,L^{-1}$. If a safety factor of 10 is applied, then this standard should be safe for the longer-term (1-year) survival of fish unless erratic variations in the concentration of the test substance occur from time to time in the test water. This long-term standard ($S_x/10 = 22\,\mu g\,L^{-1}$) incorporating a safety factor of 10 might be stated as the annual long-term average concentration of test substance in test water. However, adoption of this standard might allow higher concentrations of the test substance in test water to occur for shorter periods during the year, and there is a potential risk that these excursions in concentration of the test substance would be sufficiently great as to cause damage to the fish.

To overcome this, the 95th percentile concept has been adopted, that is, the concentration of test substance that could be safely exceeded for 5% of 1 year (i.e. 17 days per year if daily analysis of the water is performed or 2 weeks per year if weekly analysis of the water is performed).

Figure 1.7 shows a plot of the percentage of nickel in water during which the concentration of the test substance in the test water lies between particular concentration limits versus measured concentration. Thus, 19% of 365 daily samples, that is, 69 out of 365 samples, have a test substance concentration between 200 and $300\,\mu g\,L^{-1}$. Two percent of 365 daily samples, that is 7 out of 365 samples, have a test substance concentration between 800 and $900\,\mu g\,L^{-1}$. Reference to this graph shows that for 5% of the time, that is, 17 samples out of 365, the concentration of test substance is about $1000\,\mu g\,L^{-1}$, that is, $S_{95} = 1000\,\mu g\,L^{-1}$. Reference to the 17-day exposure results in Figure 1.6 gives an S_{95} value of $0.9\,mg\,L^{-1}$ ($900\,\mu g\,L^{-1}$) excluding the safety factor of 10 or $S_{95} = 90\,\mu g\,L^{-1}$ incorporating the safety factor of 10. Thus, providing the concentration of nickel in test water does not exceed $S_{95} = 90\,\mu g\,L^{-1}$ in this example for more than 5% of the exposure time, in this case 1 year, all fish will survive unharmed for 1 year in the test water.

The data quoted in Figure 1.6 refer to LC_{50} values (i.e. concentration of test substance which causes 50% mortality) of test animals. The S_1 and S_{95} values also refer to maximum permissible concentrations present in test water for 95% and 50% of a year, respectively, for the survival of fish with nil mortalities. However, we are concerned not only with mortalities, but also ill health, that is, adverse effects. A concentration of test substance at which fish do not die could be too high for them to both survive and remain in good health in order to ensure their continued existence in the sea. S_{95} adverse effect concentration, that is, LE_{50}, S and S_{95} will obviously be the LE_{50}, S and S_{95} concentrations quoted above for lethal effect. Adverse effects induce impaired reproductions, reduced rate of growth and illness.

The above discussion is concerned with nickel. However, similar considerations can apply to a range of other metals. In Table 1.1 are listed the S_x and S_{95} tile values for a range of elements from the least toxic (nickel, $S_x = 220\,\mu g\,L^{-1}$, S_{95} tile = $900\,\mu g\,L^{-1}$) to the most toxic (cadmium, $S_x = 2\,\mu g\,L^{-1}$, S_{95} tile = $6\,\mu g\,L^{-1}$; mercury, $S_x = 2\,\mu g\,L^{-1}$, S_{95} tile = $22\,\mu g\,L^{-1}$). These data are obtained by plotting the data shown in Table 1.1 in the same manner as shown in the case of nickel (Figure 1.6). The maximum safe concentrations quoted in Table 1.1 are not amended by safety factors and have not been weighted for the effects of such

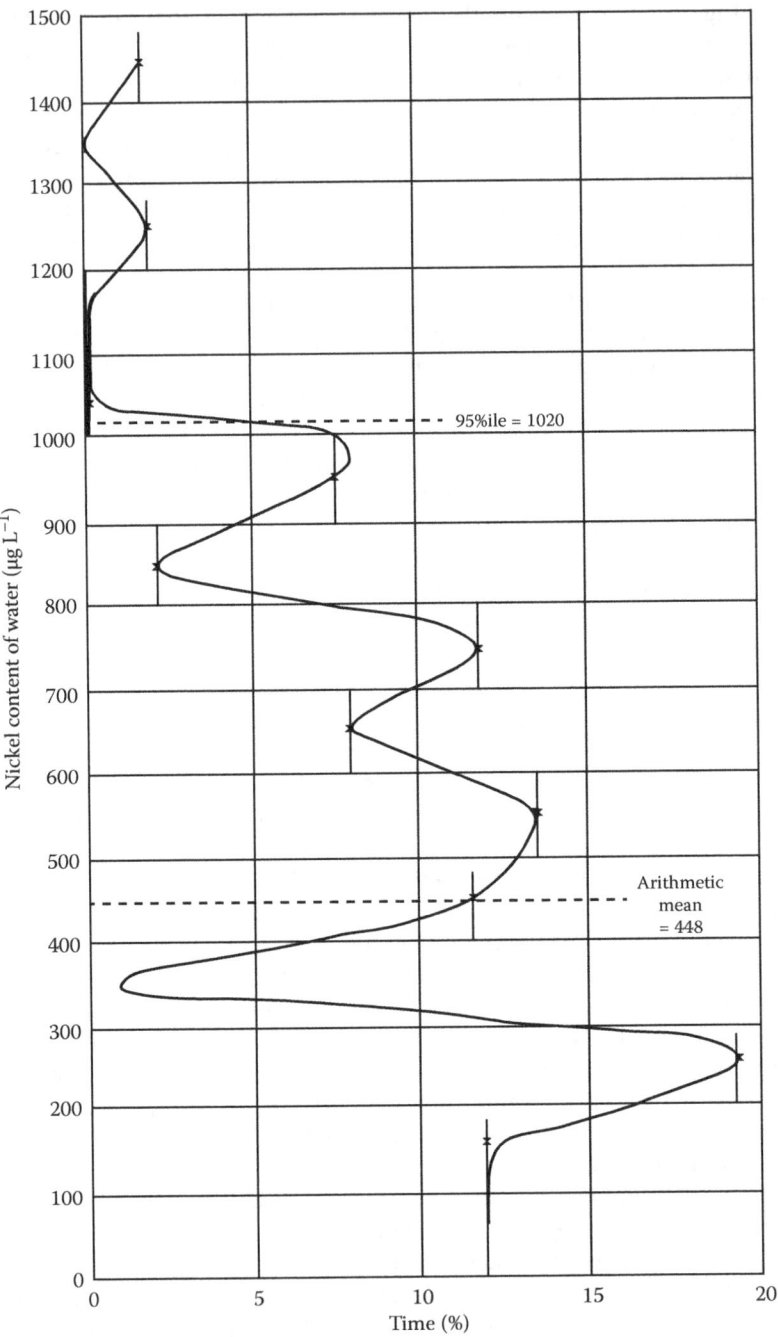

FIGURE 1.7   Ninety-fifth percentile determination of nickel in water over a 12-month period.

TABLE 1.1

S_x and S_{95} Values for Various Fish Exposed to Various Metals

Fish Species	Metal	Standard Maximum Safe Concentration Supporting Fish Life (μg L^{-1}), i.e. S_x (365 days)	Standard S_{95} (μg L^{-1}) Maximum Metal Concentration Permitted for 17 days out of 365 days	S_{95}/S_x
Non-salmonid	Ni	220	900	4.5
	Se	200	1300	6.5
	V	100	1000–1600	10–16
Salmonid	Cr	100	800	8
Non-salmonid	As	80	600	4.8
Non-salmonid	Ag	70	850	12.1
	Cr	100	1000–3000	10–30
	Zn	23	200	8.7
	Pb	20	100	5
	Cd	4	16	4
Salmonid	Cu	4	17	4.2
Non-salmonid	Hg	2	22	11
Salmonid	Cd	2	6	3

environmental factors as water hardness, pH, temperature and, in saline waters, salinity. In practice, the available data do not permit this, and the effects of experimental factors as demonstrated in short-term acute toxicity tests are extrapolated to long-term exposure.

1.1.5 Cumulative LC$_{50}$ Values

Only rarely does a water which is toxic towards fish life contain a single toxin. If the impurities are present in any appreciable amount, then as is often the case, several of them will adversely affect fish life. The following progressive dilution technique enables the commutative effect of toxins on fish to be assessed.

In these tests, fish are exposed to the water for a number of days and the fish mortality rate and pollutant concentrations are measured at daily intervals during this period. Simultaneously, caged fish are exposed to a range of dilutions of river water and the same measurements are repeated. From the results obtained, the dilution causing 50% mortality in 2 days is estimated for various fish species at each location.

Figure 1.8a–d shows test duration–percentage mortality curves obtained from fish in (A) polluted waters and (B) less polluted waters at zero dilution and 1×, 2×, 5× and 10× dilutions of river water. From these curves, the percent of mortality occurring after 2-day exposure for those polluted (A) and less polluted (B) waters can be obtained. Plots of percent mortality versus dilution enable the dilution corresponding to 50% mortality to be read off (Figure 1.9a and b). From these curves, it is seen (Figure 1.9a)

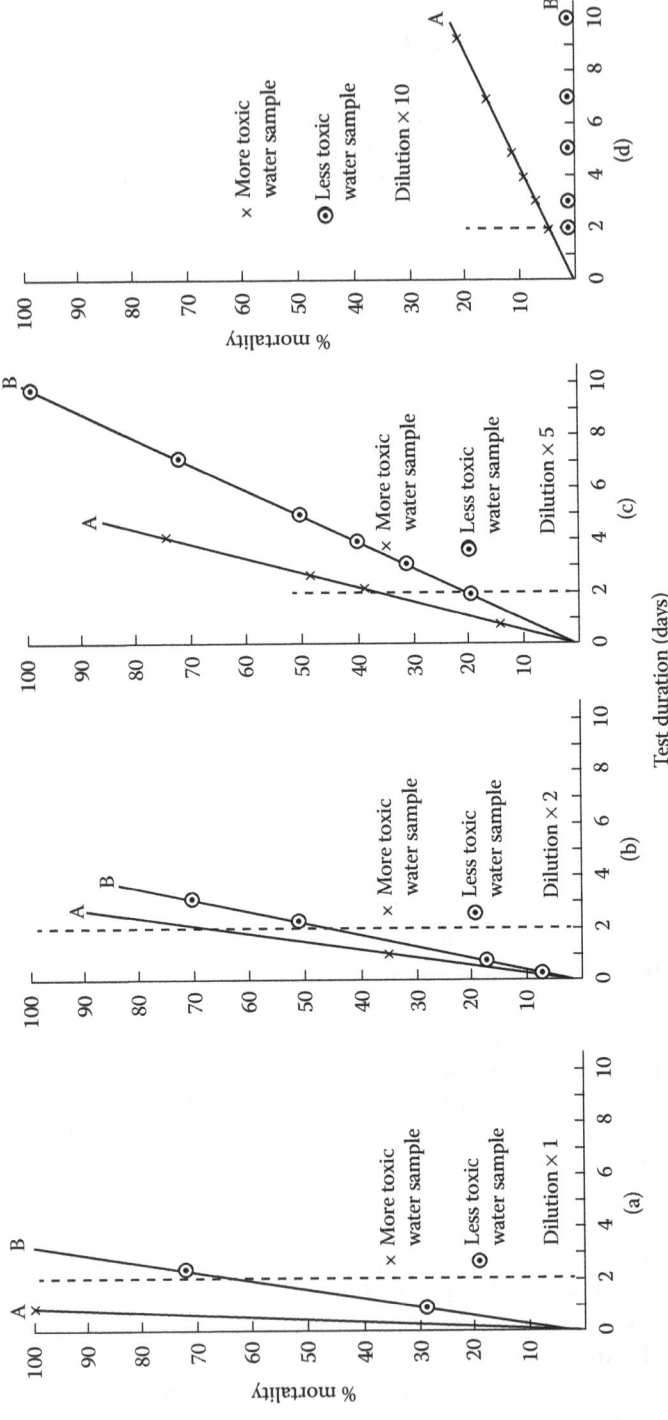

FIGURE 1.8 Test duration–percent mortality curves for a range of river water dilutions.

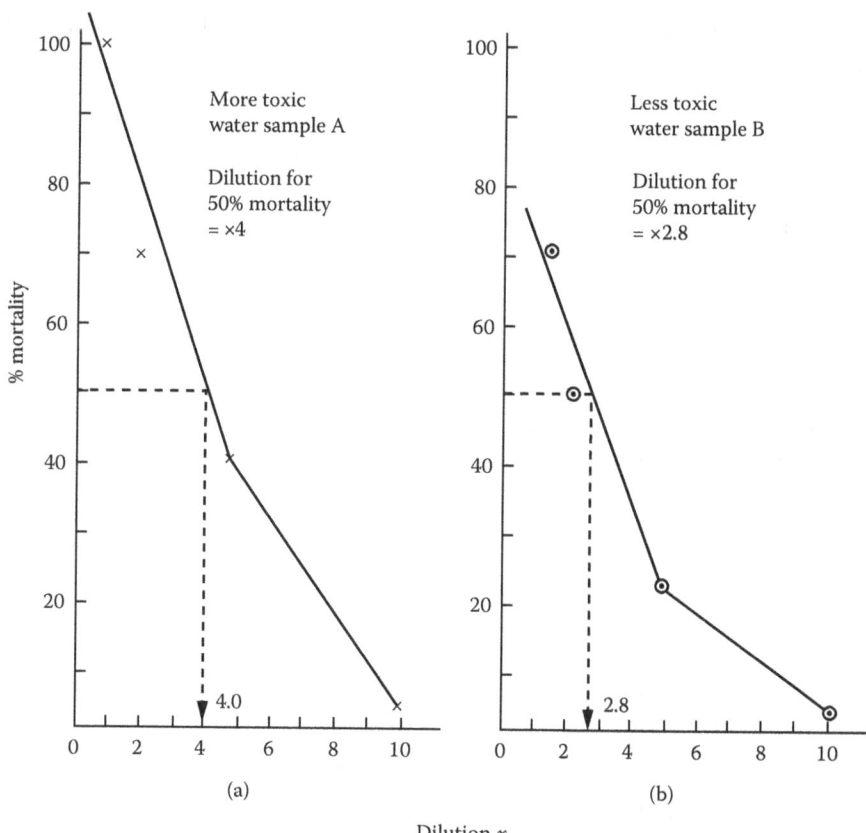

FIGURE 1.9 Dilution–percent mortality curves. (a) More polluted river water. (b) Less polluted river water.

that for the more polluted water a 50% mortality results when the original water sample has been diluted 4×, and for the relatively unpolluted water B (Figure 1.9b) only 2.8× dilution is required to achieve the same effect. The results of these studies are presented not as a concentration of pollutants in the rivers, but as cumulative fractions of the relevant laboratory-derived 2-day LC_{50} for each species and substance, the sum of which is compared with the observed toxicity at each location.

Thus, considering a simple example, if a relatively toxic water A before dilution contained $50\,mg\,L^{-1}$ zinc and $10\,mg\,L^{-1}$ copper, then the 4× dilution of this causing 50% mortality (Figure 1.9a) would contain $12.5\,mg\,L^{-1}$ zinc and $2.5\,mg\,L^{-1}$ copper; that is, the river-derived cumulative 2-day $LC_{50} = 12.5 + 2.5 = 15\,mg\,L^{-1}$. Similarly, if a relatively less toxic water B before 2.8× dilution (Figure 1.9b) contains 8 and $1.5\,mg\,L^{-1}$ of zinc and copper, respectively, then the fourfold dilution would contain 2.8 and $0.5\,mg\,L^{-1}$ zinc and copper, respectively; that is, river-derived cumulative 2-day $LC_{50} = 2.8 + 0.5 = 3.2\,mg\,L^{-1}$. If the laboratory-derived 2-day LC_{50} values for zinc and copper are 12 and $6\,mg\,L^{-1}$, respectively, that is, cumulative 2-day $LC_{50} = 18\,mg\,L^{-1}$, then the river-derived cumulative 2-day LC_{50} as a fraction of the laboratory-derived 2-day LC_{50} (i.e. cumulative proportion of laboratory-derived 2-day LC_{50}) is given by

$$\text{Polluted water A} = \frac{\text{2-day } LC_{50} \text{ river derived}}{\text{2-day } LC_{50} \text{ laboratory derived}} = \frac{15}{18} = 0.83$$

$$\text{Polluted water B} = \frac{\text{2-day } LC_{50} \text{ river derived}}{\text{2-day } LC_{50} \text{ laboratory derived}} = \frac{3.2}{18} = 0.18$$

The observed difference between river-derived and laboratory-derived cumulative LC_{50} can be ascribed to the effects of factors such as hardness, pH, temperature and dissolved oxygen prior to summation.

This approach has been applied to an assessment of the fishery status of rivers where it has been found that if the sum of proportions of 2-day LC_{50} exceeds about 0.3, then fish will not survive well enough to support activities [1–7]. Table 1.2 shows this effect for a range of river waters of different total hardness.

Accumulation factors of metal concentrations on metal content in fine sediments of the Severn and Humber Estuaries are shown in Table 1.3.

TABLE 1.2

Differences of Cumulative Proportions of 2-day LC_{50} (Laboratory Derived) vs. Fishing Status and Water Hardness

		Median Cumulative 2-day LC_{50} ($\mu g\,L^{-1}$)		
Survey	Total Hardness	Fishless	Marginal	Fish Present
1	11–20	0.45	0.42	0.05–0.25
2	100–170	0.1–0.2	0.16	0.13–0.16
3	100–300	>0.28	—	<0.28
4	134–292	>0.1	—	<0.1
5	500	>0.32	0.15–0.32	<0.25
6	70–745	0.32–2.95	0.3–0.37	0.005–0.02

TABLE 1.3

Accumulation of Metals from Humber and Severn Estuary Water into Sediment

Location	Copper	Lead	Nickel	Zinc	Arsenic	Cadmium
Severn Estuary	15,710–16,300	26,830–67,330	15,280–22,600	13,090–25,640	—	1280–3230
Humber Estuary	57,350–43,000	68,000–136,000	2130–32,000	4060–102,500	3700–37,000	800–4000

Note: Accumulation factor = $\dfrac{\mu g\,kg^{-1} \text{ dry weight in sediment}}{\mu g\,L^{-1} \text{ in water}}$

1.2 ANIMAL TISSUE–BASED ANALYSIS

Similar bioaccumulation of metal phenomena has been observed in the case of fish, and indeed, bioaccumulation has been studied not only in the whole fish, but also in individual fish organs where appreciable differences have been reported between different organs. Van Hoof and Van Son [8] have reported on the extent of bioaccumulation occurring in five different organs taken from rudd (muscle, gill, opercle, liver and kidney). Table 1.4 reports the concentration factors for four metals (zinc, copper, cadmium and chromium) for organs taken from fish exposed to different levels of the metals for various exposure times between 4 h and greater than 10 weeks. Of the various organs taken from this particular type of fish, it is seen in Table 1.5 that the highest concentration factors always occur in opercle tissue and the lowest in muscle, with the other organs being intermediate. It will not be that higher concentration factors are obtained when the exposure time is extended from 3 to 10 weeks, even though the metal concentrations of the water were lower in the 10-week test.

Figure 1.10 plots concentration factors obtained from the opercle versus test duration and concentration of zinc in the water (Table 1.3). It is seen that the concentration factor increases linearly with increasing exposure time but seems to bear an exponential relationship with metal concentration in the water. A plot of the logarithm of the metal concentration and the exposure time versus concentration factor is linear, as illustrated in Table 1.6 and Figure 1.11.

The bioaccumulation has been measured on copper and zinc in the barnacle *Balanus amphitrite* in estuary water [9]. At concentrations of $1-11\,\mu g\,L^{-1}$ copper in the water, between 39,700 and $625,700\,\mu g\,L^{-1}$ of copper were found in the barnacle tissue, giving bioaccumulation factors of 7060 and 384,310. At concentrations of $13-46\,\mu g\,L^{-1}$ zinc in the water, between 203,600 and $1,937,000\,\mu g\,kg^{-1}$ zinc were found in barnacle tissue, giving bioaccumulation factors between 10,660 and 84,600.

Langston and Zhou [10] studied the bioaccumulation of cadmium in the tellinid clam *Macoma balthica* taken at the Whitehaven Cumbria coastline. At $100\,\mu g\,L^{-1}$ cadmium in water, the clam picked up $10.150\,\mu g\,kg^{-1}$ cadmium during a 29-day exposure ($0.35\,\mu g\,Cd\,g^{-1}$ per day), giving a bioaccumulation factor of 101. A bioaccumulation factor of 68,000 has been obtained for iron in kelp (*Ecklonia radiate*) taken from harbour water [11].

There are a number of terms associated with aquatic toxicity which will be defined, and then the most important features of toxicity tests will be considered in relation to the potential complications arising from physical, chemical and biological factors.

1.2.1 TERMINOLOGY

Pollution: Change in water quality which causes a deleterious change in the biological community or makes the aesthetic character of the water unacceptable.

Contamination: Presence of a potentially harmful substance at concentrations which do not cause harm to the environment.

Toxicity test: Use of living organisms to define the poisonous effects of a substance or substances under controlled conditions.

TABLE 1.4
Interrelationship between Concentration of Metals in Water and Concentration Factor in Rudd Fish

Element	Exposure Time	Water (μg L⁻¹)	Organs (μg g⁻¹) (or mg kg⁻¹)					Concentration Factor in Organ				
			Muscle	Gill	Opercle	Liver	Kidney	Muscle	Gill	Opercle	Liver	Kidney
Zinc	>10 weeks[a]	680[a]	16.4	47.9	120.2	29.4	57.0	91	266	667	163	317
	3 weeks[a]	800[a]	24.4	101.9	195.5	104.9	151.7	28	127	244	131	189
	≤24 h[a]	1600[a]	10.5	38.6	115.3	42.5	154.6	6.5	24	72	26	97
	≤12 h[b]	7500[a,b]	6.6	51.2	90.6	34.1	92.2	0.9	6.8	12.1	4.5	12
	≤4 h[b]	18,000[a,b]	11.2	657.2	174.5	63.5	216.1	0.6	36	9.7	3.5	12
Copper	>10 weeks[a]	11[a]	0.7	5.5	12.4	6.9	6.0	64	500	11.27	627	545
	3 weeks[a]	50[b]	1.6	8.9	30.9	202	28.5	32	178	618	404	570
	<12 h[b]	250[b]	2.3	22.9	52.6	22.3	30.4	9.2	9.2	210	89	121
	<12 h[b]	1200[b]	2.2	29.3	72.1	31.1	39.0	1.8	24	60	26	32
	<12 h[b]	1600[b]	4.0	43.2	104	39.8	100	2.5	27	65	25	62
Cadmium	>10 weeks[a]	3[a]	0.3	2.6	9.5	5.0	4.2	100	868	3166	1666	1400
	3 weeks[a]	250[a]	0.41	2.5	8.7	9.6	15.7	1.6	10	35	38	55
	<12 h[b]	1100[b]	0.6	3.9	6.0	4.1	14.4	0.5	3.5	5.4	4.5	13
	<12 h[b]	4000[b]	0.5	10.4	20.7	3.8	12.8	0.12	2.6	5.2	0.95	3.2
	<12 h[b]	11,000[b]	3.2	87.9	29.2	12.3	28.2	0.29	8.0	2.65	1.1	2.5
Chromium	>10 weeks[a]	3	<0.2	<0.2	<0.2	<0.2	<0.2	<66	<66	<66	<66	<66
	3 weeks[a]	16	<2	<2	<2	<2	<2	<125	<125	<125	<125	<125
	<12 h[a]	20	0.5	4.9	8.3	5.6	10.3	125	245	415	280	515
	<12 h[b]	80	0.8	48.2	26.0	18.4	23.8	10	602	325	230	297
	<12 h[b]	145	0.6	30.6	19.6	15.2	27.8	41	211	135	105	192

[a] Concentration factor = $\dfrac{\text{Concentration in organ (mg kg}^{-1}\text{ or μg kg}^{-1} \times 1000)}{\text{Concentration in water (μg L}^{-1})}$

[b] Subacute toxicity tests, no fish mortalities; acute toxicity tests, 100% fish mortality.

TABLE 1.5

Summary of Concentration Factors Obtained for Organs Taken from Rudd Fish at Different Metal Concentrations in Water and Different Exposure Times

Exposure Time (weeks)	Metal	Concentration in Water (µg L⁻¹)	Highest Concentration Factor	Lowest Concentration Factor
3	Zn	800	244 (opercle)	28 (muscle)
	Cu	50	618 (opercle)	32 (muscle)
	Cd	250	25/38 (opercle/muscle)	1.6 (muscle)
10	Zn	180	677 (opercle)	91 (muscle)
	Cu	11	1127 (opercle)	64 (muscle)
	Cd	3	3166 (opercle)	100 (muscle)

Note: $\dfrac{\mu g\ L^{-1}\ \text{in tissue} \times 100}{\mu g\ L^{-1}\ \text{in water}}$

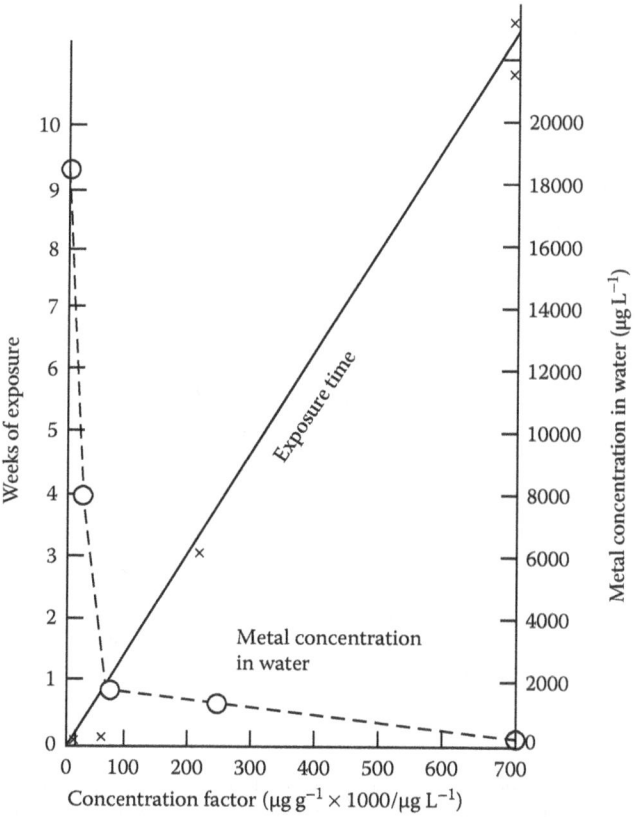

FIGURE 1.10 Plot of weeks of exposure versus concentration factor.

TABLE 1.6

Dependent of Concentration Factor Obtained for Opercle on Product of Log Concentration Factor and Exposure Time

Element	Exposure Time of Rudd (Weeks)	Concentration of Metal in Water (μg L^{-1})	(Log Concentration) Exposure Time (*a*)	Observed Concentration $\left(\dfrac{mg\ kg\ /\ 1000}{\mu g\ L^{-1}} \right)$ (*b*)	Scope (*a/b*)
Zinc	10	180	22.55	677	0.034
	3	800	8.71	244	0.036
	0.143 (1 day)	1600	0.46	72	
	0.0715 (0.5 day)	7500	0.25	12.1	
	0.024 (4 h)	18,000	0.10	9.7	
Copper	10	11	10.41	1127	0.0092
	3	50	5.10	618	0.0082
	0.0715 (0.5 day)	250	0.17	210	
	0.0715 (0.5 day)	1200	0.22	60	
	0.0715 (0.5 day)	1600	0.23	65	

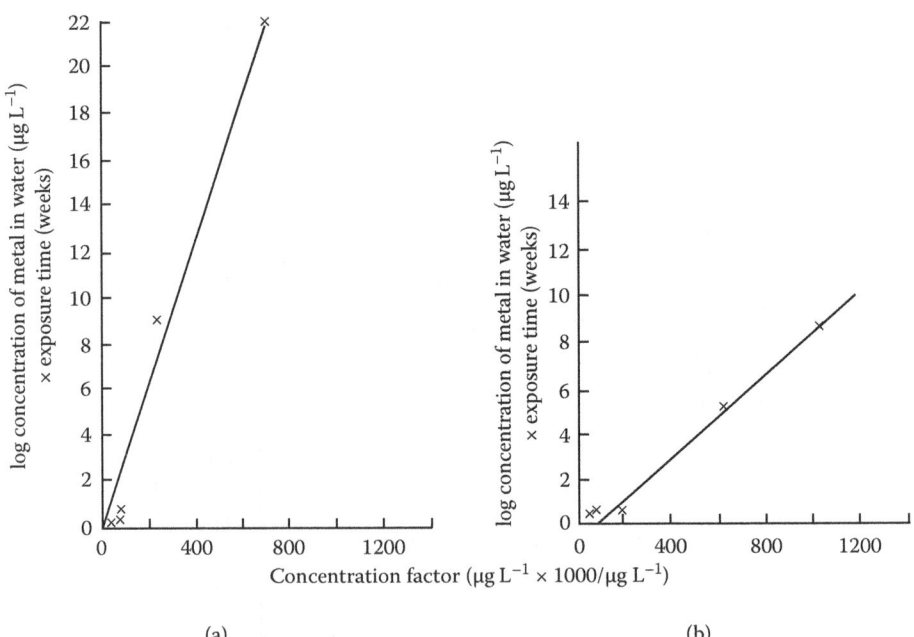

(a)   (b)

FIGURE 1.11 Linear relationship between concentration factor (μg g^{-1} in fish × 1000)/(μg L^{-1} in water) and product of concentration in water (μg L^{-1}) and exposure time (weeks). (a) Zinc. (b) Copper.

Static test: Toxicity test in which the test solution is not replaced during the test.

Semistatic test: Toxicity test in which all or most of the test solution is replaced at intervals of 12 h or more during the test.

Flow-through test: Toxicity test in which there is continuous or near-continuous replacement of the test solution.

Nominal concentration: Concentration of the test substance calculated from the mass added to a known volume of water.

Observed concentration: Concentration of the test substance measured analytically in a sample or samples of test solution recovered from the test chamber during the test.

Composite sample: Sample formed by the combination of several separate samples of test solution.

Acute toxicity: Lethal response caused by a short exposure to a substance (at the most a few days, commonly 4).

Sublethal: Having a deleterious effect but not causing mortality.

Chronic toxicity: Deleterious effects (not exclusively fatalities) resulting from prolonged exposure (i.e. more than a few days).

Controls: Group of organisms of the same species as tested which are exposed to the same test conditions as the test population but in the absence of the test substance.

LC_{50}: Medium lethal concentration, that is, the concentration which is calculated to cause the mortality of 50% of a test population.

EC_{50}: Medium effect concentration, that is, the concentration which is estimated to cause a particular response in 50% of the population.

Median period of survival (or median lethal time or median survival time): Length of time that 50% of a test population survives at a particular concentration of the substance.

1.2.2 TOXICITY EVALUATION: ANIMAL TISSUE–BASED ANALYSIS

There are several reasons for monitoring the concentration of toxic metals in creatures such as fish and shellfish.

1.2.2.1 Protection of Human Health

This applies to organisms which are harvested for food. Direct analysis of the organisms against accepted standards enables a decision to be made as to whether the organisms are acceptable for human consumption. In the UK, for example, regulations exist concerning levels of zinc, chromium, copper, nickel and arsenic in fish and shellfish, and these are based on maximum acceptable intake of these foods for 1 week. It is stated that the 90th percentile consumption of fish should not exceed 0.79 kg per week, and for shellfish 0.26 kg per week. Table 1.7 shows weekly intakes of metals by consumers observing the recommendations that would result following the consumption of fish containing different levels of total metals. For example, the weekly recommended maximum intake of chromium from fish caught in coastal waters would be about 0.3 mg, while that of arsenic would be between 14 and 17 mg.

TABLE 1.7
Weekly Intake of Total Metals by Consumers

	Concentration of Metal in Organisms (mg kg^{-1}, dry weight)						Maximum Intake (Recommended kg)	Weight (mg of Metals Consumed)
	Cu	Ni	Zn	As	Cr	Total		
Fish								
Coastal waters	0.5	0.7–1.4	4.6	14.1–16.7	0.3	20.2–23.5	0.79	17.3–18.5
Vicinity of municipal outfall	1–19	0.7–1.4	0.8–2.8	10	0.5–1.5	13.0–16.6		10.3–13.1
Remote area	1.2	0.2	2.4	0.5–1.5	0.2	4.5–5.5		3.5–4.3
Shellfish								
Vicinity of municipal outfall	0.4–2.4	2.2–6.5	0.3–09	10	1–10	13.9–29.8	0.26	3.6–7.7
Remote area	1	3	0.4	0.5–1.5	0.8	5.7–6.7		1.5–1.7

1.2.2.2 Protection of Animal Species

Biomagnification and bioaccumulation of metals and organics by fish and creatures other than fish (e.g. crustaceans and molluscs) in non-saline and saline waters will now be considered.

Biomagnification is an increase in concentration of a toxicant moving along a food chain and has been observed for organochlorine pesticides, which occur at progressively higher concentrations along the food chain.

Two competing factors operate a *bioaccumulation*, namely the rate of uptake of metals or organics and the rate of loss. These govern whether there is a net decrease or increase in toxicant content of the water creatures [1–7].

Bioaccumulation in fish and other creatures is greatest in the following circumstances:

1. When body weight is lowest, that is, just after spawning or in younger and smaller creatures
2. During periods of low rate growth
3. In waters of low salinity
4. In waters of higher temperature
5. In the absence of competing metals; for example, bioaccumulation is greater in soft waters than in hard waters
6. When the species is nearer the water surface

Thus, rates of bioaccumulation are greater in creatures of low body weight and rate of growth in surface waters of low salinity and hardness which are at a relatively high temperature.

As a consequence of all these factors having an influence on the extent of bioaccumulation, the ratio between the reported concentration of metal in water ($\mu g\ L^{-1}$) and in animal or plant life ($\mu g\ kg^{-1}$ dry weight) (i.e. bioaccumulation factor = $\mu g\ kg^{-1}$ in plant or animal/$\mu g\ L^{-1}$ in water) is by no means constant.

Due to the bioaccumulation, there is an increase in concentration of a toxicant in a particular animal or plant species with time, and this has been extensively observed. Metals added to fresh or tidal water tend to be removed by absorption onto particulate matter or by chemical transformation into an insoluble form. Thus, sediment concentrations are normally higher than those of the overlying water. At the primary production level, macrophytes rooted in these metals.

1.2.2.3 Bioaccumulation of Organic Compounds

Bioaccumulation factors of 159, 66 and 17, respectively, have been obtained in rainbow trout, channel catfish and bluegills exposed to carbendazin [12]. Seawater lampreys were exposed to water containing $50–485\ \mu g\ L^{-1}$ kepone for 10 days and gave an average bioaccumulation of 1900 [13]. Striped bass (*Morone saxatilis*) exposed to $100\ \mu g\ L^{-1}$ molinate (Ordram) in water for 1 day gave a bioaccumulation factor of 25.3 [14]. The crustaceans *Daphnia pulex* and *Palaemon paucideus*, upon exposure to $1000\ \mu g\ L^{-1}$ fenitrothion for 1–3 days, gave maximum bioaccumulation factors of 76 and 6, respectively [15].

1.2.2.4   Bioaccumulation of Organometallic Compounds

The data in Table 1.8 illustrate bioaccumulation of tetramethyl lead present in water at a concentration of $3.46\,\mu g\,L^{-1}$ in rainbow trout. The concentration factor ranges from 124 after a 1-day exposure to 800–900 after a 7-day exposure.

Monitoring of bioaccumulation of fresh and tidal waters as trends in spatial monitoring has two purposes:

1. Macroscale, that is the identification of potentially unknown areas of elevated concentration and assessment of the extent of the zone of contamination.
2. Monitoring of bioaccumulation in fresh and tidal waters as trends in time. These need to be maintained to identify trends in contamination, especially near effluent discharges, in order to identify stability, improvement or deterioration in contaminated levels.

Special and time monitoring programmes of the types discussed above will also give information needed to assess the risk to top predators in a particular ecosystem.

The design of such programmes [16] is typified by the US Mussel Watch Program, which takes into account the following factors:

Species studied: *Mytilus edulis* mussel was used in this programme as this creature had already been studied for factors affecting accumulation.
Time of year: Late winter was chosen as metal content is stable during that period (i.e. avoiding post-spawning maximum).
Size or age: Dominant size of population sampled to avoid effects of age and size.
Position on shore: Collected on rocky shores to avoid contamination by soft sediments at level on shore exposed for approximately 6 h each tidal cycle, that is 3–4 h after high tide.

TABLE 1.8
Accumulation of Tetramethyl Lead in Rainbow Trout

Exposure (days)	Weight of Fish (g)	Fish Alive or Dead	Water Averaged ($\mu g\,L^{-1}$)	Fish Wet Weight ($\mu g\,g^{-1}$)	Concentration Factors
1	0.1211	Dead	3.6	0.43	124
2	0.3661	Dead		1.08	312
	0.7982	Dead		2.00	578
3	0.4116	Dead		1.32	382
	0.6300	Dead		2.09	604
7	1.3045	Alive		2.94	850
	1.5466	Alive		3.23	934
	0.8100	Alive		2.25	650
	0.4926	Alive		1.73	500

Note: Concentration factor = concentration of Me_4Pb in fish ($\mu g\,kg^{-1}$)/concentration of Me_4Pb in water ($\mu g\,L^{-1}$).

Sample size: Minimum 25 animals to allow statistical assessment.

Sampling: Transported alive in polyethylene bags regularly drained from free water. Placed in clean water for 24 h prior to analysis to ensure that gut contents are eliminated. Analysis on homogenised individual animals and shell dimensions recorded.

This scheme is designed to detect a 10% change in metal concentrations in *Mytilus* mussels with a confidence of 90%.

In one such study, mussels from a clean environment were suspended in cages at several locations in the Firth of Forth. A small number were removed periodically, homogenised and analysed for methylmercury. The rate of accumulation of methylmercury was determined by dividing this by mussel filtration rate, and the total concentration of methylmercury in the seawater was calculated.

The methylmercury concentration in caged mussels increased from low levels ($<0.01 \, \mu g \, g^{-1}$) to $0.06–0.08 \, \mu g \, g^{-1}$ in 150 days, giving a mean uptake rate of $0.4 \, ng \, g^{-1}$ daily; that is a 10 g mussel accumulated 4 ng daily. The average percentage of total mercury in the form of methylmercury increased from less than 10% after 20 days to 33% after 150 days. This may be compared with analysis of natural intertidal mussels from the area in which the proportion of methylmercury was higher in mussels of lower ($<10 \, \mu g \, g^{-1}$) than of higher total mercury concentrations.

Davies et al. [17] calculated the total methylmercury concentration in the seawater as $0.06 \, \mu g \, L^{-1}$, that is 0.1%–0.3% of the total mercury concentration, as opposed to less than 5–32 ng L^{-1} methylmercury found in Minamata Bay, Japan. The bioaccumulation factor $\mu g \, kg^{-1} / \mu g \, L^{-1}$ of methylmercury in mussels ranged from 17 (1-day exposure) to 1333 (150-day exposure).

A potentially valuable consequence of this type of bioassay is that estimates of the relative abundance of methylmercury may be obtained at different sites by the exposure of 'standardised' mussels, as used in their experiment, in cages for controlled periods of time, and by the comparison of the resultant accumulations of methylmercury.

REFERENCES

1. G. Manice, V.M. Brown, J. Gardiner and J. Yates, Proposed environmental quality standards for List II substances in water, chromium, Technical Report TR207, Water Research Centre, Marlow, Bucks, UK, 1984.
2. D.J.H. Phillips, *Quantitative Aquatic Biological Indications*, Applied Science Publishers, London, 1980.
3. A. Bohn, *Marine Pollution Bulletin*, 1975, 6, 87.
4. M. Leatherland and J.D. Burton, *Marine Biological Association*, 1974, 54, 457.
5. F. Prosi, *Schmermetallbleastung in den Sedimenten der Elsenz und ihre Austrilkung auf Unische Organismen*, Dissertation, University of Heidelberg, 1977.
6. J.J. Davis, R.W. Perkins, R.F. Palmer, W.C. Hanson and J.F. Clive, Radioactive materials in aquatic and terrestrial organisms exposed to reactor effluent water, in 2nd UN International Conference on the Peaceful Uses of Atomic Energy, 1968.
7. Y.K. Chau, P.T.S. Wong, G.A. Bengert and V. Kramer, *Analytical Chemistry*, 1979, 51, 186.
8. F. Van Hoof and M. Van Son, *Chemosphere*, 1981, 10, 1127.
9. A.C. Anil and A.B. Wagh, *Marine Pollution Bulletin*, 1988, 19, 177.

10. W.J. Langston and M. Zhou, *Marine Environmental Research*, 1987, 21, 225.
11. H.W. Higgins and D.J. Mackey, *Australian Journal of Marine and Freshwater Research*, 1987, 38, 307.
12. D.U. Palawski and C.O. Knowles, *Environmental Toxicology and Chemistry*, 1986, 5, 1039.
13. J. Mallett and N.G. Barron, *Archives of Environmental Contamination and Toxicology*, 1988, 17, 73.
14. R.S. Tjeerdema and D.G. Crosby, *Aquatic Toxicology*, 1987, 9, 305.
15. Y. Takimoto, M. Okshima and J. Miyamoto, *Ecotoxicology and Environmental Safety*, 1987, 13, 126.
16. C.F. Baydon, *Journal of Marine Biological Association*, 1977, 57, 675.
17. L.M. Davies, W.C. Graham and S.M. Pirie, *Marine Chemistry*, 1979, 7, 11.

2 Determination of Metals in Fish

2.1 ARSENIC

The arsine generation spectrophotometric procedure [1] for the determination of arsenic has been applied to the determination of arsenic in fish. The sample is first digested with a mixture of nitric, sulphuric and perchloric acids. Spiked samples of crayfish gave a 98%–100% recovery of arsenic by this procedure.

Agemian and Thomson [2] have described a semi-automated atomic absorption spectrometric method for the determination of arsenic in wet homogenised fish tissue. A combination of nitric, perchloric and sulphuric acids is used to dissolve high-fat fish tissues at 140°C–180°C in a glass tube. Extracts are then analysed by reduction to arsine with sodium borohydride, followed by atomic absorption spectrophotometry with quartz tube atomisation. Average recoveries of arsenic(III), arsenic(V), *p*-arsalinic acid, benzene arsenic acid, methyl arsenic and triphenylarsine oxide through this procedure were between 90% and 102%. Arsenic found in a National Bureau of Standards (NBS) standard bovine liver (Standard Reference Material [SRM] 1577) reference sample was 500 ± 5 mg kg^{-1} against a certified value of 550 ± 5 mg kg^{-1}. Arsenic levels found in fish samples ranged from 0.26 to 0.44 mg kg^{-1}, determined with a coefficient of variation of 7%–15%.

Brooke and Evans [3] described two methods for the digestion of fish samples prior to determination of arsenic down to 0.02 mg kg^{-1} by hydride generation atomic absorption spectrometry.

The first method involved separation of the inorganic arsenic by distilling it from 6.6 N hydrochloric acid. The second method involved chelation and extraction of inorganic arsenic after sample dissolution in sodium hydroxide solution, with subsequent back extraction and oxidation. In both methods, the arsenic concentration is measured, after hydride generation, by atomic absorption spectrophotometry with atomisation in a flame-heated silica tube: in the first method, the solution contains arsenic(III), and in the second, the solution contains arsenic(V). Results obtained by both methods are in agreement over a range of samples. The distillation method is favoured for reasons of efficiency and economy in time.

Recoveries of inorganic arsenic in spiked fish samples were, respectively, 80%–94% and 75%–88% for arsenic(III) and arsenic(V). For a sample of herring containing 7.5 mg kg^{-1} arsenic, the inorganic arsenic content obtained by hydrochloric acid digestion was 0.03 mg kg^{-1} and the arsenic content obtained by sodium hydroxide digestion was 0.04 mg kg^{-1}.

Total arsenic recovery, that is, inorganic plus organic content, was 96.14%. Thus, a high proportion of the arsenic in herring was in organic form.

The semi-automated inductively coupled argon plasma–atomic emission spectrometric technique [4] has been applied to the determination of arsenic in nitric–perchloric–sulphuric acid digests of fish.

Beauchemin et al. [5] identified and determined the arsenic species present in dogfish muscle reference material (DORM-1). The arsenic species present were identified using electron impact mass spectrometry, thin layer chromatography and high-performance liquid chromatography–inductively coupled plasma spectrometry. Determination was done by the latter technique and graphite furnace atomic absorption spectrometry. Arsenobetaine was the major arsenic species in the methanol–water fraction (84% of the total arsenic).

Arsenic(III) and arsenic(V), monomethyl arsenic acid, dimethyl arsenic acid and arsenocholine constituted 4.5% of the total arsenic.

The total arsenic content was 18.7 mg kg^{-1}. The detection limit obtained in this method was 0.3 ng arsenic.

Paul [6] used radiochemical neutron activation analysis with retention on manganese dioxide to determine arsenic in biological materials. Although this method provided very high and reproducible yields and detection limits at low microgram per kilogram levels, counting geometry uncertainties may arise from unusual distribution of arsenic in the radiochemical neutron activation analysis, and arsenic detection limits may not be optimal significant retentions of other elements.

Paul [6] described an alternate radiochemical neutron activation analysis procedure for the determination of arsenic after digestion of samples in nitric acid and perchloric acid. As(III) is extracted from 2 M sulphuric acid solution into a solution of zinc diethyldithiocarbamate in chloroform. Counting of ^{76}As allows quantitation of arsenic.

Addition of arsenic as trace solution prior to dissolution allows correction for chemical yield and counting geometries, further improving reproducibility. The hydrated manganese dioxide and solvent extraction procedures for arsenic were compared through analysis of SRMs (bovine liver), 1547 (peach leaves) and 1575 (pine needles). Both gave arsenic results in agreement with certified values with comparable reproducibility. However, the solvent extraction method yields a factor of three improvement in detection limits and is less time consuming than the hydrated manganese dioxide method.

2.2 CADMIUM

Blood and Grant [7] determined cadmium in fish tissue using flameless atomic absorption spectrometry at 228.7 nm using a tantalum ribbon.

Fish samples were digested with mixtures of nitric, sulphuric and perchloric acids for 2 h at 74°C and the digest diluted with concentrated nitric acid for 15 min at 80°C–90°C, followed by addition of 1 ml of 10% hydrogen peroxide and heating for an additional 15 min.

Mean recoveries obtained on the NBS SRM 1577 reference bovine liver sample with an authenticated cadmium content of 0.27 ± 0.04 mg kg^{-1} were 85% and 90%. The higher mean levels (13.1–5.6 mg kg^{-1}) of cadmium in wet blue gill tissue were found in kidney, gut, heart, gill and liver, and the lowest levels (0.14–1.7 mg kg^{-1}) in muscle, skin and bone.

In a series of papers, Sperling [8–10] studied the application of flameless graphite furnace atomic absorption spectrometry to the determination of cadmium in complex matrices resulting from the digestion of fish and other biological materials. Organic material in the sample is destroyed before atomisation by digestion with ammonium peroxydisulphate, thereby avoiding loss of volatile cadmium which would occur in ignition methods at temperatures exceeding 420°C [8]. Cadmium was then extracted from the digest with a saturated solution of ammonium pyrrolidone dithiocarbamate in carbon tetrachloride [9,10] and the cadmium in the lower layer determined by flameless graphite furnace atomic absorption spectrometry.

Poldoski [11] used a molybdenum- and lanthanum-treated pyrolitically coated graphite tube for the graphite furnace atomic absorption spectrometric determination of cadmium at 228.8 nm in nitric acid and perchloric acid digests of fish tissue. Molybdenum and lanthanum help reduce chemical interferences and interference from uncompensated background signals during analyte atomisation.

Digestions were carried out on 0.6 g of dry fish using 10 ml of concentrated nitric acid and 2 ml perchloric acid. After digestion was complete, the residue was dissolved in 10 ml of 0.2% w/v nitric acid and stored in Nalgene bottles.

Cadmium spiking recovery experiments were carried out on fish tissue samples and on an authenticated reference sample (NBS SRM 1577 bovine liver) under specified conditions of analysis. The determination cadmium content on NBS SRM reference fish sample (0.31 ± 0.05 mg kg^{-1}) is in good agreement with the nominal value (0.27 ± 0.04 mg kg^{-1}). In addition, 0.038 mg kg^{-1} of cadmium was found in whole catfish and blue gill fish, and average analytical recoveries on the fish samples were 91%–107%. Down to 0.2 pg of cadmium in the injected portion of sample could be determined. Cadmium contents determined by this technique were in good agreement with those obtained by anodic scanning voltammetry.

2.3   COBALT

Kuriyama and Kuroda [12] applied their combined ion exchange spectrophotometric procedure to the simultaneous determination of cobalt and vanadium in cutlass fish. In this procedure, the sample is dry ashed at 420°C, and the ash (ca. 0.5 g) is decomposed with a mixture of perchloric, nitric and hydrofluoric acids, and is finally taken up in hydrochloric acid. The metals are absorbed by anion exchange on an Amberlite CG 400 (SCN–) column from a dilute ammonium thiocyanate–hydrochloric acid solution. The adsorbed vanadium and cobalt are separated chromatographically by elution with 12 mol L^{-1} hydrochloric acid and 2 mol L^{-1} perchloric acid, respectively. Both fractions of vanadium and cobalt are subsequently purified by anion exchange from 0.1 mol L^{-1} hydrochloric acid – 3 vol% hydrogen peroxide for vanadium and 6 mol L$^-$ hydrochloric acid for cobalt. Vanadium and cobalt in the effluents are determined spectrophotometrically with 4-(2-pyridylazo) resorcinol. A 98.2% recovery of cobalt was obtained by this procedure in the presence of appreciable excesses of elements also likely to occur in the sample, namely magnesium, calcium, aluminium, iron, copper, nickel and zinc.

2.4   COPPER

Spark source mass spectrometry, employing copper 65 and copper 63 stable isotopes, has been used to study the uptake of copper by Harvey [13]. An attractive feature of this method is that both the natural copper content of the fish organs and the concentrations of added tracers are determined on the same sample, making two measurements of isotopic ratio – one before and one after the addition of the standard ^{63}Cu spike. Both of these isotopes constitute ideal tracers, being readily available at low cost and being free from radiation hazards. A spark source mass spectrometer is an ideal way of carrying out isotopic ratio measurements.

In this procedure, the wet fish sample was weighed before and after vacuum freeze drying, and then transferred to a low-temperature asher to remove organic matter. The residue was then digested in 0.5 mol L^{-1} hydrochloric acid–30% hydrogen peroxide, ascorbic acid being added to destroy residual hydrogen peroxide, and copper is extracted from the solution with a carbon tetrachloride solution of dithizone. This extract was then evaporated directly onto graphite prior to spark source mass spectrometric evaluation.

The range of copper content found in various organs, such as heart, liver, skin and muscle, ranged between 0.22 and 3 µg kg^{-1}.

2.5   LEAD

The molybdenum- and lanthanum-treated pyrolitically coated graphite furnace atomic absorption spectrometric method described by Poldoski [14] under Cadmium earlier in this section has also been applied to the determination of lead in fish tissue, lead results were obtained from spiking recovery experiments, carried out in fish tissue samples and on an authenticated reference sample (NBS SRM 1577 bovine liver) under specified conditions of analysis. It is seen that the determined lead content on NBS SRM 1577 (0.33 ± 0.01 mg kg^{-1}) is in good agreement with the nominal value (0.34 ± 0.08 mg kg^{-1}). Average analytical recoveries on the fish samples are 91%–93%. Down to 4 pg lead in the injected portion of sample can be determined. Lead contents obtained by this procedure are in good agreement with those obtained by anodic scanning voltammetry. Using this method, 0.26 mg kg^{-1} of lead was found in each of the whole catfish and blue gill samples.

Pagenkopf et al. [15] also employed graphite furnace atomic absorption spectrometry to determine lead in fish. They were able to determine down to 0.15 µg kg^{-1} lead in fish. In this procedure, 1–5 g of fish muscle tissue was removed by dissection and freeze-dried by a ThermoVac lypholiser. Approximately 1 g of the dried tissue was weighed and then digested in a mixture of 7.0 ml of concentrated nitric acid and 5.0 ml of concentrated perchloric acid. The solutions were slowly heated until all foaming had stopped and dissolution was achieved. At this point, the temperature was increased so as to reduce the volume to about 1 ml. This was accompanied by copious fuming of perchloric acid. The maximum temperature was 88°C. The colourless samples were then transferred to 100 ml volumetric flasks and diluted to volume.

To overcome problems of contamination and non-reproducibility in the determination of low levels of lead in fish, Harms [16] devised a method of sample pretreatment and enrichment, in which sample decomposition was performed in a closed system based on Mattinson's two-bottle system, and followed by the addition of purified nitric acid, followed by neutralisation with ammonia and extraction with dithizone/toluene solution. After back extraction into aqueous hydrochloric acid, the aqueous phase was subjected to measurement of Pb-203 activity (recovery control), followed by electrothermal atomic absorption spectrometry for determination of stable lead. Samples of fish muscle containing less than 0.5 µg kg^{-1} could be analysed by this procedure.

May and Brumbaugh [17] used an ammonium dihydrogen phosphate matrix modifier and a modified L'vov platform to overcome matrix interference effects in the determination of lead in fish tissue. The 283.3 deadline was used. They defined required graphite furnace atomic absorption spectrometric conditions to obtain maximum improvement in the slope ratio. Precisions were between 0.8% and 1.7% for fish samples.

Fish sample digestions were performed in polytetrafluoroethylene (PTFE)-capped glass pressure reaction vessels, in which the sample was digested with concentrated nitric acid at 70°C for 48 h. The digestates were then made up to 50 ml with 1% hydrochloric acid. This procedure did not yield complete digestion, as lipids are not destroyed and remain as a floating white solid in the digestate.

2.6 MERCURY

Various workers have discussed the application of cold vapour atomic absorption spectrometry for the determination of mercury in fish [18–30]. Various digestion procedures have been used, including concentrated nitric acid in a Teflon-lined sealed bomb [18] or glass tube [21], mixtures of nitric acid and sulphuric acid [27,28], 50% hydrogen peroxide [25,29] and sulphuric acid–potassium permanganate [24] in open tubes. Wickbold combustion procedures have also been used [30].

Hendzel and Jamieson [28] digested 0.1–0.5 g of fish sample with 5 ml of 1:1 v/v nitric acid–sulphuric acid in a glass digestion tube at 180°C until white fumes appeared. After reduction with a reagent comprising hydroxylamine sulphate and stannic chloride, the elemental mercury was swept off with a stream of air and estimated at 253.7 nm by cold vapour atomic absorption spectrometry.

Louie [27] used concentrated hydrochloric acid–nitric acid–sulphuric acid in open tube digestions, followed by a cold vapour atomic absorption spectrometry, to determine down to 1.01 mg kg^{-1} mercury in fish tissue. He claimed that 3 g of fish was completely digested at 85°C–100°C within 30 min. Using this procedure, Louie [27] obtained a mercury content on an NBS Albacore Tuna Research Material 50 Reference Sample of 0.94 ± 0.05 mg kg^{-1} against a certified value of 0.95 ± 0.01 mg kg^{-1}. Levels found in various fish samples ranged between 0.1 and 0.4 mg kg^{-1}.

Davidson [25] used digestion on a hot plate with 4:1 50% sulphuric acid–hydrogen peroxide to digest tissue prior to the determination of mercury at 253.7 nm by cold vapour atomic absorption spectrometry.

Nine replicate samples of NBS reference tuna (Research Material 50) were analysed by the 50% hydrogen peroxide method to determine the repeatability of the method. The mean and standard deviations were 1.00 and $0.02\,\mu g^{-1}$ dry mass, respectively, against the reported value of the mean standard deviation of $0.95 \pm 0.1\,\mu g^{-1}$ and indicated that 80%–90% of the mercury content is present as methylmercury. Between 0.24 and $1.11\,mg\,kg^{-1}$ of mercury was found in pike and lake trout samples by this method.

Konishi and Takahashi [29] have described a method for the determination of inorganic mercury in fish in the presence of organic mercury. This is based on the fact that hydrogen peroxide oxidatively liberates inorganic mercury from organic substances in strong alkali, and reduces it to the metallic state without decomposing organic materials concomitantly present. The metallic mercury, vaporised with a nitrogen stream, is trapped by gold amalgamation, and then released for electrothermal atomisation atomic absorption spectrometry. The detection limit is 1 ng of inorganic mercury, and the coefficient of variation for 40 ng of inorganic mercury is 2.8%. A 92% recovery of mercury was obtained in this procedure.

Jones and Nickless [31] converted inorganic mercury in fish samples to its methyl derivative using 2,2′-dimethyl-2-silapentane-5-sulphonate as a reagent, prior to the determination of inorganic mercury in benzene extracts of the reaction product by gas chromatography. The highest yield was obtained by digesting the fish sample at 100°C with 5 N nitric acid in the presence of sodium nitrite and then extracting with benzene. Between 2.8 and $8.6\,mg\,kg^{-1}$ mercury was found in fish samples by this method.

Thomas et al. [32] described a rapid pyrolytic procedure to determine the total mercury content in fish. A weighed amount of homogenised fish tissue is combusted in a flowing air stream at 900°C, and then over copper oxide at 850°C to ensure complete combustion. Elemental mercury vapour is expelled into the carrier stream and, after passing through silver oxide absorbent traps to remove possible interfering gases, is detected and measured in an ultraviolet photometer at 253.5 nm. Relative error is approximately ±10% for inorganic and organic mercury over a linear response range of $0.05-3.0\,mg\,kg^{-1}$.

Thomas et al. [32] compared mercury contents obtained by this method with those based on the gas chromatographic method involving the conversion of inorganic mercury to methylmercury, and with determination of total mercury by a sulphuric acid–potassium permanganate acid digestion method.

The total mercury determinations are lower; the pyrolysis method and the gas chromatographic method give results that are in reasonably good agreement.

Nitric acid–perchloric acid digestion in a Teflon autoclave bomb has been used to prepare digests of finely powdered freeze-dried fish [33]. The extract was ultraviolet light irradiated to complete fish sample degradation prior to determination of mercury by using a gold disc electrode. Results obtained compared well with total mercury contents obtained by neutron activation analysis.

Uthe et al. [19] found that mercury determinations in fish by digestion–flameless atomic absorption spectrometry were only slightly lower than those obtained by neutron activation analysis, but had a poorer precision.

Svasankara-Pillay et al. [34] determined mercury fish samples by neutron activation analysis. As a further check, the samples were wet ashed at 120°C–160°C with sulphuric and perchloric acids in the presence of an accurately known amount of mercury carrier. A preliminary precipitation as mercury sulphide is followed by further purification, and electrodeposition or precipitation as mercuric oxide to isolate mercury. The radioactivities due to mercury-196 and mercury-197 are then measured by scintillation. The errors in this method are 5% at the 2 mg kg^{-1} mercury level and 15% at the 0.01 mg kg^{-1} level. Standard deviations are, respectively, less than 5% at the 5 mg kg^{-1} level and less than 17% at the 0.01 mg kg^{-1} level. Fish samples contain both organic and inorganic mercury, predominantly organic. Svasankara-Pillay et al. [34] showed that freeze drying of homogenised fish samples caused a 16%–39% loss of organic mercury compounds, but did not cause any loss of inorganic mercury. Similarly, low-temperature ashing caused an 81%–98% loss of mercury from fish. Exposure of fish samples to X-rays or neutrons before mercury analysis, to convert volatile organomercury compounds to inorganic mercury, reduced mercury losses to 4.5%–16.4%, but did not eliminate them. Low-temperature (60°C) oven drying caused up to 72% losses of volatile mercury from fish. As a consequence of these findings, Svasankara-Pillay et al. [34] decided that it was good practice, before analysis, not to preprocess fish samples to limit their bulk or reduce their water content, and not to store samples in containers that absorb mercury onto their surfaces. The procedure they adopted was to keep the samples frozen until use. They were then homogenised using a blender or a grinder made of stainless steel or borosilicate glass. The portion of sample for neutron activation analysis was then vacuum sealed in a polyethylene bag.

Lo et al. [35] digested wet fish with concentrated sulphuric–nitric acids until white fumes appeared, and then added excess potassium permanganate, sodium chloride and hydroxylamine hydrochloride to reduce mercury. Mercury in the digest was then preconcentrated into a small volume of lead diethyldithiocarbamate dissolved in chloroform. The chloroform was then allowed to evaporate in an ampoule, and then the ampoule was sealed for neutron activation analysis and subsequent gamma spectrometry of the sensitive mercury-197 peak. As well as reducing the detection limit of 1 μg kg^{-1} fish, the preconcentration has the additional advantage of overcoming interferences by sodium-24 and bromine-82, which commonly occur in fish samples. Recoveries of 95% of mercury in fish samples were obtained by this procedure.

This method was applied to the determination of total mercury to a range of different fish taken in Lake Erie. Values between 0.21 and 0.79 mg kg^{-1} were obtained. Determinations of total mercury obtained by neutron activation analysis were, in general, higher than those obtained by other methods, such as atomic absorption spectrometry and pyrolysis.

2.7 SELENIUM

The semi-automated atomic absorption method [36] for the determination of arsenic in fish has been applied to the determination of selenium. Average recoveries of selenium(IV), selenium(VI), selenourea, selenomethonine and selenocysteine by this method were between 91 ± 10% and 100 ± 1%.

Selenium found in NBS standard bovine liver (SRM 1577) reference sample was 1100 ± 100 mg kg^{-1} against a certified value of 1020 ± 10 mg kg^{-1}. Selenium levels found in a range of fish samples ranged from 0.308 to 0.548 mg kg^{-1}, determined with a coefficient of variation of 4.5%–6.0%.

The semi-automated inductively coupled plasma–atomic emission spectrometric technique described by Goulden et al. [4] has also been applied to the determination of selenium in nitric–sulphuric–perchloric acid digests of fish.

2.8 TIN

The flameless atomic absorption spectrometric method [37] has also been applied to the determination of tin in fish. Between 0.4 and 6.6 mg of tin was reported in homogenised fish samples.

2.9 VANADIUM

The combined ion exchange spectrophotometric procedure [85], described in Section 2.3 for cobalt, has been applied to the determination of vanadium in cutlass fish. A recovery of 9.63% vanadium was obtained by this procedure.

Cation exchange chromatography followed by neutron activation analysis has been used [86] to determine down to 30 µg kg^{-1} vanadium in fish.

2.10 MULTIELEMENTS

2.10.1 ATOMIC ABSORPTION SPECTROMETRY

Various workers have discussed the application of this technique to the determination of elements in fish tissue digests [38–43]. Elements determined include cadmium, lead, copper, manganese, zinc, chromium and mercury [38]; cadmium, zinc, lead, copper, nickel, cobalt and silver [39]; copper, zinc, cadmium, nickel and lead [41]; lead, cadmium, copper and zinc [42]; and lead and cadmium [43].

Various digestion systems have been studied for the decomposition of fish samples prior to analysis, including digestion with nitric acid–perchloric acid [39,40] and nitric acid–hydrogen peroxide [40–42], all in open tubes, or decomposition with nitric acid in a closed Teflon-lined bomb [38].

2.10.2 NITRIC ACID–SULPHURIC ACID DIGESTION

Agemian et al. [41] have reported a simple and rapid digestion method for the simultaneous acid extraction of chromium, copper, zinc, cadmium, nickel and lead from high-fat fish tissue. Samples are digested with nitric (5 ml of 16 N) and sulphuric (5 ml of 36 N) acids at 150°C in a modified aluminium hot block. After digestion, acid extracts of the sample are analysed by direct-flame atomic absorption spectrometry for copper, zinc and chromium. The other three elements, cadmium, nickel and lead, are concentrated by chelation with ammonium tetramethylene dithiocarbamate, followed by solvent extraction with isobutyl methylketone, and determined by flame atomic absorption spectrometry.

Detection limits in whole fish tissue are 0.02 mg kg^{-1} (cadmium), 0.05 mg kg^{-1} (nickel), 0.1 mg kg^{-1} (lead) and 0.2 mg kg^{-1} (chromium, copper and zinc). Recoveries through the whole analytical procedure ranged from 90% to 110%. Precisions were in the range of 9.1%–12.1% (cadmium), 5%–15% (nickel and copper), 4.3%–17.0% (lead), 3.9%–6.7% (zinc) and 7.9%–15% (chromium).

2.10.3 NITRIC ACID–PERCHLORIC ACID DIGESTIONS [39]

To carry out this digestion, 0.5–3 g of freeze-dried fish ball mill ground sample is digested in a silica flask with 10–20 ml of concentrated nitric acid, and then 5–10 ml of 1:1 nitric-perchloric acid to dryness. The residue is dissolved in dilute hydrochloric acid–nitric acid and adjusted to pH 8 with ammonia. This solution is extracted with a 0.02% solution of dithizone in chloroform. Metals are then back extracted from the organic phase with 2 mol L^{-1} hydrochloric acid prior to atomic absorption spectrometry. Using this method, the following values (mg kg^{-1}) were obtained for an NBS reference kale sample (nominal values in parentheses): cadmium, 0.9 (0.84); zinc, 29.9 (31.8); lead, 2.6 (3.2); copper, 4.2 (4.9); and cobalt, 0.05 (0.056). Concentrations (mg kg^{-1}) of metals found in whale tissues were as follows: zinc, 26–103; lead, 0.45–1.37; copper, 1.2–7.6; nickel, 0.17–0.60; cobalt, 0.07–0.38; silver, 0.02–0.04; and cadmium, not detected. Kale brought from Iceland contained the following concentrations (mg kg^{-1}): zinc, 39; lead, 0.89; copper, 2.6; nickel, 0.34; cobalt, 0.14; silver, 0.04; and cadmium, not detected.

2.10.4 NITRIC ACID–HYDROGEN PEROXIDE DIGESTIONS

Van Hoof and Van Son [40] worked on fish samples that had been calcined at 450°C prior to digesting the ash in 2.5 ml of 14 N nitric acid–30% hydrogen peroxide. Elements determined included copper, zinc, cadmium and chromium. Low recoveries of at least some of these elements would be expected under these conditions.

Borg et al. [42] digested 10 mg of freeze-dried fish livers with concentrated nitric acid at 50°C for 2 h in quartz tubes, and then slowly raised the temperature to 110°C over 18 h. Hydrogen peroxide (30%) was added to the cooled samples, which were again heated to 110°C for 6 h to digest fats completely. When made up to a standard volume, this digest was used for the determination of copper, lead, cadmium and zinc by graphite furnace atomic absorption spectrometry.

2.10.5 NITRIC ACID BOMB DIGESTION

Ramelow et al. [38] determined cadmium, lead copper, manganese, zinc and chromium in wet fish by digestion of a 0.5–1.0 g sample with 2–3 ml of concentrated nitric acid in a Teflon-lined bomb at 150°C for 1.5 h. Elements were determined in the digest by flame atomisation or graphite furnace atomisation atomic absorption spectrometry.

A typical example of whole white bream gave values for cadmium, lead, copper, manganese, zinc and chromium of 0.04, 0.61, 1.11, 0.51, 10.6 and 0.58 mg kg^{-1}.

Ramelow et al. [38] digested fish samples with concentrated nitric acid in a Teflon-lined bomb for 1.5 h at 150°C prior to the determination of mercury by reduction to elemental mercury with stannous chloride and determination by cold vapour atomic absorption spectrometry.

2.10.6   HYDRIDE GENERATION ATOMIC ABSORPTION SPECTROMETRY

Welz and Melcher [44] decomposed fish tissue with nitric–sulphuric–perchloric acids in a Teflon-lined bomb to decompose arsenic, selenium and mercury. Nitric acid alone gave low recoveries for arsenic and selenium but quantitative recovery for mercury. Finally, determination of down to 0.3 mg kg^{-1} arsenic, 0.2 mg kg^{-1} selenium and 0.005 mg kg^{-1} mercury was carried out by hydride generation and cold vapour atomic absorption spectrometry.

2.10.7   INDUCTIVELY COUPLED PLASMA–ATOMIC EMISSION SPECTROMETRY

Sakai and May [45] used inductively coupled plasma–atomic emission spectrometry, atomic absorption spectrometry and hydride generation atomic absorption spectrometry to determine cadmium, arsenic, boron, chromium, mercury, molybdenum, nickel, lead and selenium in common carp. The highest concentrations found were arsenic, 1.5 mg kg^{-1}; boron, 20 mg kg^{-1}; cadmium, 0.27 mg kg^{-1}; chromium, 2.2 mg kg^{-1}; mercury, 2.9 mg kg^{-1}; molybdenum, 3.6 mg kg^{-1}; nickel, 2.2 mg kg^{-1}; lead, 2.3 mg kg^{-1}; and selenium, 5.5 mg kg^{-1}.

Low et al. [46] evaluated microwave-assisted digestion conditions for the determination of various tilapia fish samples by inductively coupled plasma–mass spectrometry.

A Plackett–Burman design was carried out as a multivariate strategy to investigate the main effects of the following parameters on microwave-assisted nitric acid digestion: microwave irradiation time, ramp time, digestion temperature, microwave power limit and the addition of hydrogen peroxide or hydrochloric acid. The most significant microwave setting parameters (radiation time, ramp and temperature) were further evaluated by the response surface methodology under the Box–Behnken design, while others were kept constant. The influences of different parameters vary according to metal element; thus, the working conditions were established as a compromise within an optimum region found for each targeted element, which ensures quantitative recoveries and time efficiency. The compromised conditions are as follows: ramp to 185°C in 10.5 min, and then hold for 14.5 min with 1600 W (50%) of microwave power, using a reagent mixture composed of 2.5 ml of nitric acid, 0.5 ml of hydrochloric acid and 7.0 ml of water. Good agreements were demonstrated between measured and certified values.

Ababneh and Al-Mohammed [47] have used inductively coupled plasma–optical emission spectrometry to determine mercury, cadmium, lead, arsenic, nickel vanadium, aluminium, barium and silver in canned tuna.

In samples of canned tuna fish, the mean concentrations (mg kg^{-1}) found were as follows: total Hg (0.06–0.57), Cd (0.01–0.63), lead (0.04–0.24) and Ag (0.02). The data in this study compared well with data obtained from similar studies carried out in different parts of the world. Few samples had the mercury and cadmium levels slightly exceeding the Codex Committee on Food Additives and Contaminants draft guidelines. However, the estimated weekly intakes of these metals showed that there was no health risk associated with the consumption of the analysed canned tuna samples.

2.10.8 DIFFERENTIAL PULSE ANODIC STRIPPING VOLTAMMETRY

Adeljou et al. [48] used this technique to determine selenium, copper, lead and cadmium in fish tissues. Detection limits were in the microgram per kilogram range. Samples were first digested with concentrated nitric acid and 80% magnesium nitrate solution, and then dry ashed at 500°C. The ash was dissolved in boiling 6 mol L^{-1} hydrochloric acid. This solution was analysed for selenium on a hanging mercury drop polarographic analyser, and copper, lead and cadmium were determined in the anodic scanning voltammetry mode using the peaks appearing at −0.20, −0.5 and −0.7 V versus standard electrode (SCE), respectively. Results (mg kg^{-1}) obtained by this method for crayfish are in good agreement with certified values (reference values in parentheses): selenium, 0.17 (0.16); copper, 3.46 (3.10); lead, 0.48 (0.48); and cadmium, 0.10 (0.05). Relative standard deviations in determinations of selenium, copper, lead and cadmium were 12%, 5%, 15% and 20%, respectively.

2.10.9 NEUTRON ACTIVATION ANALYSIS

This technique has been applied to the determination in fish of cobalt, chromium, selenium, silver, rubidium, nickel and zinc [49], and aluminium, gold, bromine, calcium, chlorine, cobalt, chromium, copper, iron, iodine, potassium, magnesium, manganese, sodium, rubidium, scandium, vanadium and tungsten [50].

2.10.10 SECONDARY ION MASS SPECTROMETRY AND X-RAY SPECTROMETRY

This technique has been used [51] to provide simultaneous morphological and chemical identifications in historical sections of fish, molluscs and crustaceans.

The concentrations of metals found in whole fish and in fish organs from a variety of sources are summarised in Table 2.1. It can be observed that a wide range of concentrations occur in fish or in organs and that, in general, the highest concentrations of metals are found in fish organs rather than whole fish tissue, and this is particularly so for cadmium and lead.

A wide variety of analytical methods have been employed for the determination of metals and non-metals in fish (Table 2.2). The major techniques employed (reviewed in Table 2.2) include electrothermal atomic absorption spectrometry and its cold vapour and hydride generation variants, inductively coupled plasma–atomic emission spectrometry and neutron activation analysis. These methods have adequate sensitivity for environmental testing, that is, below 1 mg kg^{-1}, and in some cases below 0.1 mg kg^{-1}. Other techniques employed for particular elements or groups of elements include gas chromatography (mercury), high-performance liquid chromatography (arsenic) and anodic scanning voltammetry (mercury, selenium, copper, lead and cadmium).

TABLE 2.1
Metal Content of Whole Fish and Fish Organs

Reported Values (mg kg^{-1}) (dry weight)

Element		Whole Fish Tissue			Fish Organs			Compounds
		Minimum Reported Value	Fish Type	Maximum Reported Value	Fish Type	Minimum Reported Value	Organ	Maximum Reported Value	Organ	Maximum Value in Whole Creature
As	Inorganic	0.02	Herring and haddock	0.44	Smel					
	Total	1.1	Herring	2.9	Tuna					
Cd		0.02	Sardine	0.17	Horse mackerel	0.038	Gill	9.5	Opercle	64
								10.9	Skin	
								9.0	Liver	
								7.1	Kidney	
Cr		0.10	Grey mullet	2.2	Rainbow trout	0.8	Muscle	23.7	Kidney	12.6
								26.0	Opercle	
								18.4	Liver	
Cu		0.39	Flathead	3.46	Crayfish	0.6	Gill	48	Perch liver	13.9–17.9
		0.53	Rainbow trout	2.18	Sardine			62	White fish liver	

(Continued)

TABLE 2.1 (*Continued*)
Metal Content of Whole Fish and Fish Organs

Reported Values (mg kg⁻¹) (dry weight)

| Element | Whole Fish Tissue | | | Fish Organs | | | | Compounds |
	Minimum Reported Value	Fish Type	Maximum Reported Value	Fish Type	Minimum Reported Value	Organ	Maximum Reported Value	Organ	Maximum Value in Whole Creature
Pb	0.12	Striped mullet	1.36	Grey mullet	0.12	Muscle	36	Kidney	26.5
Ni	0.15	Rainbow trout	0.2	Rainbow trout	0.34	Trout	1.9	Kidney	9.5
Se	0.19	Shark	0.55	Coho salmon					
Ag	0.02	Whale meat	0.04	Trout					
	0.04	Trout							
Zn	6.3	Sardine	39	Trout	12.6	Liver	150	Liver	2.5–3.8
		Striped mullet					120	Opercle	
							57	Kidney	

TABLE 2.2

Methods for the Determination of Metals and Non-Metals in Fish

Determined	Technique	Detection Limit (mg kg^{-1} Unless Otherwise Stated)	Reference
Arsenic	Hydride (AsH$_3$) generation atomic absorption spectrometry	0.3	[52,53]
		0.002	[54,55]
	High-performance liquid chromatography with inductively coupled plasma–atomic emission spectrometric detection	0.3 ng of As absolute	[56]
Cadmium	Flameless graphite furnace atomic absorption spectrometry	0.2 pg in portion (0.6 pg of fish per 10 extracts)	[57–60]
	Atomic absorption spectrometry	—	[61,62]
Copper	Spark source mass spectrometry, atomic absorption spectrometry	—	[62,63]
Lead	Atomic absorption spectrometry	—	[61,62]
	Graphite furnace atomic absorption spectrometry	0.15	[64,65]
Mercury	Cold vapour atomic absorption spectrometry	0.005	[52]
		0.01	[66–69]
	Conversion to methylmercury–gas chromatography	—	[70]
	Derivativisation with 2,2^1-dimethyl-2-silapentane-5-sulphonate–gas chromatography	—	[71]
Mercury	Anodic scanning voltammetry using gold disk electrode	—	[72]
	Neutron activation analysis	—	[73,74]
Selenium	Hydride generation atomic absorption spectrometry	0.2	[52]
	Inductively coupled plasma–atomic emission spectrometry	—	[75]
Tin	Flameless atomic absorption spectrometry	—	[76]
Vanadium	Cation exchange chromatography neutron activation analysis	0.03	[77]
Zinc	Atomic absorption spectrometry	—	[62]
Co, V, Pb	Ion exchange spectrophotometric method	0.4 pg metal (injected portion)	[78]
Cr, Cu, Zn, Cd, Ni, Pb	Atomic absorption spectrometry	0.02 (Cd)	[79]

(Continued)

TABLE 2.2 (*Continued*)
Methods for the Determination of Metals and Non-Metals in Fish

Determined	Technique	Detection Limit (mg kg^{-1} Unless Otherwise Stated)	Reference
		0.05 (Ni)	
		0.1 (Pb)	
		0.1 (Cr, Cu, Zn)	
Cd, Zn, Pb, Cu, Ni, Co, Hg		—	[80]
Cd, Pb, Cu, Mn, Zn, Cr, Hg			[81]
Cd, As, B, Cr, Hg, Mo, Ni, Pb, Se	Inductively coupled plasma–atomic emission spectrometry	—	[82]
Se, Cu, Pb, Cd	Differential pulse anodic stripping voltammetry	—	[83]
Co, Cr, Se, Ag, Rb, Ni, Zn	Neutron activation analysis	—	[84]
As	Hydride generation	0.3	[85]
Se	Atomic absorption spectrometry	0.2	
	Cold vapour atomic mercury	0.005	[86]

REFERENCES

1. W.A. Maher, *Analyst*, 1983, 108, 939.
2. H. Agemian and R. Thompson, *Analyst*, 1980, 105, 902.
3. P.J. Brooke and W.H. Evans, *Analyst*, 1981, 106, 574.
4. P.D. Goulden, D.H.J. Anthony and K.D. Austen, *Analytical Chemistry*, 1981, 53, 2027.
5. B. Beauchemin, M.E. Bednaz, S.J. Berman, J.W. McLaren, K.W. Siu and R.E. Sturgeon, *Analytical Chemistry*, 1988, 60, 2209.
6. R.L. Paul, *Analytical Chemistry*, 2011, 83, 152.
7. E.R. Blood and G.C. Grant, *Analytical Chemistry*, 1975, 47, 1438.
8. K.R. Sperling, *Fresenius Zeitschrift für Analytische Chemie*, 1977, 287, 23.
9. K.R. Sperling, *Fresenius Zeitschrift für Analytische Chemie*, 1980, 301, 294.
10. K.R. Sperling, *Fresenius Zeitschrift für Analytische Chemie*, 1982, 310, 254.
11. J.E. Poldoski, *Analytical Chemistry*, 1980, 52, 1147.
12. T. Kuriyama and R. Kuroda, *Analyst*, 1982, 107, 505.
13. B.R. Harvey, *Analytical Chemistry*, 1978, 50, 1866.
14. J.E. Poldoski, *Analytical Chemistry*, 1980, 52, 1147.
15. G.K. Pagenkopf, D.R. Neuman and R. Woodruff, *Analytical Chemistry*, 1972, 44, 2248.
16. S.U. Harms, *Fresenius Zeitschrift für Analytische Chemie*, 1985, 323, 53.
17. T.W. May and W.G. Brumbaugh, *Analytical Chemistry*, 1982, 54, 1032.
18. W. Holak, B. Krinitz and J.C. Williams, *Journal of the Association of Official Analytical Chemists*, 1972, 55, 741.

19. J.F. Uthe, F.A.J. Armstrong and K.C. Tam, *Journal of the Association of Official Analytical Chemists*, 1971, 54, 866.
20. J.G. Saha and Y.W. Lee, *Bulletin of Environmental Contamination and Toxicology*, 1972, 7, 301.
21. G. Cumont, *Chim Analyst*, 1971, 53, 634.
22. A. Fabbrini, G. Modi, L. Signorelli and G. Simiani, *Bollettino Laboratorie Chimico*, 1971, 22, 339.
23. V. Lidums, *Chemia Scipta*, 1972, 2, 159.
24. J.F. Uthe, F.A.J. Armstrong and M.P. Stainton, *Journal of the Fisheries Research Board, Canada,* 1970, 27, 805.
25. J.W. Davidson, *Analyst*, 1979, 104, 683.
26. S.L. Tong and W.K. Leow, *Analytical Chemistry*, 1980, 52, 581.
27. H.W. Louie, *Analyst*, 1983, 108, 1313.
28. M.R. Hendzel and D.M. Jamieson, *Analytical Chemistry*, 1976, 48, 926.
29. T. Konishi and H. Takahashi, *Analyst*, 1983, 108, 827.
30. E. Kunkel, *Fresenius Zeitschrift für Analytische Chemie*, 1972, 258, 337.
31. P. Jones and G. Nickless, *Journal of Chromatography*, 1974, 89, 201.
32. R.J. Thomas, R.A. Hagstrom and E.J. Kuchar, *Analytical Chemistry*, 1972, 44, 512.
33. I. Gustarsson and J. Giolimowski, *Science of the Total Environment*, 1981, 22, 85.
34. K.K. Svasankara-Pillay, C.C. Thomas, J.A. Sondel and C.M. Hyche, *Analytical Chemistry*, 1971, 43, 1419.
35. J.M. Lo, J.C. Wei, M.H. Yan and S.I. Yeh, *Journal of Radioanalytical Chemistry*, 1982, 72, 571.
36. H. Agemian and R. Thomson, *Analyst*, 1980, 105, 902.
37. S. Dogan and W. Haerdi, *International Journal of Environmental Analytical Chemistry*, 1980, 8, 249.
38. G. Ramelow, M.A. Ozkan, G. Tuncel, C. Saydam and T.I. Balkas, *International Journal of Environmental Analytical Chemistry*, 1978, 5, 125.
39. H. Armannsson, *Analytica Chimica Acta*, 1979, 110, 21.
40. F. Van Hoof and M. Van Son, *Chemosphere*, 1981, 10, 1127.
41. H. Agemian, D.P. Sturtevant and K.D. Austen, *Analyst*, 1980, 105, 125.
42. H. Borg, A. Edin, K. Holm and E. Skold, *Water Research*, 1981, 15, 1291.
43. K.R. Sperling, *Fresenius Zeitschrift für Analytische Chemie*, 1988, 332, 565.
44. B. Welz and M. Melcher, *Analytical Chemistry*, 1985, 57, 427.
45. M.K. Sakai and W. May, *Science of the Total Environment*, 1988, 74, 199.
46. K.W. Low, S.M. Zain and M.R. Abas, *International Journal of Environmental Analytical Chemistry*, 2012, 92, 1161.
47. F.A. Ababneh and I.F. Al-Mohammed, *International Journal of Environmental Analytical Chemistry*, 2013, 93, 755.
48. S.B. Adeljou, A.M. Bond and H.C. Hughes, *Analytica Chimica Acta*, 1983, 148, 59.
49. R.A. Greig and J. Jones, *Archives of Environmental Contamination and Toxicology*, 1976, 4, 420.
50. R.M. Awadallah, A.E. Mohamed and S.S. Gabr, *Journal of Radioanalytical and Nuclear Chemistry Letters*, 1985, 95, 145.
51. C. Chassard-Bouchard, *Analytica Chimica Acta*, 1987, 195, 307.
52. B. Welz and M. Melcher, *Analytical Chemistry*, 1985, 57, 427.
53. H. Agemian and M. Thomas, *Analytical Chemistry*, 1980, 105, 902.
54. D.T. Heggie, Study of reservoirs, fluxes and pathways in an Alaskan fjord, PhD dissertation, University of Alaska, 1977.
55. P.J. Brook and W.H. Evans, *Analyst*, 1981, 106, 574.
56. A.R. Lima, C. Curtis, D.E. Hammermeister, I.E. Markee, C. Northcott and L.T. Brocke, *Archives of Environmental Contamination and Toxicology*, 1984, 13, 595.

57. G.W. Holcombe, G.L. Phipps and J.T. Fiant, *Ecotoxicology and Environmental Safety*, 1977, 7, 400.
58. J.D. Meisner and W.Q. Hum, *Bulletin of Environmental Contamination and Toxicology*, 1987, 39, 898.
59. A.L. Hilmy, N.A. El-Domiaty, A.Y. Dabees and H.A.A. Latife, *Comparative Biochemistry and Physiology*, 1987, 86c, 263.
60. J. Vostal, *Mercury in the Environment*, CRC Press, Cleveland, OH, 1972.
61. F.L. Harrison, K. Watness, D.A. Nelson, J.E. Miller and A. Calabrese, *Estuaries*, 1987, 10, 78.
62. M. Willis, *Archiv für Hydrobiologie*, 1988, 112, 299.
63. R. Schgal and A.B. Saxena, *International Journal of Environmental Studies*, 1987, 29, 157.
64. M.D. Knittel, Heavy Metal Stress and Increased Susceptibility of Steelhead Trout (*Salino Gairneri*) to Yersinia Poakeri Infection, in J.G. Eaton, R.R. Parrish and A.C. Hendricks (eds), *Aquatic Toxicology*, American Society for Testing Materials, Philadelphia, PA, 1980, p. 321.
65. V.M. Snarski, *Environmental Pollution (Series A)*, 1982, 28, 219.
66. D.G. Dixon and J.W. Hilton, *Journal of Fisheries and Biology*, 1981, 19, 509.
67. S. Hatakeyama, *Ecotoxicology and Environmental Safety*, 1988, 49, 77.
68. J.G. Nemcsok and G.M. Hughes, *Environmental Pollution*, 1988, 49, 77.
69. E.P. Sheerban, *Hydrobiological Journal*, 1979, 13, 75.
70. C.G. Ingersoll and R.W. Winner, *Environmental Contamination and Toxicology*, 1982, 1, 321.
71. J.R. Sinley, J.P. Geoltl and P.H. Davies, *Bulletin of Environmental Contamination and Toxicology*, 1974, 12, 193.
72. A.V. Nebecker, M.A. Cairns and C.M. Wise, *Environmental Contamination and Toxicology*, 1984, 3, 151.
73. M. Ahsanullah, M.C. Mohley and P. Rankin, *Australian Journal of Marine and Freshwater Research*, 1988, 39, 33.
74. M. De. N. Guidici, L. Migliore and S.M. Guirino, *Hydrobiologia*, 1987, 146, 63.
75. D. Lee (ed.), *Metallic Contaminants and Human Health*, Academic Press, New York, 1972.
76. L. Friberg, M. Piscator and G. Nordberg, *Cadmium in the Environment*, CRC Press, Cleveland, OH, 1971.
77. J. McCaull, *Environment*, 1971, 13, 3.
78. D.B. Laurie, M.M. Joselow and A.A. Browder, *Annals of Internal Medicine*, 1972, 76, 307.
79. J.F. De and L.G. Solbe, *Water Research*, 1974, 8, 389.
80. J.G. Eaton, *Water Research*, 1973, 7, 1723.
81. J.G. Eaton, J.M. McKin and G.W. Holcombe, *Bulletin of Environmental Contamination and Toxicology*, 1978, 19, 95.
82. J.S. Alabaster and R. Lloyd (ed.), *EIFAC Water Quality Criteria for Freshwater Fish*, UN Food and Agriculture Organisation/Butterworths, London, 1980.
83. L. Friberg, T. Kjellstrom, G. Nordberg and M. Piscator, *Cadmium in the Environment*, EPA 650 (3–75–049), Environmental Protection Agency, Washington, DC, 1975.
84. S.W. Reeder, A. De Mayo and N.C. Taylor, *Guidelines for Surface Water Quality*, Vol. 1, *Inorganic Chemical Substances: Cadmium*, Environment, Canada, Ottawa, 1979.
85. T. Kuriyama and R. Kuroda, *Analyst*, 1982, 107, 505.
86. A.J. Blokety, V.A. Medira, C. Falcone and E.P. Rack, *Analytical Chemistry*, 1979, 51, 178.

3 Determination of Organic Compounds in Fish

The occurrence of organic compounds in whole fish tissue is summarised in Table 3.1. The main classes of compounds so far investigated are alicyclic and aromatic hydrocarbons and polyaromatic hydrocarbons (Table 3.1); various chlorinated compounds, including chlorinated aliphatics and aromatics, polychlorinated biphenyls (PCBs) and chlorinated insecticides (Table 3.2); and a compound which is causing great environmental concern, 2,3,7,8-tetrachlorodibenzopdioxin (Table 3.3).

Polyaromatic hydrocarbons are present in the exhaust gases of most vehicles operating on heavy hydrocarbon fuels. The total concentration of these found in fish is in the range of 0.1–5 mg kg^{-1} (Table 3.1), and certainly, concentrations in fish at the higher end of this range give cause for environmental concern in the effect on not only the fish, but also consumers of that fish. The maximum permitted World Health Organization (WHO) level of polyaromatic hydrocarbons in drinking water is, for example, 0.2 µg L^{-1} (six compounds: fluoranthene, benzo[d]fluoranthene, benzo[k] fluoranthene, benzo[a]pyrene, benzo[ghi]perylene and indeno[1,2,3-ed]-pyrene).

As in the case of metals, certain organic substances tend to concentrate in the organs of fish. For example, the concentration of PCBs found in the liver of longnose gar (1.11–3.7 mg kg^{-1}) is appreciably higher than that found in whole fish tissue (0.5–1.0 mg kg^{-1}).

Rogers and Hall [19] determined polychlorobiphenyls in starry flounder (*Platichthys stellatus*) and in muscle, bone and liver in polluted sites. Polychlorobiphenyls have been found in the flesh of starry flounders (*Platichthys flesus*) caught in the Elbe Estuary, German Bight [16] and San Francisco Bay [17].

O'Connor and Pizza [18] studied the pickup by tissues and elimination routes for polychlorobiphenyls in striped bass (*Morone saxatilis*) from the Hudson River. In a single close study, measurable quantities of polychlorobiphenyls were detected in tissues 6 h after dosing and peaked 1–2 days after dosing.

Approximately 53% of the administered dose was eliminated by the fish within 120 h. Polychlorobiphenyl burdens on the fish increased with successive doses of polychlorobiphenyls.

3.1 HYDROCARBONS

Farrington et al. [21] used column chromatography and thin-layer chromatography to isolate hydrocarbons, arising from marine contamination, in fish lipids. The hydrocarbon extracts were then examined to select those that could be determined by gas

TABLE 3.1
Concentrations of Hydrocarbons (mg kg^{-1} Dry Weight) Occurring in Environmental Fish Samples

Compound	Tuna	Trout	White Fish
Pristine	2.4 [1]		
Methylcyclohexane		0.002 [2]	
Ethylcyclohexane		0.001 [2]	
Propylcyclohexane		0.002 [2]	
Benzene		0.008 [2]	
Toluene		0.008 [2]	
m/p-Xylene		0.005 [2]	
Methyl-3-methylbenzene		0.04 [2]	
1,3,5-Trimethylbenzene		0.05 [2]	
1-Methyl-1,4-propylbenzene		0.01 [2]	
2-Ethyl-1,4-dimethylbenzene		0.01 [2]	
2-Methylcyclopentanol		0.090 [2]	
4-Ethyl-1,2-dimethylbenzene		0.001 [2]	
4-Methylindan		0.002 [2]	
Naphthalene		0.001 [2]	
2-Methylnaphthalene			0.001–0.006 [3]
Biphenyl			0.001–0.014 [3]
C$_2$ naphthalenes			0.017–0.18 [3]
Acetnaphthalene			0.043–0.27 [3]
Acenaphthalene			0.007–0.039 [3]
Dibenzothiophene			0.021–0.27 [3]
Phenanthacene			0.002–2.7 [3]
Methyldibenzothiophenes			0.17 [4]
Fluoranthene			0.004–1.8 [3]
Phenanthro[4,5-bcd] thiophene			0.016–0.078 [3]
Pyrene			0.004–1.5 [3]
Benzone[b]naphtho[2,1-d] thiophene			0.006 [3]
Benz[a]anthracene			0.004–0.022 [3]
Chrysene			0.003–0.061 [3]
Benzo[e]pyrene and benzo[a]pyrene			0.014 [3]
Pyrylene			0.001–0.007 [3]
Phenanthrene plus anthracene			0.008 [3]

TABLE 3.2

Concentration of Organic Chlorine Compounds (mg kg^{-1} Dry Weight) Occurring in Environmental Fish

	Perch	Trout	Pike	Cod	Salmon	White Bass	Whale	Herring
Bromochloromethane		0.008 [2]						
Pentachlorobenzene				0.001–0.002 [2]				
Hexachlorobenzene				0.08–0.17 [8]				
Octachlorostyrene				1.3–4.2 [8]				
PCB 1242			0.89 [5]					
PCB 1254			1.01 [5]					
PCB 1260			0.48 [5]					
PCB				2.2 [6]				
DDE				2.2 [6]				
DDD				0.59–8.0 [6] 0.47–7.5 [6]				
Mirex	0.3–0.33 [9]	0.05–0.36 [12]	0.05 [5]		0.09–0.33 [7]	0.06–0.43 [7]	1.25–7.4 [10]	
Polychloro-2-(chlormethyl sulphonamide) diphenylether (Eulan WA)								
Dieldrin							0.007–0.04 [10]	
Toxaphane				1.1 [11] (liver)			0.4–1.0 [11]	

TABLE 3.3

Concentrations (mg kg⁻¹ Dry Weight) of 2,3,7,8-Tetrachlorodibenzo-*p*-Dioxin Occurring in Environmental Fish

Lake Trout	Carp	Ocean Herring	Rainbow Trout	Edible Fish	Catfish	Buffalo Fish	Predator Fish	Bottom-Feeding Fish
<0.004–0.014 [13]	0.001–0.094 [14]	<0.001–<0.01 [14]	0.031–0.038 [14]	0.48 [15]	<0.01 [15]	0.015–0.23 [15]	0.04–0.05 [15]	0.077 [15]
0.054–0.058 [14,20]								

chromatography–mass spectrometry, combinations of spectrophotometric methods or wet chemistry. As a screening method, gas chromatography was shown to be fairly accurate and precise for hydrocarbons boiling in the range 287°C–450°C and of suitable polarity.

Farrington et al. [22] also described a gas chromatographic method for the determination of hydrocarbons such as petroleum cuts, fuel oil and lubricating oils in tuna meat and cod liver lipid extracts. The cod liver oil samples used in this study were spiked with known amounts of various commercial hydrocarbon products and subjected to analysis in order to check analytical recoveries. In one recovery procedure, the cod liver oil sample was refluxed with methanolic 1 N potassium hydroxide to saponify esters. An ether extract of the digest was prepared, the ether was removed, and the residue dissolved in chloroform. This extract was subjected to thin-layer chromatography and gas chromatography. Further cod liver oil samples were chromatographed on a column consisting of layers of alumina and silica gel, elution of hydrocarbons being carried out with a 5% solution of benzene in pentane. Solvent was then removed from the eluate *in vacuo* and the residue dissolved in a small volume of carbon disulphide. This extract was then gas chromatographed to give the distribution of hydrocarbons present.

Table 3.4 gives the results of naturally occurring pristine determination in unspiked cod liver oil and tuna meat samples analysed by different methods. Agreement and precision are good for cod liver oil, but not so good for tuna meat.

In a sample of cod liver [21–23] spiked with a known amount of crude oil, the measured concentration of petroleum hydrocarbons is in fair agreement with the actual concentration spiked to the sample (373–426), 372 mg kg⁻¹ lipid.

Law [24] and Chesler et al. [25] have discussed the application of gas chromatography–mass spectrometry to the determination of traces of hydrocarbons in fish. Chesler et al. [25] used dynamic headspace analysis of an alkaline digest of the sample to prepare extracts for gas chromatography.

TABLE 3.4
Reproducibility of Pristine Analyses

Analyst	Sample	Technique	Pristine Concentration (mg kg⁻¹)
Cod liver	Sample A	Saponification-TLC, GC	30.1
		Saponification-TLC, GC	35.8 ± 1.6
		Saponification-CC	39.4 ± 2.3
		No saponification-CC	36.4 ± 1.6
		No saponification-CC, GC	7.3 ± 1.0
			Mean = 35.7 ± 3.5
Cod liver	Sample B	No saponification-CC, GC	
		1	38.7 (n-C_{14})
		2	37.1 (n-C_{14})
			39.7 (n-C_{28})
			Mean = 39.0 ± 1.2
Tuna meat		Saponification-TLC, GC	3.3
		Saponification-CC, CG	2.0
		Saponification-CC, GC	2.0
			Mean = 2.4 ± 1.5

Note: TLC = thin-layer chromatography; CC = column chromatography; GC = gas chromatography.

3.2 POLYAROMATIC HYDROCARBONS

Vassilaros et al. [26] have described a method for determining polyaromatic hydrocarbons down to 0.2 µg kg⁻¹ in fish tissue. The analytical procedure includes the following steps: aqueous alkaline digestion; acidification of the digestate with glacial acetic acid; extraction with methylene chloride; liquid–liquid partitioning with water and then 10% potassium hydroxide solution; adsorption chromatography on basic alumina using hexane, benzene and chloroform sequentially; gel permeation chromatography on BioBeads with methylene chloride; capillary gas chromatography using nitrogen and sulphur-specific detectors; and gas chromatography–mass spectrometry. An average recovery of 72% of spiked ¹⁴C-labelled anthracene was obtained.

Figure 3.1 shows capillary gas chromatograms obtained for the polyaromatic hydrocarbon (PAH) and polycyclic aromatic sulphur heterocycle (PASH) fractions of a brown bullhead catfish, obtained from Table 3.5. Retention indexes for some of the compounds identified heavily polluted Black River, Ohio.

The fraction obtained from fish in Black River consisted primarily of polyaromatic hydrocarbons ranging from 2-methyl-naphthalene to benzo(*ghi*)perylene, with fairly high levels of polycyclic aromatic sulphur heterocycles. The major components are acenaphthylene, dibenzofuran, fluorine, phenanthrene, fluoranthene and pyrene. The lower detection limit for this sample was about 0.5 µg kg⁻¹.

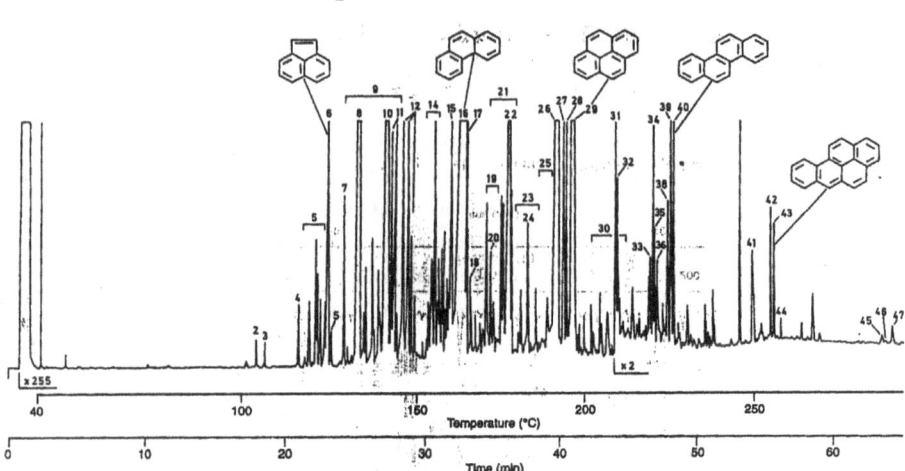

FIGURE 3.1 Capillary gas chromatogram of PAH/PASH fraction from Black River bull-head catfish. See Table 3.5 for peak identifications and text for chromatographic conditions.

TABLE 3.5
PAH and PASH Determined in Fish Samples Taken in Black River

Peak No.	Retention Index	Compound Name
1	200	Naphthalene
2	220	2-Methylnaphthalene
3	223	1-Methylnaphthalene
4	236	Biphenyl
5		C_2-Naphthalenes
6	246	Acenaphthylene
7	252	Acenaphthene
8	258	Dibenzofuran
9		C_3-Naphthalenes
10	269	Fluorene
11		Methylbiphenyls or methylacenaphthenes
12		Methyldibenzofurans
13		C_2-Biphenyls
14		Methylfluorenes
15	295	Dibenzothiophene
16	300	Phenanthrene
17	301	Anthracene
18		Naphtho[2,3-*b*]thiophene
19		Methyldibenzothiophenes
20	316	1-Phenylnaphthalene

(Continued)

TABLE 3.5 (*Continued*)
PAH and PASH Determined in Fish Samples Taken in Black River

Peak No.	Retention Index	Compound Name
21		Methylphenanthrenes
22	321	4*H*-Cyclopenta[*def*]phenanthrene
23		C$_2$-Dibenzothiophenes
24	333	2-Phenylnaphthalene
25		C$_2$-Phenanthrenes
26	334	Fluoranthene
27	347	Acephenanthrylene
28	348	Phenanthro[4,5-*bcd*]thiophene
29	351	Pyrene
30		Methylfluoranthenes and methylpyrenes
31	366	Benzo[*a*]fluorene
32	369	Benzo[*b*]fluorene
33	389	Benzo[*b*]naphtho[2,1-*d*]thiophene
34	389	Benzo[*ghi*]fluoranthene
35	391	Benzo[*c*]phenanthrene
36	392	Benzo[*b*]naphtho[1,2-*d*]thiophene
37	395	Benzo[*b*]naphtho[2,3-*d*]thiophene
38	396	Cyclopenta [*cd*] pyrene
39	398	Benz[*a*]anthracene
40	400	Chrysene
41		Benzofluoranthenes
42	452	Benzo[*e*]pyrene
43	453	Benzo[*a*]pyrene
44	457	Perylene
45	493	Indeno[1,2,3-*cd*]pyrene
46		Dibenzanthracenes
47	500	Benzo[*ghi*]perylene

3.3 PHTHALATE ESTERS

Giam et al. [27] determined phthalate esters in amounts down to $5\,\mu g\,kg^{-1}$ in marine fish using capillary column gas chromatography with an electron capture detector. Chlorinated insecticides and chlorinated biphenyls interfere in this chromatographic analysis and consequently have first to be removed by column chromatography on water-deactivated Florisil.

The tissue was macerated with acetonitrile, and then diluted with ethylene chloride–petroleum ether (1:5) and extracted with salt water. The dried organic phase was concentrated and diluted with iso-octane and subjected to clean-up in a Florisil column. Elution of the Florisil column with 6%, 15% and then 20% diethylether in petroleum ether provided three fractions containing, respectively, (1) chlorinated insecticides and chlorinated biphenyls, (2) diethyl hexyl phthalate and dibutyl phthalate, and (3) dibutyl phthalate.

Extreme precautions are necessary in this procedure to avoid contamination due to phthalates present as impurities in commonly used laboratory materials; for example, aluminium foil contains 300 mg kg^{-1} phthalate.

Between 2 and 20 µg kg^{-1} of diethyl hexyl phthalate was found by this procedure in various types of fish taken from the Gulf of Mexico.

3.4   NEUTRAL PRIORITY POLLUTANTS

An extraction procedure [28] utilising sonication with acetonitrile and clean-up using aminopropyl or C-18 bonded silica phases prior to gas chromatography has been applied to the determination of priority pollutant extracts. These included DDE, DDD and dichlorodiphenyltrichloroethane (DDT) occurring in the 4–18.6 mg kg^{-1} concentration range.

3.5   FLURIDONE

A method has been described [29] for the simultaneous determination of underivatised fluridone herbicide and its major metabolite, 1-methyl-3-(4-hydroxyphenyl)-5-(3-trifluoromethyl)phenyl-4(1H)-pyridinone, in fish and crayfish tissues by liquid chromatography. Compounds are extracted with methanol, followed by evaporation, acid hydrolysis to release conjugated residues, liquid–liquid partitioning, purification using Florisil Sep-Pak and liquid chromatographic analysis. In the absence of interfering peaks, the detection limit was 0.04–0.05 mg kg^{-1} for either compound. Overall recoveries of fluridone and its metabolite averaged 84% and 70% in edible crayfish tissues, 74% and 67% in inedible crayfish tissues, 111% and 103% in edible fish tissues and 109% and 76% in inedible fish tissues, that is, slightly high recoveries.

3.6   GLUCOSE

Takase et al. [30] have described a biocomposite glucose biosensor for real-time blood glucose monitoring of fish.

Polymers have a structure similar to that of living organisms and are thus used to make metallic materials more compatible with the living body. These workers focused on three widely used biocompatible polymers, 2-methacryloyloxyethyl phosphorlycholine (MPC) polymers, polypyrroles and polyurethanes, to achieve biocompatibility of a glucose biosensor. The developed glucose biosensor has a Pt–Ir wire (0.178 mm) as the working electrode and Ag/AgCl paste as the reference electrode. The biosensor was first coated with Nafion to prevent coexisting substances such as ascorbic acid and uric acid from interfering with the sensor output current, and then glucose oxidase (GOx) was fixed on top of the Nafion layer along with biocompatible polymers. The sensor was inserted into the fish eyeball interstitial sclera fluid, which contains low levels of proteins and correlates well with the glucose levels in the whole blood. Those three sensors were tested for durability, and sensors coated with MPC (Nafion/GOx/MPC sensor) proved to be most durable: the sensor output current maintained 93% output for 15 h in standard glucose solution, whereas the output current of the other sensors decreased more rapidly over time. A one-point calibration method was used to calibrate the sensor output current. As a result, 24 h of wireless monitoring was successfully achieved.

3.7   ALKYL PHENOLS

Bulukin et al. [31] used an enzymatic peroxide biosensor to expose rapid screening of alkyl phenol in a fish bile screen-printed electrode coupled with peroxides immobilised by glutaraldehyde cross-linking. The sensor was optimised with regard to factors such as immobilisation procedures, substrate selectivity and matrix effects. The biosensor was used for the analysis of fish bile samples from Atlantic cod (*Godus morhua L.*) exposed in the laboratory during a 2-week period to different petroleum-related compounds. The biosensor could distinguish between bile samples of fish exposed to water containing high concentrations of alkyl phenols and those of the control group.

3.8   MICROCYSTINE

Suchy and Berry [32] showed that microcystins are widespread cyanobacterial toxins in freshwater systems, and have been linked to both acute and chronic health effects. A great number of studies suggest that microcystine can bioaccumulate in food webs. Although several methods, such as enzyme-linked immunosorbent assay (ELISA)–liquid chromatography–mass spectrometry, have been developed for analysis of microcystin in water, extraction for subsequent analysis of the toxin from a biological matrix such as fish is impeded owing to covalent binding of toxins and activities of these cellular targets, that is, protein phosphatises. As an alternative approach, chromatographic methods for analysis of a unique marker, namely 2-methyl-3-methyloxy-4-phenyl/butanoic (the product of Lemieux oxidation of microcystin), have been previously developed and shown to measure total bound and unbound microcystine. Application of this method has, however, been limited by poor recovery of the analyte. Suchy and Berry [32] improved recovery by the use of solid-phase microextraction. The 2-methyl-2-methoxy-4-phenyl butanoic acid analogue 4-phenyl-butanoic acid and oxidised microcystine were used to develop methods, specifically a method employing post-oxidation methyl esterification, followed by headspace solid-phase microextraction recovery of 2-methyl-2-methyl-4-phenyl butanoic acid was developed and applied to fish tissue analysis. This method showed a high linearity for both water and fish tissue spiked with microcystin and an improved limit of quantitation of 140 ng g^{-1}.

Evaluation of field samples was performed by solid-phase microextraction–gas chromatography–mass spectrometry. This method reveals microcystine (particularly in the bound form) that is not detected by other methods. These results indicate that the method proposed by Suchy and Berry [32] provides improved detection capabilities for microcystin in biological materials and will improve our ability to understand the bioaccumulation process.

3.9   MALACHITE CRYSTAL RESIDUES

Xu et al. [33] described a rapid, easy-to-use trace-level direct competitive ELISA for detection of total residual malachite green, crystal violet and their corresponding primary metabolites, leucomalachite and leucocrystal violet, in a fishery.

The monoclonal antibodies (mAbs) malachite green and crystal violet were prepared using carboxyl-malachite green cationised bovine serum albumin conjugates. The linear range for the quantitative detection of total malachite green, crystal violet and their primary metabolite was between 0.15 and 4.0 ng ml^{-1}, with a half maximal concentration of 0.56 ± 0.04 ng ml^{-1}.

The anti-malachite green mAbs exhibited 98% cross-reactivity to crystal violet, less than 0.1% cross-reactivity with leucomalachite green and crystal violet and no cross-reactivity with chloramphenicol, enrofloxacin, sulfadiazine and tetracycline. Application of the direct competitive (dc)–ELISA in fish samples gave a limit of detection of 0.37 ng g^{-1}. The improved total detection lead to a recovery of $74.60 \pm 8.38\%$ at 0.5 ng g^{-1} and $87.47 \pm 12.83\%$ at 2.0 ng g^{-1} that was better than those of existing techniques. The dc-ELISA showed total malachite green in 7 out of 44 field fish samples that were confirmed with liquid chromatography–tandem mass spectrometry (LC-MS/MS). The easy-to-use, inexpensive and rapid dc-ELISA for the detection of malachite green, crystal violet and their corresponding primary metabolites holds promise for field applications.

3.10   B$_2$ VITAMINS

A sensitive method has been described [34] for the determination of riboflavin, flavin mononucleotide and flavin adenine dinucleotide in serum of different fish using reversed-phase high-performance liquid chromatography with fluorimetric detection. Trichloroacetic was used to isolate B$_2$ vitamins from the serum, and an aliquot of this solution was analysed by high-performance liquid chromatography using a Zorbax-NH$_2$ column with methanol–0.2 mol L^{-1} phosphate buffer $(1 + 9)$ as the mobile phase. The detection limits of riboflavin, flavin mononucleotide and flavin adenine dinucleotide in the serum were 4.89, 9.13 and 73.1 ng ml^{-1}, respectively.

Samples were prepared for analysis either by acid hydrolysis in 1 N hydrochloric acid at 120°C for 1 h or by enzyme hydrolysis at pH 4.2. Proteins were then precipitated with trichloroacetic acid to provide a solution ready for high-performance liquid chromatography. Typical concentrations in fish sera were riboflavin, 0.31–0.37 µg ml^{-1}, and flavin adenine dinucleotide, 0.25–0.44 µg ml^{-1}. Recoveries were in the range 92%–97%.

3.11   VITAMIN E (TOCOPHEROL)

Shultz [35] has described a high-performance liquid chromatographic method using fluorimetric detection, which permitted the simultaneous determination of four tocopherol isomers. The four tocopherols were completely separated during 15 min on a 5 µm silica gel column with a silanised stationery phase. Fluorescence detection, with excitation at 206 nm and measurement at 340 nm, permitted recoveries averaging 95% from spiked samples, based on measurement of the peak areas.

3.12   SQUOXIN (1,1′-METHYLENE-2-NAPHTHOL)

Kuigamagi et al. [36] developed a method for the determination of 0.1 mg kg^{-1} residues of this piscicide in fish using derivatisation gas chromatography and spectrophotometric methods. Fish were homogenised by grinding with dry ice. The pulverised mixture was poured into a plastic bag, which was lightly sealed and placed in a −10°C freezer overnight to allow the carbon dioxide to sublime. The samples were stored in this condition prior to extraction. A benzene extract of the fish was treated with diazomethane to produce the methylated derivative of the piscicide. After concentration, the residue was dissolved in acetonitrile and then extracted with hexane prior to Florisil clean-up and gas chromatography, or conversion to a Diazo Blue B complex and spectrophotometric evaluation at 552 nm.

3.13   GEOSMIN AND 2-METHYLISOBORNEOL

Martin et al. [37] have described a method for isolating and quantifying 2-methylisoborneol in fish flesh. Samples of channel catfish flesh were cooked in a microwave oven under a nitrogen stream, and the condensate trapped at −80°C. Hexane extracts of the condensate were then concentrated to 100–200 µl and analysed by gas chromatography. Concentrations as low as 5 ng methylisoborneol per gram of fish were detectable.

About 80% of the organoleptically off-flavoured fish had elevated 2-methylisoborneol concentrations (5.0–815.5 ng g^{-1} 2-methylisoborneol), but other musty odorants (geosmin and pyrazine isomers) were not detected. On average, 2-methylisoborneol levels in channel catfish from off-flavouring ponds were approximately 34 and 28 times higher than those in mud and water, respectively.

Lovell et al. [38] showed that a high proportion of catfish samples taken in ponds in Alabama had an off-flavour due to the presence of geosmin. Persson [39] investigated threshold odour concentrations of geosmin and 2-methylisoborneol in fish.

3.14   EULAN WA (POLYCHLORO-2(CHLOROMETHYL SULPHONAMIDE) DIPHENYLETHERS)

Wells and Cowan [40] described a gas–liquid chromatographic method for determining this moth-proofing agent in fish tissue down to 0.005 mg kg^{-1}. Eulan was extractively methylated using tetrabutylammonium ion, which forms ion pairs with sulphonamide at pH 10–12. This ion pair was subsequently back-extracted into the organic phase and methylated using methyl iodide. The methyl derivatives of the Eulan were quantified by gas–liquid chromatography using electron capture detection.

The coefficient of variation ($p > 0.05$) for perch liver was 4.82 mg kg^{-1} ± 5.6% for between-batch extractions.

To prepare the fish digest, 5 g of fish tissue is ground with sufficient anhydrous sodium sulphate to obtain a free-flowing powder. The sample is then extracted for 2 h with 100 ml of hexane with a Soxhlet extractor. The extract is then passed down a column containing acidic and basic alumina and eluted with hexane to remove chlorinated insecticides, and then with diethylether–glacial acid to remove Eulan. Addition of tetrabutylammonium hydroxide and methyl iodide to the second extract methylates the Eulan ready for gas chromatographic analysis.

Concentrations of Eulan were determined in perch muscle and liver, with coefficients of variation of 38% and 5.6%, respectively, at the 0.2 and 4.8 mg kg^{-1} levels.

Concentrations of Eulan WA in perch liver (4.5–55 mg L^{-1}) are appreciably higher those that found in the whole fish tissue (0.30–0.33 mg kg^{-1}).

3.15   α,α,α-TRIFLUORO-4-NITRO-*m*-CRESOL

This pesticide has been determined gas chromatographically in fish in amounts down to 0.01 mg kg^{-1} [41]. The fish sample is homogenised with hexane-ethylether (7:3). The phenol is extracted into 0.1 mol L^{-1} sodium hydroxide, back extracted into hexane-ether (7:3) after acidification and methylated with diazomethane. The product formed is analysed by gas chromatography at 140°C with the use of a glass column (1.8 m × 4 mm) packed with 3% of OV-1 on 80- to 100-mesh Chromosorb W, with nitrogen as carrier gas (60 ml min^{-1}) and electron capture detection.

3.16   CHLORINATED ALIPHATIC HYDROCARBONS

Various workers have discussed the application for gas chromatography to the determination of chlorinated aliphatics in fish [42–48]. Compounds that have been determined include trichloroethane, tetrachloroethene, tetrachloroethylene, chloroform and carbon tetrachloride [49,50,55], and 1,1,1-trichloroethane, trichloroethylene, perchloroethylene, 1,1,1,2-tetrachloroethane, 1,1,2,2-tetrachloroethane, pentachloroethane, hexachloroethane, pentachlorobutadiene, hexachlorobutadiene, chloroform and carbon tetrachloride [51–54].

Hela and Papadopoulos [56] described a gas chromatographic method for the determination of organochlorine insecticide and polychlorobiphenyls in such samples from location in western Greece. An experimental design was employed to estimate linear toxicity in results.

Luckas and Harms [16,57] measured the characteristic level of chlorinated hydrocarbons and trace metals in the coastal waters of the North Sea and the Baltic Sea. Compounds such as hexachlorobenzene, octachlorostyrene, hexachlorocyclohexane (HCH) isomers, *p,p'*-DDT and its metabolites and PCBs and heavy metals such as mercury, cadmium and lead were determined in a selected flatfish species (flounder, *Platichthys flesus L.*).

The sampling network covered the outer estuaries of the rivers Weser and Elbe, the German Bight, the Danish North Sea coast and coastal regions of the southwestern Baltic. Organochlorine compounds were determined by high-resolution glass capillary gas chromatography electron capture detector after sample pretreatment and clean-up.

3.17   HEXACHLOROBENZENE

Residues of hexachlorobenzene in fish have been determined at the microgram per kilogram level using gas chromatography combined with mass spectrometry [58].

3.18   TOXAPHENE

Hughes and Lee [59] have used gas chromatography to determine toxaphene in fish samples.

Musial and Uthe [60] describe a simple procedure using a combination of chromatography and fuming nitric-concentrated sulphuric acid clean-up followed by capillary gas chromatography for the estimation of toxaphene residues in marine fish. Wet weight concentrations were $1.1 \, mg \, kg^{-1}$ in cod liver and $1.0–0.4 \, mg \, kg^{-1}$ in herring fillet from the Gulf of St Lawrence and Halifax, respectively.

3.19   POLYCHLORINATED NITROBENZENES

Procedures have been described for the separation and identification of polychlorinated nitrobenzene compounds in fish [61]. The method comprised liquid chromatographic separation in silica gel and selective fractionation on Florisil, followed by gas chromatography, with mass spectroscopy and mass fragmentography as methods of identification. In all, 12 separate nitro compounds were detected, together with other polychlorinated pesticide derivatives.

3.20   POLYCHLORINATED STYRENES

Polychlorinated styrenes have been determined in fish [62–64]. Steinwandter and Zimmer [63], used analysis by gas chromatography, mass spectroscopy and mass fragmentography. Liquid chromatography identified 14 isomeric polychlorinated styrenes in Rhine fish using negative ion chemical ionisation mass spectrometry. Ramdahl et al. [64] determined polychlorinated styrenes in fish samples. The fish muscle and liver samples were homogenised and extracted with equal parts of cyclohexane and isopropanol. Concentrated extracts were treated with concentrated sulphuric acid before gas chromatography–mass spectrometry analysis. Separation of chlorinated styrenes was achieved using gas chromatography columns of fused silica with injector and detector temperature at 250°C, and temperature programming at 60°C for 2 min, then 4°C per minute to 250°C. All eight possible chlorostyrenes with fully chlorinated aromatic nuclei were identified in cod samples by this method.

3.21   1,1-DICHLORODIMETHYL SULPHONE

Lindstrom and Schubert [65] and others [66–72] used gas chromatography combined with multistage mass spectrometry and direct-inlet multistage mass spectrometry to determine 1,1-dichlorodimethyl sulphone in aquatic organisms in waters. The results obtained using the multistage mass spectrometry technique are reported, using homogenised liver extracts from flounders. The method is shown to be sensitive, selective and rapid, and does not require selective workup procedures.

3.22   CHLOROPHENOLS

Stark [66] has described a gas chromatograph method for the determination of pentachlorophenol as the trimethylsilyl ether in amounts down to 0.5 mg kg^{-1} in fish. Rudling [67] determined pentachlorophenol in fish and water by an electron capture gas chromatographic method. In this method, a sample fish tissue (1 g) in water is transferred to a centrifuge tube with 5 ml of water, 1 ml of 6 mol L^{-1} sulphuric acid and 5 ml of 1:5 isopropyl alcohol-hexane. The tube is centrifuged and cooled in ethanol–solid carbon dioxide. The organic layer is decanted and extracted with 0.1 mol L^{-1} Na$_2$B$_4$O$_7$. Hexane (0.5 ml) and acetylation reagent (pyridine [2 ml] plus acetic anhydride [0.8 ml]) (40 μl) are added to the combined aqueous extracts. The hexane phase is analysed by gas chromatography on a glass column (1 m × 1.5 mm) packed with 5% of QF-1 on Varaport 30 (100 to 120 mesh).

Renberg [68] has used an ion exchange technique for the determination of chlorophenols and phenoxyacetic acid herbicides in fish tissue. The sample (5 g) is homogenised in a mixture of hexane and acetone. The liquid is dropped into a separatory funnel containing 1.0 mol L^{-1} hydrochloric acid (5 ml). The funnel is shaken and the upper phase transferred into a centrifuge tube. Sodium sulphate (100–300 mg) is added to bind any water present. After centrifugation, the extract is transferred into a weighed flask, the sodium sulphate is rinsed with diethylether (1 ml) and the solvents gently evaporated on a water bath in a nitrogen stream. The flask is reweighed and the fat content calculated; the fat which originates from the fish samples is then dissolved in benzene. After treatment with ion exchange resin, the chlorophenols in the extract are converted to their methylethers using diazomethane and determined by gas chromatography. Detection limits for a 10 g fish sample are 0.1–1 μg kg^{-1}.

Hoben et al. [69] have described a gas chromatographic technique for determining 0.1 ppb pentachlorophenol in fish tissues. Confirmation of the identity with hexane of the chlorophenol was provided by gas chromatography–mass spectrometry. Then pentachlorophenol is extracted from the acidified sample and then reextracted into a borax solution. It is then acetylated with n-hexane containing acetic acid anhydride and pyridine. The resulting acetate is analysed by gas chromatography using an electron pentachlorophenol acetate detector.

This extract procedure gave 83%–91% recovery of pentachlorophenol from fish. The method was used successfully to determine pentachlorophenol at the 0.15–3 mg kg^{-1} level in fish. The identity of chlorophenol was confirmed at the 0.15–3 mg kg^{-1} level in fish. Confirmation of the identity of the chlorophenols was established by combined gas chromatographic–mass spectrometric analysis.

Thin-layer chromatography and gas chromatography have been used to determine microgram levels of pentachlorophenol, trichlorophenol and 2,4-dichlorophenyl in fish tissue [76].

3.23 POLYCHLORINATED BIPHENYLS

Sackmauer et al. [71] determined PCBs in water and fish by thin-layer chromatography on silica gel plates impregnated with 8% paraffin oil. As a mobile phase, a mixture of acetonitrile, acetone, methanol and water (20:9:20:1) was used. For detection, a solution of silver nitrate and 2-phenoxyethanol was used, followed by irradiation with UV light. The detection sensitivity for Aroclor 1242 is 0.5–1.0 µg.

Szelewski et al. [72] claim that some loss of PCB homologues occurs during the chromium trioxide extraction of fish tissue. The biphenyl-free PCB extract is then perchlorinated using antimony pentachloride at 200°C. Following acidification and toluene extraction, the aqueous phase remaining is extracted with hexane, and this extract is passed down an anhydrous sodium sulphate microcolumn and concentrated in a Kuderna-Danish evaporator prior to gas chromatography. Comparison of the gas chromatograms of the polychlorobiphenyls thus obtained with those obtained for the perchlorinated product of an authentic sample of PCB enables identification to be made of the types of PCBs in the sample extract.

Tausch et al. [73] determined polychlorobiphenyls in fish taken from the Danube River, by capillary gas chromatography and mass spectrometry. Altogether, 40 separate peaks representing various PCB isomers were identified; the total PCB concentration in the sample amounted to 6.1 mg kg^{-1}, with much lower amounts of pesticide residues and DDT breakdown products.

Szelewski et al. [79] have also determined PCBs in fish samples.

Gaskin et al. [80] have described a gas chromatographic method for determining DDT, dieldrin and PCBs in the organs of whales and dolphins. Total DDT in blubbers ranged from 1.25 to 7.4 mg kg^{-1}, dieldrin in blubber from 0.007 to 0.04 mg kg^{-1} and PCB in blubber from 0.69 to 5.0 mg kg^{-1}.

Ludke and Schmitt [81] have reported on pesticide and polychlorobiphenyl concentrations found in fish in the US National Pesticide Monitoring Program.

3.24 CHLORINATED INSECTICIDES

Gas chromatography has been extensively used for the determination of chlorinated insecticides in extracts of fish tissue. Solvent extraction and clean-up procedures are summarised in Table 3.6.

Concentrations of α-benzene hexachloride (BHC), β-BHC, aldrin, heptachlor, heptachlor expoxide, α-endosulphan, β-endosulphan, α-chlordane and γ-chlordane have been determined in samples of 13 commercially significant fish species caught in the northwest American Gulf [82]. Concentrations of all of these in fish tissue were below the analytical detection limit of 1 µg kg^{-1} wet weight. DDT was the most prevalent organochlorine pesticide, with average concentrations in fish ranging from 1 to 28 µg kg^{-1}. Dieldrin was detected in about 25% of fish species examined at 1–4 µg kg^{-1} wet weight. Total DDT and endrin residues in fish caught in an insecticide-sprayed lake were 5–72 and 3–67 µg kg^{-1}, respectively.

Levels of DDT and polychlorobiphenyl have been determined in immature cod species (*Gadus morhua*) liver and herring (*Clupea larengus*) muscle. DDT levels in herring muscle between 1979 and 1986 were 0.3–2.2 mg kg^{-1}, reducing to

TABLE 3.6
Extraction and Clean-Up Procedures Used in the Determination of Chlorinated Insecticides in Fish

Extraction Solvent	Sample Clean-Up	Gas Chromatography	Recovery %	Reference
Miscellaneous Chlorinated Insecticides				
Mix fish with granular sodium sulphate and sand and Soxhlet extract with *n*-hexane	—	Electron capture	*p,p'* DDE 97.8	[83]
			p,p' DDT 91.2	[89]
			p,p' TDE 94.6	
			p,p' DDT 89.6	
			Dieldrin 89.3	
Fish mixed with deactivated Florisil and extracted in a column with 1:4 v/v dichloromethane-hexane	—	Electron capture		
Blend fish with sodium sulphate and solid carbon dioxide; extract with cyclohexane	Florisil column	—	95–100	[84]
Dieldrin				
Light petroleum–acetonitrile partitioning or direct homogenisation	Florisil column and capture detection	Flame ionisation	Dieldrin 100	[85]
Photodegradation products of endrin	Florisil column	Flame ionisation of silylated or acetylated derivatives		[86]
Dieldrin and polychlorobiphenyls	—	Gas chromatography–mass spectrometry		[87]
BHC isomers, DDE, DDT hexachlorobenzene. Homogenised fish mixed with standard anhydrous sodium sulphate extracted with light petroleum; extract concentrated prior to clean-up on Celite-oleum column	Celite-oleum column or Florisil for aldrin (unstable on Celite-oleum)	Electron capture	BNH isomers 93–103 DDT, DDE 90–93	[88]

0.010–0.017 mg kg⁻¹ in 1988. Polychlorobiphenyl levels in cod liver between 1979 and 1986 were 0.3–3.7 mg kg⁻¹, reducing to 0.013–0.19 mg kg⁻¹.

Data have been obtained on the concentrations of aldrin, endrin, endosulphan, heptachlor, heptachlor epoxide, lindane and the DDT group in black bullhead, bleak, chub, common carp, eel and tench collected in Italian rivers.

Aldrin, lindane, heptachlor and endosulphan were detected in less than 20% of the fish examined, whereas dieldrin was found in almost all the fish studied. Total DDT group residue concentrations in fish were between 17 and 153 µg kg⁻¹, depending on the river. Other pesticides were at lower concentrations (up to 39 µg kg⁻¹) [76–81].

Fingerling rainbow trout (*Salmo gairdneri*) exposed for up to 4 days to 10 µg L⁻¹ of aminocarb in water at pH 4.6–8.2 picked up a 9.1 mg kg⁻¹ aminocarb level in fish tissue in the first 6 h of exposure. At pH 8.2, whole body aminocarb increased to 12 mg kg⁻¹ in 1 h and remained elevated until the fish died in 72 h.

Luckas et al. [90] have described a method for determining PCBs and chlorinated insecticides in fish by the simultaneous use of electron capture gas chromatography and derivatisation gas chromatography. The method is based on the different stabilities of chlorinated insecticides and PCBs towards magnesium oxide in a microreactor. Extracts of samples are injected twice, first into a regular gas chromatograph and then into a gas chromatograph equipped with a microreactor for derivatisation. A basic chromatogram and a derivatisation chromatogram are obtained, and the combination of the two chromatograms provided a satisfactory solution.

Chemical derivatisation of sample extracts is very convenient. The extracts containing insecticides and PCBs, after the first injection into the gas chromatograph, are treated with derivatisation reagents, the insecticides being converted into derivatives, while the PCBs remain unchanged.

Luckas et al. [90,91], as a result of these considerations, developed a microreactor gas chromatographic technique in which derivatisation is carried out *in situ*. Preheated magnesium oxide affects the rapid quantitative dehydrochlorination of saturated DDT metabolites to the corresponding DDT olefins [91]. The derivatisation products immediately obtained in the gaseous phase by means of the microreactor (with nitrogen as the carrier gas and magnesium oxide as the catalyst) are comparable with the products of chemical derivatisation with an alkali in the liquid phase, and substances that are stable to treatment with alkali are also not decomposed in the microreactor. Two gas chromatographs, with an all-glass system and an electron capture detector, were used. One chromatograph was equipped with a microreactor for the derivatisation gas chromatography. Luckas et al. [90] extracted the fish with *n*-hexane and cleaned up with sulphuric acid.

A basic chromatogram of an extract of fish showed peaks due to γ-HCH and DDT metabolites, but the background suffers from interference from peaks of PCBs. After derivatisation, the peaks of γ-HCH and the saturated DDT metabolites disappeared.

The saturated DDT metabolites (*p,p'*-DDT, *p,p'*-DDD and *o,p*-DDT) are converted quantitatively into the corresponding DDT olefins (*p,p'*-DDE and *o,p*-DDE). The main peak in the derivatisation gas chromatogram represents the sum of *p,p'*-DDT and *p,p'*-DDE from the basic chromatogram and is often sufficient for the determination of the total DDT content. The content of PCBs can be calculated in the derivatisation gas chromatogram without interference effects due to the saturated DDT metabolites.

Norheim and Okland [74] studied the distribution of down to $0.001\,mg\,kg^{-1}$ persistent chlorinated insecticides, such as DDT, PCBs and hexachlorobenzene, in cod samples taken along the Norwegian coast.

In their analytical procedure, a $0.5\,g$ sample of fish was digested with $6\,ml$ of concentrated sulphuric acid for $4\,h$ at $60°C$. The cooled digest was shaken with $1\,ml$ of hexane, and this extract examined by electron capture gas chromatography. Recoveries of 95% of added hexachlorobenzene and octachlorostyrene were obtained in spiking experiments. Good inter-laboratory comparison was obtained in determination of hexachlorobenzene, octachlorostyrene and pentachlorobenzene, present at $0.001–4\,mg\,kg^{-1}$ levels in fish.

Neeley [75] determined PCBs in fish and water in Lake Michigan, while Frederick [76] measured by gas chromatography the comparative uptake of PCBs and dieldrin by the white sucker. Jan and Malservic [77] determined PCBs and polychlorinated terphenyls in fish using acid hydrolysis of the fish tissue and destructive clean-up of the extract. Olsson et al. [78] studied the seasonal variation of PCB.

3.25   POLYCHLORINATED DIBENZO-*p*-DIOXINS AND POLYCHLORINATED

Polychlorinated dibenzo-*p*-dioxins, polychlorinated dibenzofurans and other substituted PCBs are three structurally and toxicologically related families of anthropogenic chemicals that have, in recent years, been shown to have the potential to cause serious environmental contamination. These substances are trace-level components or by-products in several large-volume and widely used synthetic chemicals, principally PCBs and chlorinated phenols, and they can also be produced during combustion processes and by photolysis. In general, polychlorinated dibenzo-*p*-dioxins, polychlorinated dibenzofurans and non-*ortho* PCBs are classified as highly toxic substances, although the toxicities are dramatically dependent on the number of positions of the chlorine substituents. About 10 individual members of a total of 216 polychlorinated dibenzo-*p*-dioxins, polychlorinated dibenzofurans and non-*ortho* PCBs are among the most toxic man-made or natural substances to a variety of animal species. The toxic hazards posed by these chemicals are exacerbated by their propensity to persist in the environment and to bioaccumulate readily, and although the rate of metabolism and elimination is strongly species dependent, certain highly toxic isomers have been observed to persist in the human body for more than 10 years. Work on the determination of these classes of compounds in fish is discussed below.

Lamparski et al. [92] have developed a procedure for the determination of $10–100\,\mu g\,kg^{-1}$ quantities of 2,3,7,8-tetrachloro-dibenzo-*p*-dioxin in fish. The technique involves digestion with alcoholic potassium hydroxide and extraction of the matrix with hexane, followed by a series of absorbent and chemically modified adsorbent liquid column chromatographic clean-up steps involving the use of silica–sulphuric acid, alumina and silica–silver nitrate. A final 'residue polishing' step via elevated-temperature reversed-phase high-performance liquid chromatography is applied prior to detection by multiple-ion-mode gas chromatography–mass spectrometry. Using ^{13}C-labelled 2,3,7,8-tetrachlorodibenzo-p-dioxin as an internal standard and carrier. The procedure has been validated for rainbow trout from approximately 10 to 100 mg kg^{-1} 2,3,7,8-tetrachlorodibenzo-p-dioxin up to 0.48 mg kg^{-1} of this substance has been detected in edible fish. Relative to this range, the 2,3,7,8-tetrachlorodibenzo-p-dioxin recovery is 75% ± 25%, and the precision of a single determination at the 95% confidence level (2 σ) is ±20% relative to a 50 ng kg^{-1} concentration.

Results obtained by this procedure in spiking experiments in which 10–100 ng kg^{-1} quantities of 2,3,7,8-tetrachlorodibenzo-p-dioxin were added to trout gave recoveries of between 80% and 120%. Other anthropogenic compounds in the trout did not interfere.

Phillipson and Puma [93] identified chlorinated methoxybiphenyls as contaminants in fish and recognised them as potential interferences in their gas chromatography–mass spectrometric method for the determination of chlorinated dibenzo-p-dioxins. These workers used a sample digestion clean-up procedure similar to that used by Lamparski et al. [92], involving digestion of the fish with alcoholic potassium hydroxide, followed by hexane extraction, extraction with concentrated sulphuric acid and clean-up on alumina and Florisil.

Gas chromatography–mass spectrometry of the solvent extracts of carp samples revealed the presence of a group of xenobiotics (Cl_3, Cl_4 and Cl_5 ring-substituted methoxybiphenyls). The compounds when gas chromatographed in the region of Cl_3, Cl_4 and Cl_5 polychlorinated dibenzo-p-dioxins and produced intense molecular ions having the same nominal masses and chlorine isotopic abundances as those observed in the molecular ion clusters from trichloro-, tetrachloro- and pentachlorodibenzo-p-dioxins. 3,3′,4′,5-Tetrachloro-4-methoxybiphenyl – and 2,3′,4,4′-tetrachloro-3-methoxybiphenyl – provided model compounds came through the polychlorinated dibenzo-p-dioxin clean-up procedure, and had gas chromatographic and mass spectrometric properties consistent with those of the residues recovered from the fish. The finding of chlorinated methoxybiphenyls as contaminants in fish, combined with the potential for their molecular ions to be mistaken for those from polychlorinated dibenzo-p-dioxins, indicates a need for reappraisal of reported identifications of polychlorinated dibenzo-p-dioxin residues in environmental samples by selected ion monitoring gas chromatography–mass spectrometry methods based on monitoring exclusively for the isotopic molecular ions from polychlorinated dibenzo-p-dioxins by gas–liquid chromatography and low-resolution mass spectrometry. It might be necessary to examine a sufficient segment of the mass spectrum of the suspect residue to rule out the presence of polychloromethoxybiphenyl (by the absence of fragment ion clusters resulting from losses of CH_3, CH_3CO and $CH_3CO=2$ Cl from the M$^+$ cluster) and other potential interferences not yet observed.

Paasivirta et al. [94] reported on the occurrence of polychlorinated aromatic ethers in salmon and fish liver oil. They described an improved clean-up technique which separated polychlorinated diphenylethers, anisoles and veratroles, phenoxyanisoles and biphenylanisoles, and dioxins and dibenzofurans into different fractions. Compounds from all these groups of substances were detected.

Mitchum et al. [95] determined 10–30 µg kg^{-1} levels of 2,3,7,8-tetrachlorodibenzo-p-dioxin in various fish samples by a procedure involving the use of a capillary column gas chromatograph interfaced directly to an atmospheric pressure ionisation mass spectrometer. Isolation of 2,3,7,8-tetrachlorodibenzo-*p*-dioxin from tissue samples was accomplished via multistep high-pressure liquid chromatography on a silica gel incorporating a stable label isotope dilution. PCBs were found not to interfere with 2,3,7,8-tetrachlorodibenzo-*p*-dioxin analysis at the low nanogram per kilogram levels. No mention is made of potential interference by chlorinated methoxy biphenyls as discussed by Phillipson and Puma [93]. Again, ethanolic potassium hydroxide was used to digest fish samples, followed by hexane extraction and clean-up on charcoal column.

The 2,3,7,8-tetrachloro-*p*-dioxin contents obtained by this procedure, found in fish samples taken from the Arkansas River, ranged from 7 µg kg^{-1} (catfish and buffalo fish) through 230 µg kg^{-1} (predator fish) up to 480 µg kg^{-1} (edible fish flesh).

Harless et al. [96] also used high-resolution gas chromatography–mass spectrometry to determine 2,3,7,8-tetrachlorodibenzo-*p*-dioxin in fish tissues.

Lawrence et al. [97] detailed the equipment and procedure for the determination of dioxins in fish tissues and sediment samples involving isolation and extraction by acid digestion, gel permeation chromatography, trisodium phosphate treatment, microalumina chromatography and carbon-fibre column chromatography, followed by determination of tetrachlorodibenzo-*p*-dioxins by gas chromatography–electron capture detection screening, and confirmation by high-resolution gas chromatography–mass spectrometry.

Clement et al. [98] compared results obtained by various techniques for the determination of chlorinated dibenzo-*p*-dioxins and chlorinated dibenzofurans in fish in a round-robin experiment involving gas chromatography–mass spectrometry, as well as mass spectrometry–mass spectrometry and high-resolution mass spectrometry. They commented that analysis of these complex samples was difficult without extensive clean-up procedures. However, determination by gas chromatography and high-resolution mass spectrometry was possible, as was with gas chromatography and low-resolution mass spectrometry.

Taguchi et al. [99] used high-resolution mass spectrometry to determine polychlorinated dibenzo-*p*-dioxins and polychlorinated dibenzofurans in fish. They used mixtures of the solid aromatic hydrocarbons, coronene, tetraphenylcyclopentadiene and decacyclene, as lock masses in the analysis of polychlorinated dibenzo-*p*-dioxins and polychlorinated dibenzofurans with between four and eight chlorine atoms. Enhanced sensitivity, easier control of the lock-mass concentration in the ion source and greater resolution between lock mass and sample ion signals were obtained for this system compared with the conventional perfluorokerosene lock-mass system. Both systems exhibited a trend of decreasing signal strength with increasing chlorine substitution, and stronger signals obtained for polychlorinated dibenzofurans than for polychlorinated-dibenzo-*p*-dioxins. Crummett [100] has reviewed methods for the determination of polychlorinated dibenzo-*p*-dioxins and polychlorinated dibenzofurans in fish.

3.26 MIREX (DECHLORANE, C$_{10}$H$_{12}$)

Markin et al. [101] have discussed the possible confusion between mirex and PCBs in the analyses of crabs, shrimp, fish and fish products. In their method, the samples were thoroughly scrubbed to remove mud, algae and other residues; they were ground whole and mixed in a Waring blender to make a composite sample. Samples were prepared and analysed on a whole body basis as received. A 20 g subsample of the composite was removed and analysed as follows. The homogenised sample was extracted with a mixture of hexane and isopropanol, and the extract subjected to a concentrated sulphuric acid clean-up. The sulphuric acid destroys dieldrin, endrin and organophosphorus insecticides, but the improvement in sensitivity by this clean-up was considered more than adequate compensation for the loss of these other insecticides. The final extract was cleaned up on a Florisil column, and concentrated to the desired level for analysis. If PCBs were suspected in the first analysis, their presence usually being indicated by a series of characteristic peaks, the sample was reprocessed to separate the PCBs from the insecticides as described by Armour and Burke [102], Gaul and Cruz-La Grange [103] and Markin et al. [104]. After concentrating to the appropriate volume, the extracts from both methods of clean-up were chromatographed on a gas chromatograph equipped with dual electron capture detection. Each sample was analysed on two different columns: the first column was a mixture of 1.5% OV-17 and 1.95% QF-1 on Gas Chrom Q. The second column was 2% DC-200 on Gas Chrom Q. Level detection was 0.001 mg kg^{-1} for DDT and its metabolites, 0.005 mg kg^{-1} for mirex and 0.01 mg kg^{-1} for Aroclor 1260.

Markin et al. [101] said that they found mirex in only a minority of the samples they analysed, contrary to results obtained by earlier workers. All samples containing mirex were from around Savannah, Georgia, an area with a history of concentrated mirex use that is among the most extensive in the United States. The recovery of mirex in only 12% of the samples, all from one area, could indicate that mirex is not so general or widespread a contaminant of seafood as are PCBs and DDT. This does not correspond to earlier seafood studies [105–110,110a], which reported that mirex occurred much more frequently and densely in many of these same collection sites. Probably the reason for the discrepancy between their study and earlier studies is the confusion of Aroclor 1260 with.

Laseter et al. [111] also used a gas chromatography–mass spectrometry system to determine mirex in Lake Ontario fish samples. Quantitative analyses employing reconstructed mass chromatograms with the mirex (perchloropentacyclo)decane or dodecachlorooctahydro-1,3,4-metheno-2H-cyclobuta[cd]pentalene) base peak (m/e 272) and gas chromatography with electron capture and Hall electrolytic detectors provided concentration values ranging from 0.15 to 0.33 mg kg^{-1} fresh weight of tissue.

Onuska et al. [112] analysed lake trout and lamprey samples from Lake Ontario, to provide a rapid, quantitative procedure for the determination of mirex. Statistically insignificant differences were found, based on the correlation of results from linear regression analysis, for the determination of mirex by selected ion monitoring mass spectrometric data and electron capture detection. The single-ion detection procedure is rapid and capable of screening and quantitating mirex levels as low as 500 fg. The multiple-ion detection technique extends the determination to include degradation products of mirex. It does not suffer from previously reported interferences and can be used for a wide range of samples.

3.27 NITROGEN BASES

3.27.1 TRIMETHYLAMINE [113]

Amines and basic substances that are not rendered non-reactive by formaldehyde are released at room temperature from an extract using potassium hydroxide and extracted into toluene. The toluene phase is separated and dried, and picric acid reagent is added in order to form a coloured complex with trimethylamine. The absorbance of this solution at 410 nm is a measure of the concentration of trimethylamine. Other tertiary bases, not completely rendered non-reactive by formaldehyde under the conditions of the method, may also give coloured complexes, but normally this interference is small.

3.27.2 METHYLAMINES [113]

The volatile amines in the sample are extracted by means of perchloric acid. The perchloric acid is made alkaline with sodium hydroxide, and the amines liberated are steam distilled into hydrochloric acid. The concentration of amines, including trimethylamine, is determined in the hydrochloric acid solution by gas–liquid chromatography.

3.27.3 TRIMETHYLAMINE OXIDE [113]

Trimethylamine oxide in an extract of the sample is reduced with titanium(III) chloride to trimethylamine, which is then determined. The value obtained for the trimethylamine content of the sample before reduction is subtracted from this value to give the trimethylamine oxide.

3.27.4 HYPOXANTHINE [113]

Hypoxanthine is extracted from the sample by macerating it with perchloric acid. After neutralisation, the extract is treated with an enzyme that converts the hypoxanthine quantitatively into uric acid, which is determined by measuring its absorbance at 290 nm. Both the enzyme and the extract absorb at this wavelength, and so it is necessary to carry out measurements on blanks. The method is not specific for hypoxanthine, as similar results will be given by xanthine, which, although not normally present, should be looked for when unusual samples are being examined.

3.28 PHOSPHORUS-CONTAINING COMPOUNDS

Murray [114] has described a gas chromatographic method for the determination in fish tissue of triaryl phosphate esters (1 mol S-140, tricresyl phosphate, cresol diphenyl phosphate). These substances are used commercially as lubricating oil and plastic additives, hydraulic fluids and plasticisers. The method involves extraction from the sample, hydrolysis and measurement of the individual phenols by gas chromatography as the trimethyl derivatives. The lower detection limit was about 3 mg kg^{-1} of fish.

TABLE 3.7

Methods for the Determination of Organic Compounds in Fish

Determined	Type of Sample	Technique	Detection Limit (mg kg^{-1} Unless Other Stated)	Reference
Hydrocarbons, aliphatic and aromatic	Fish	Gas chromatography with mass spectrometric detector	—	[119–121]
Pristine	Tuna, cod, liver, fish lipids	Gas chromatography, column and thin-layer chromatography	—	[122]
Polyaromatic hydrocarbons	Flounder bile	Derivativisation spectrofluorimetry	5 mg L^{-1}	[123]
	Fish	Capillary gas chromatography	0.5	[124]
	Fish	Gas chromatography with mass spectrometric detector	µg kg^{-1}	[125]
Chloroaliphatic compounds	Fish	Gas chromatography	—	[126–132]
Trichloroethylene tetrachloroethylene	Fish		—	[133,134]
1,1,1-trichloroethane, trichloroethylene, perchloroethylene, 1,1,1,2-tetrachloroethylene, 1,1,2,2-tetrachloroethylene, pentachloroethylene, hexachloroethylene, pentachlorobutadiene, hexachlorobutadiene, chloroform, carbon tetrachloride	Fish		—	[135–138]
Haloparaffins	Fish	Gas chromatography–mass spectrometry	—	[139]
Organic halogen compounds	Fish	Extraction–neutron activation analysis	—	[140]
Pentachlorophenol	Fish	Gas chromatography	0.15–0.3	[141]
	Fish	Conversion to silyl derivative–gas chromatography	—	[142]

(Continued)

TABLE 3.7 (*Continued*)
Methods for the Determination of Organic Compounds in Fish

Determined	Type of Sample	Technique	Detection Limit (mg kg^{-1} Unless Other Stated)	Reference
	Fish	Gas chromatography with electron capture detection	—	[143]
Pentachlorophenol trichlorophenol	Fish	Gas chromatography	—	[144]
2,4-Dichlorophenol chlorophenols	Fish	Ion exchange chromatography conversion to methyl esters, gas chromatography	0.00–0.001	[145]
PCBs	Fish, white sucker	Gas chromatography	—	[146,147]
PCBs	Biota	Photoactivated luminescence	—	[148]
Methyl sulphone substituted polychlorobiphenyls	Adipose liver grey seals	Mass spectrometry	—	[149]
Polychlorinated dibenzo-*p*-dioxins	Fish	Gas chromatography	Low ng kg^{-1}	[150]
	Fish	Gas chromatography with mass spectrometric detector	Low ng kg^{-1}	[151–154]
2,7,7,8-Tetrachlorodibenzo-*p*-dioxin	Fish, trout	High-performance liquid	10^{-5} to 10^{-4} (10–100 ng kg^{-1})	[155]
	Fish	High-resolution electron impact mass spectrometry, also chemical ionisation atmosphere pressure mass spectrometry	ng kg^{-1} level	[156–158]
	Plasma blood	Collection on C$_{18}$ bonded cartridge	—	[159]
Dibenzofurans	Fish	Gas chromatography–mass spectrometry	—	[154–157]

(Continued)

TABLE 3.7 (*Continued*)

Methods for the Determination of Organic Compounds in Fish

Determined	Type of Sample	Technique	Detection Limit (mg kg^{-1} Unless Other Stated)	Reference
	Fish	High-resolution electron impact mass spectrometry, also chemical ionisation atmospheric pressure	ng kg^{-1}	[156]
Chlorinated insecticides	Fish	Mass spectrometry, gas chromatography	0.001	[155–167]
	Fish	Derivativisation–gas chromatography	—	[169]
DDE, DDT, DDD		Fish extracts		
DDT dieldrin	Whale	Gas chromatography	—	[168]
Mirex (C$_{10}$H$_{12}$)	Lake fish	Gas chromatography–mass spectrometry	—	[170–176]
	Trout, lamprey		50 gf absolute	[175]
	Fish	Gas chromatography with electron capture detector	0.005 µg kg^{-1}	[176–179]
Phenoxy acetic acid herbicides	Fish	Ion exchange chromatography, conversion to methyl esters, gas chromatography	0.0001–0.001	[145]
Pyrimylphos methyl *O*-(2-dimethylamino)-6-methyl-4-pyriminyl-*O,O*-dimethyl phosphorothioate	Fish	Gas chromatography	—	[180]
Dursban	Fish	Gas chromatography	0.01	[181]
Acephate methamidophos	Fish		—	[182]
Fenthion	Fish		0.01	[183]
Fluridone	Crayfish	Liquid chromatography	0.04–0.05	[184]

(*Continued*)

TABLE 3.7 (*Continued*)

Methods for the Determination of Organic Compounds in Fish

Determined	Type of Sample	Technique	Detection Limit (mg kg⁻¹ Unless Other Stated)	Reference
Atrazine, terbutryn	Fish	Assay based on inhibition of photosynthetic electron transport in spinach thylalkaloids by target compounds with spectrophotometric detection of redox dye	—	[185]
Triaryl phosphate	Fish	Gas chromatography	3	[186]
Polychloronitrobenzenes	Fish	Gas chromatography–mass spectrometry	—	[187]
Methylamine, trimethylamine oxide	Fish	Gas chromatography	—	[188]
Hypoxanthine oxide, 1,1-dichloromethyl sulphone	Flounders	Gas chromatography–mass spectrometry	—	[189]
Polychlorostyrene	Fish		—	[190–192]
Hexachlorobenzene, octachlorostyrene	Fish eggs	Gas chromatography	1	[193]
Phthalate esters	Marine fish	Gas chromatography with electron capture detector	0.5	[194]
Squoxin (1,1-methylene 2-napthol)	Fish	Derivativisation then gas chromatography	—	[195]
Eulan WA (polychloro-2-chloromethyl sulphamido) diphenyl esters	Fish	Methylation–gas chromatography	0.005	[196]
α,α-Tri-fluoro 4-nitro-*m*-cresol	Fish		0.01	[197]

Deutsch et al. [115] determined dursban insecticide in fish. After a preliminary clean-up, the extract is chromatographed on a column packed with 3% Carbowax 20M on Gas-Chrom (60–80 mesh), which gives excellent separation of dursban from other organophosphorus insecticides. Both thermionic and flame photometric detectors are satisfactory. Recoveries range from 75% to 105% (i.e. 90 ± 15%) depending on the nature of the sample. This procedure will detect as little as 0.5 ng of dursban, corresponding to a level of 0.01 mg kg^{-1} in a 10 g sample of fish.

Pyrimylphos methyl (O-(2-diethylamino)-6-methyl-4-pyrimidyl-O,O-dimethyl phosphorothioate) insecticide has been determined in fish by a procedure involving gas chromatography of a hexane–acetone extract of the sample [116].

Lores et al. [117] described a method for the determination of fenthion in amounts down to 0.01 mg kg^{-1} in fish. The method involved solvent extraction followed by a silica gel clean-up procedure, and then determination by gas–liquid chromatography with thermionic detection. The clean-up procedure required relatively little time. Recovery of fenthion exceeds 85% with adequate sensitivity.

Szeto et al. [118] have described a gas chromatographic method for the determination of acephate and methamidophos residues in fish.

3.29 MISCELLANEOUS ORGANIC COMPOUNDS

Other methods for the determination of organic compounds in fish are reviewed in Table 3.7. Gas chromatographic methods of identifying the separated compounds predominate. Direct chemical ionisation mass spectrometry has been used to identify and determine polychlorobiphenyls and methyl sulphone–substituted polychlorobiphenyls, and high-resolution electron impact mass spectrometry, as well as chemical ionisation atmospheric pressure mass spectrometry, to identify and determine 2,3,7,8-tetrachlorobenzo-p-dioxins and other dioxins (also dibenzofurans). High-performance liquid chromatography has also been employed to examine dioxins as these, due to low volatility, are not amenable to gas chromatographic techniques.

REFERENCES

1. J.W. Farrington, J.M. Teal, J.G. Quinn, T. Wade and K.A. Burns, *Bulletin of Environmental Contamination and Toxicology*, 1973, 10, 129.
2. M.H. Hiatt, *Analytical Chemistry*, 1983, 55, 506.
3. D.L. Vassilaros, P.W. Stoker, G.M. Booth and M.L. Lee, *Analytical Chemistry*, 1982, 54, 106.
4. J.W. Poldoski, *Analytical Chemistry*, 1980, 52, 1147.
5. K.L.E. Kaiser, *Science*, 1974, 185, 523.
6. R.J. Ozretich and W.P. Schroeder, *Analytical Chemistry*, 1986, 58, 2041.
7. J.L. Laseter, I.R. De Leon and P.C. Remele, *Analytical Chemistry*, 1978, 50, 1169.
8. G. Norheim and E.M. Okland, *Analyst*, 1980, 105, 990.
9. N. Ichinse, K. Adachi and G. Schwedt, *Analyst*, 1985, 110, 1505.
10. D.E. Gaskin, G.J.D. Smith, P.W. Arnold, M.V. Louisy, R. Frank, M. Moldrinet and J.W. McWade, *Journal of Fisheries Research Board, Canada*, 1974, 31, 1235.
11. C.J. Musial and J.F. Uthe, *International Journal of Environmental Analytical Chemistry*, 1983, 14, 117.
12. K.R. Sperling, *Fresenius Zeitschrift für Analytische Chemie*, 1982, 310, 254.
13. L.L. Lamparski, T.J. Nestrick and R.H. Stehl, *Analytical Chemistry*, 1979, 51, 1453.
14. L.M. Smith, D.L. Stalling and J.L. Johnson, *Analytical Chemistry*, 1984, 56, 1830.
15. R.K. Mitchum, G.F. Moier and W.A. Korfmacher, *Analytical Chemistry*, 1980, 52, 2278.

16. B. Luckas and U. Harms, *International Journal of Environmental Analytical Chemistry*, 1987, 29, 215.
17. R.B. Spies and D.W. Rice, *Marine Biology*, 1988, 98, 191.
18. J.M. O'Connor and J.C. Pizza, *Estuaries*, 1987, 10, 68.
19. T.H. Rogers and K.J. Hall, *Water Pollution Research Journal of Canada*, 1987, 22, 197.
20. Private communication.
21. J.W. Farrington, J.M. Teal, J.G. Quinn, T. Wade and K.A. Burns, *Bulletin of Environmental Contamination and Toxicology*, 1973, 10, 132.
22. J.W. Farrington, J.M. Teal, J.G. Quinn, T. Wade and K.A. Burns, *Bulletin of Environmental Contamination and Toxicology*, 1973, 10, 129.
23. G.C. Medeiros and J.W. Farrington, Page 6, in *Marine Pollution Monitoring (Petroleum)*, Special Publication 409, US National Bureau of Standards, Gaithersburg, MD, 29.
24. R.J. Law, *Analytical Proceedings (London)* 1982, 19, 248.
25. S.N. Chesler, B.H. Gump, H.S. Hertz, W.E. May and S.A. Wise, *Analytical Chemistry*, 1978, 50, 805.
26. D.L. Vassilaros, P.W. Stoker, G.M. Booth and M.L. Lee, *Analytical Chemistry*, 1982, 54, 106.
27. C.S. Giam, H.S. Chan and G.S. Neff, *Analytical Chemistry*, 1975, 47, 2225.
28. R.J. Ozretich and W.P. Schroeder, *Analytical Chemistry*, 1986, 58, 2041.
29. S.D. West and W. Day, *Journal of the Association of Official Analytical Chemists*, 1986, 69, 856.
30. M. Takase, E. Takahashi, M. Murata, H. Ohnuki, K. Hibi, H. Ren and H. Endo, *International Journal of Environmental Analytical Chemistry*, 2013, 93, 125.
31. E. Bulukin, G. Bagni, G. Jonsson, T. Baussant and M. Mascini, *International Journal of Environmental Analytical Chemistry*, 2006, 86, 1039.
32. P. Suchy and J. Berry, *International Journal of Environmental Analytical Chemistry*, 2014, 92, 1443.
33. H. Xu, ZX Chan, L. Guo, J. Zhang, W. Lai, Z.P. Aguilar, H. Wei and Y. Xiong, *International Journal of Environmental Analytical Chemistry*, 2013, 93, 959.
34. N. Ichinose, K. Adachi and G. Schwedt, *Analyst*, 1985, 110, 1505.
35. H. Shultz, *Fresenius Zeitschrift für Analytische Chemie*, 1985, 320, 725.
36. U. Kiigemagi, J. Burnard and L.C. Terriere, *Journal of Agriculture and Food Chemistry*, 1975, 23, 717.
37. J.F. Martin, C.P. McCoy and W. Greenleaf, *Canadian Journal of Fisheries and Aquatic Science*, 1987, 44, 909.
38. H.T. Lovell, Y.Y. Lelona, C.E. Boyd and M.S. Armstrong, *Transition of American Fisheries Society*, 1986, 115, 484.
39. P.E. Persson, *Water Research*, 1980, 14, 1113.
40. D.E. Wells and A.A. Cowan, *Analyst*, 1981, 106, 862.
41. J.L. Allen and J.B. Sills, *Journal of the Association of Official Analytical Chemists*, 1974, 57, 387.
42. K. Itoh, M. Chikuma and H. Tanaka, *Fresenius Zeitschrift für Analytische Chemie*, 1988, 330, 600.
43. S.A. Sinex, A.Y. Cantillo and G.R. Helz, *Analytical Chemistry*, 1980, 52, 2342.
44. A.Y. Cantillo, S.A. Sinex and G.R. Helz, *Analytical Chemistry*, 1984, 56, 33.
45. N.H. Suhr and C.O. Ingasmello, *Analytical Chemistry*, 1968, 38, 730.
46. R.E. Sturgeon, J.A.H. Desauliniers, S.S. Berman and D.S. Russell, *Analytica Chimica Acta*, 1982, 134, 283.
47. N.R. Mc Quaker, P.D. Kluckner and G.N. Chang, *Analytical Chemistry*, 1979, 51, 888.
48. J.N. Walsh and R.A. Howie, *Mine Management*, 1980, 43, 967.
49. A.J. Murray and J.P. Riley, *Analytica Chimica Acta*, 1973, 65, 261.
50. A.J. Murray and J.P. Riley, *Nature (London)*, 1973, 242, 37.

51. A.A. Deetman, P. Demeulemeester, M. Garcia, G. Hauck, J.I. Hollies, D. Krockenberger, D.E. Palin, H. Prigge, L. Rohrschneider and L. Schmidhammer, *Analytica Chimica Acta*, 1976, 82, 1.

52. B. Parejko and R. Keler, *Bulletin of Environmental Contamination and Toxicology*, 1975, 14, 480.

53. J. Solomon, *Analytical Chemistry*, 1979, 51, 186.

54. I.R. DehLeon, M.A. Maberry, E.B. Overton, C.K. Raschke, R.C. Remele, C.F. Steele, V.L. Warren and J.C. Laester, *Journal of Chromatography Science*, 1980, 18, 85.

55. M.H. Hiatt, *Analytical Chemistry*, 1983, 55, 506.

56. D.E. Hela and V.D. Papadopoulos, *International Journal of Environmental Analytical Chemistry*, 2013, 93, 1676.

57. B. Luckas and U. Harms, *International Journal of Environmental Analytical Chemistry*, 1987, 29, 215.

58. J.L. Johnson, D.L. Stalling and J. Hogen, *Bulletin of Environmental Contamination and Toxicology*, 1974, 11, 393.

59. R.A. Hughes and G.F. Lee, *Environmental Science and Technology*, 1973, 7, 934.

60. C.J. Musial and J.F. Uthe, *International Journal of Environmental Analytical Chemistry*, 1983, 14, 117.

61. H. Steinwandter, *Fresenius Zeitschrift für Analytische Chemie*, 1987, 326, 139.

62. D.W. Kuchl, H.L. Kopperman, G.D. Veith and G.E. Glass, *Bulletin of Environmental Contamination and Toxicology*, 1976, 16, 127.

63. H. Steinwandter and L. Zimmer, *Fresenius Zeitschrift für Analytische Chemie*, 1983, 316, 705.

64. T. Ramdahl, G.E. Carlberg and P. Kolsaker, *Science of the Total Environment*, 1986, 48, 147.

65. K. Lindstrom and R. Schubert, *Journal of High Resolution Chromatography and Chromatography Communications*, 1984, 7, 68.

66. A. Stark, *Journal of Agriculture and Food Chemistry*, 1969, 17, 871.

67. L. Rudling, *Water Research*, 1970, 4, 533.

68. L. Renberg, *Analytical Chemistry*, 1974, 46, 459.

69. H.J. Hoben, S.A. Ching, L.J. Casarett and R.A. Young, *Bulletin of Environmental Contamination and Toxicology*, 1976, 15, 78.

70. M. Sackmasserova-Vennigerova and M. Uhnak, *Vodni Hospodarstvi, Series B* 1981, 31, 133.

71. O.M. Sackmauer, O. Pal'Usova and A. Szokolay, *Water Research*, 1977, 11, 551.

72. M.J. Szelewski, D.R. Hill, S.J. Spiegel and E.C. Tifft, *Analytical Chemistry*, 1979, 51, 2405.

73. H. Tausch, G. Stehlik and H. Widlidal, *Chromatographia*, 1931, 41, 403.

74. G. Norheim and E.M. Okland, *Analyst (London)*, 1980, 105, 990.

75. W.B. Neeley, *Science of the Total Environment*, 1977, 7, 117.

76. L.L. Frederick, *Fisheries Research Board, Canada*, 1975, 32, 1705.

77. J. Jan and S. Malservic, *Bulletin of Environmental Contamination and Toxicology*, 1978, 6, 772.

78. M. Olsson, B. Jenson and L. Rentergard, *Ambio*, 1978, 7, 66.

79. M.J. Szelewski, D.R. Hill, S.J. Spiegel and E.C. Tifft, *Analytical Chemistry*, 1979, 51, 2405.

80. D.E. Gaskin, G.J.D. Smith, P.W. Arnold, M.V. Louisy, R. Frank, M. Moldrinet and J.W. McWade, *Fisheries Research Board, Canada*, 1974, 31, 1235.

81. J.L. Ludke and C.J. Schmitt, *Proceedings of the 3rd USA-USSR Symposium on the Effect of Pollutants upon Aquatic Ecosystems*, Duluth, MN, 1980, p. 97.

82. A.A.Z. Douabul, H.T. Al-Saad, Z. Al-Obaidy and H.N. Al-Rekabi, *Water, Air, and Soil Pollution*, 1987, 35, 187.

83. R. Frank, A.F. Armstrong, R.G. Boeleus, H.H. Braun and C.N. Douglas, *Pesticides Monitoring Journal*, 1974, 7, 165.
84. R.J. Hesselberg and J.L. Johnson, *Bulletin of Environmental Contamination and Toxicology*, 1972, 7, 115.
85. J. Simal, J. Crous Vidal, A. Maria-Chareo Arias, M.A. Boado, R. Diaz and D. Vilas, *Analytical Bromat (Spain)*, 1971, 23, 1.
86. A.S.Y. Chau, *Journal of the Association of Official Analytical Chemists*, 1972, 55, 519.
87. D.W. Keum, *Analytical Chemistry*, 1977, 49, 521.
88. H. Sackmauerova, O. Pal'Usova and A. Szokolay, *Water Research*, 1977, 11, 537.
89. R.E. Langlois, A.R. Stemp and B.J. Liska, *Milk Food Technology*, 1954, 27, 202.
90. B. Luckas, H. Pscheidl and P. Haberland, *Journal of Chromatography*, 1978, 147, 41.
91. B. Luckas, H. Pscheidl and D. Haberland, *Nahrung* 1976, 20, K-K2.
92. L.L. Lamparski, T.J. Nestrick and R.H. Stehl, *Analytical Chemistry*, 1979, 51, 1453.
93. D.W. Phillipson and B.J. Puma, *Analytical Chemistry*, 1980, 52, 2328.
94. J. Paasivirta, J. Tohanen and J. Sakkeli, *Chemosphere*, 1986, 15, 1429.
95. R.K. Mitchum, G.F. Moier and W.A. Korfmacher, *Analytical Chemistry*, 1980, 52, 2278.
96. R.L. Harless, E.O. Oswald, R.G. Lewis, A.E. Dupey, D.D. McDaniel and H. Tai, *Chemosphere*, 1982, 11, 193.
97. J. Lawrence, F. Onuska, R. Wilkinson and B.K. Afghan, *Chemosphere*, 1986, 15, 1085.
98. R.E. Clement, B. Babbie and V. Taguchi, *Chemosphere*, 1986, 15, 1147.
99. V.Y. Taguchi, E.J. Reiner, D.T. Wang, O. Meres and B. Hallas, *Analytical Chemistry*, 1988, 60, 1429.
100. W.B. Crummett, *Chemosphere*, 1983, 12, 429.
101. G.P. Markin, J.C. Hawthorne, H.L. Collins and J.H. Ford, *Pesticides Monitoring Journal*, 1974, 7, 139.
102. J.A. Armour and J.A. Burke, *Journal of the Association of Official Analytical Chemists*, 1970, 53, 761.
103. J. Gaul and P. Cruz-La Grange, *Separation of Mirex* and *PCB's in Fish*, Laboratory Information Bulletin, Food and Drug Administration, New Orleans District, 1971.
104. G.P. Markin, J.H. Ford, J.H. Spence, J. Davies and C.D. Loftis, *Environmental Monitoring for the Pesticide Mirex*, USDA APHIS 81-83, US Department of Agriculture, Washington, DC, November 1972.
105. P.H. Butler, *Biological Science*, 1969, 19, 889.
106. M.D. McKenzie, Fluctuations in the abundance of the blue crab and factors affecting mortalities, South Carolina Wildlife Resources Division Technical Rep. No. 1, South Carolina Department of Natural Resources, Columbia, 1970.
107. R.K. Mahood, M.D. McKenzie, D.P. Middough, S.J. Davies and D. Spitsbergen, *A Report on the Co-Operative Blue Crab Study*, in South Atlantic States, Project No. 2-79-R-1, 2-81-R-1, 2-82-R-1, US Department of the Interior Bureau of Commercial Fisheries, US Fish and Wildlife Service, Falls Church, VA.
108. K.L.E. Kaiser, *Science*, 1974, 185, 523.
109. L.M. Reynolds, *Research Review*, 1971, 34, 27.
110. A.S.Y. Chau and W.K. Wilkinson, *Personal Communications: Pesticide Analytical Manual*, Vols. I and II, Department of Health, Education and Welfare, Food and Drug Administration, Washington, DC, 1971.
110a. L.D. Sawyer, *Journal of the Association of Official Analytical Chemists*, 1971, 56, 1015.
111. J.L. Laseter, I.R. DeLeon and P.C. Remele, *Analytical Chemistry*, 1978, 50, 1169.
112. F.I. Onuska, M.E. Comba and J.C. Coburn, *Analytical Chemistry*, 1980, 52, 2272.
113. Y.K. Chau, P.T.S. Wong, G.A. Bengert and O. Kramer, *Analytical Chemistry*, 1979, 51, 186.
114. D.A.J. Murray, *Fisher Research Canada*, 1975, 32, 457.

115. M.L. Deutsch, W.E. Westlake and F.A. Gunther, *Agricultural and Food Chemistry*, 1970, 18, 178.
116. L.H. Zakitis and E.M. McCray, *Bulletin of Environmental Contamination and Toxicology*, 1982, 28, 334.
117. E.M. Lores, J.C. Moore, J., Knight, J. Forester, J. Clark and P. Moody, *Journal of Chromatographic Science*, 1985, 23, 124.
118. S.Y. Szeto, J. Yee, M.J. Brown and P.C. Oloffs, *Journal of Chromatography*, 1982, 240, 526.
119. R.J. Law, *Analytical Proceedings (London)*, 1982, 19, 248.
120. S.N. Chesler, B.H. Gump, H.S. Hertz, W.E. May and S.A. Wise, *Analytical Chemistry*, 1978, 50, 805.
121. J.W. Farrington, J.M. Teal, J.G. Quinn, T. Wade and K.A. Burns, *Bulletin of Environmental Contamination and Toxicology*, 1973, 10, 129.
122. I. Ariese, S.J. Kok, M. Verback, G. Py, C.D. Hoorning, C. Govifer, N.H. Velhorst and J.W. Hafstraat, *Analytical Chemistry*, 1993, 65, 1100.
123. D.L. Vassilaros, P.W. Stoker, G.M. Booth and M.L. Lee, *Analytical Chemistry*, 1982, 54, 106.
124. D.A. Birkholz, A.T. Coutts and S.E. Hrudey, *Journal of Chromatography*, 1988, 449, 251.
125. J.L. Johnson, D.L. Stalling and J. Hogen, *Bulletin of Environmental Contamination and Toxicology*, 1974, 11, 393.
126. K. Itoh, M. Chikuma and H. Tanaka, *Fresenius Zeitschrift für Analytische Chemie*, 1988, 330, 600.
127. A.Y. Cantillo, S.A. Sinex and G.R. Helz, *Analytical Chemistry*, 1984, 56, 33.
128. N.H. Suhr and C.O. Ingasmello, *Analytical Chemistry*, 1968, 38, 730.
129. R.E. Sturgeon, J.A.H. Desauliniers, S.S. Berman and D.S. Russell, *Analytica Chimica Acta*, 1982, 134, 283.
130. N.R. McQuaker, P.D. Kluckner and G.N. Chang, *Analytical Chemistry*, 1979, 51, 888.
131. J.N. Walsh and R.A. Howie, *Mine Management*, 1980, 43, 967.
132. S.A. Sinex, A.Y. Cantillo and G.R. Helz, *Analytical Chemistry*, 1980, 52, 2342.
133. A.J. Murray and J.P. Riley, *Analytica Chimica Acta*, 1973, 65, 261.
134. A.J. Murray and J.P. Riley, *Nature (London)*, 1973, 242, 37.
135. A.A. Deetman, P. Demeulemeester, M. Garcia, G. Hauck, J.I. Hollies, D. Krockenberger, D.E. Palin, H. Prigge, L. Rohrschneider and L. Schmidhammer, *Analytica Chimica Acta*, 1976, 82, 1.
136. R. Parejko and R. Keller, *Bulletin of Environmental Conamination and Toxicology*, 1975, 14, 480.
137. J. Solomon, *Analytical Chemistry*, 1979, 51, 186.
138. I.R. De Leon, M.A. Maberry, E.B. Overton, C.K. Raschke, P.C. Remele, C.F. Steele, V.L. Warren and J.L. Laester, *Chromatographic Science*, 1980, 18, 85.
139. M.H. Hiatt, *Analytical Chemistry*, 1983, 55, 506.
140. G. Linde, J. Gether and E. Steinnes, *Ambio*, 1976, 5, 180.
141. H.J. Hoben, S.A. Ching, J.J. Casarett and R.A. Young, *Bulletin of Environmental Contamination and Toxicology*, 1976, 15, 78.
142. A. Stark, *Journal of Agriculture and Food Chemistry*, 1969, 17, 871.
143. L. Rudling, *Water Research*, 1970, 4, 533.
144. M. Sackmasserova-Vennigerova and M. Uhnak, *Vodni Hospodarstvi, Series B*, 1981, 31, 133.
145. L. Renberg, *Analytical Chemistry*, 1974, 46, 459.
146. H. Tausch, G. Stehlik and H. Widlidal, *Chromatographia*, 1981, 41, 403.
147. W.B. Neeley, *Science of the Total Environment*, 1977, 7, 117.
148. A. Jones, *Analytical Chemistry*, 1994, 66, 1264.
149. H.R. Buser, D.R. Zook and C. Rappe, *Analytical Chemistry*, 1992, 64, 1176.

150. D.W. Phillipson and B.J. Puma, *Analytical Chemistry*, 1980, 52, 2328.
151. R.K. Mitchum, G.F. Moier and W.A. Korfmacher, *Analytical Chemistry*, 1980, 52, 2278.
152. J. Lawrence, F. Onuska, R. Wilkinson and B.K. Afghan, *Chemosphere*, 1986, 15, 1085.
153. R.L. Harless, E.O. Oswald, R.G. Lewis, A.E. Dupey, D.D. McDaniel and H. Tai, *Chemosphere*, 1982, 11, 193.
154. R.E. Clement, B. Babbie and V. Taguchi, *Chemosphere*, 1986, 15, 1147.
155. L.L. Lamparski, T.J. Nestrich and R.H. Shehl, *Analytical Chemistry*, 1979, 51, 1453.
156. L.M. Smith, D.L. Stalling and J.L. Johnson, *Analytical Chemistry*, 1984, 56, 1830.
157. V.Y. Taguchi, E.J. Reiner, D.T. Wang, O. Meresz and B. Hallas, *Analytical Chemistry*, 1988, 60, 1429.
158. W.B. Crummett, *Chemosphere*, 1983, 12, 429.
159. R.D. Chang, W.M. Jarman and J.A. Hennings, *Analytical Chemistry*, 1993, 65, 2420.
160. G. Norheim and E.M. Okland, *Analyst*, 1980, 105, 990.
161. G.P. Markin, J.C. Hawthorne, H.L. Collins and J.H. Ford, *Pesticides Monitoring Journal*, 1974, 7, 139.
162. R. Frank, A.F. Armstrong, R.G. Boeleus, H.H. Braun and C.N. Douglas, *Pesticides Monitoring Journal*, 1974, 7, 165.
163. R.J. Hesselberg and J.L. Johnson, *Bulletin of Environmental Contamination and Toxicology*, 1972, 7, 115.
164. J. Simal, J. Crous Vidal, A. Maria-Chareo Aria, M.A. Moado, R. Diaz and D. Vilas, *Analytical Bromat (Spain)*, 1971, 23, 1.
165. A.S.Y. Chau *Journal of the Association of Official Analytical Chemists*, 1972, 55, 519.
166. D.W. Keum, *Analytical Chemistry*, 1977, 49, 521.
167. H. Sackmauerova, O. Pal'Usova and A. Szokolay, *Water Research*, 1977, 11, 537.
168. D.E. Gaskin, G.J.D. Smith, P.W. Arnold, M.V. Louisy, R. Frank, M. Moldrinet and J.W. McWade, *Fisheries Research Board*, Canada, 1974, 31, 1235.
169. R.J. Ozretich and W.P. Schroeder, *Analytical Chemistry*, 1986, 58, 2041.
170. K.L.E. Kaiser, *Science*, 1974, 185, 523.
171. L.M. Reynolds, *Research Review*, 1971, 34, 27.
172. A.S.Y. Chau and W.J. Wilkinson, *Personal Communications: Pesticide Analytical Manual*, Vols. I and II, Department of Health, Education and Welfare, Food and Drug Administration, Washington, DC, 1971.
173. E.J. Bonelli, *Analytical Chemistry*, 1972, 44, 603.
174. J.L. Laseter, I.R. De Leon and P.C. Remele, *Analytical Chemistry*, 1978, 50, 1169.
175. F.I. Onuska, M.E. Comba and J.C. Coburn, *Analytical Chemistry*, 1980, 52, 2272.
176. G.P. Markin, J.C. Hawthorne, H.L. Collins and J.H. Ford, *Pesticides Monitoring Journal*, 1974, 7, 139.
177. J.A. Armour and J.A. Burke, *Journal of the Association of Official Analytical Chemists*, 1970, 53, 761.
178. J. Gaul and P. Cruz-La Grange, *Separation of Mirex and PCB's in Fish*, Laboratory Information Bulletin, Food and Drug Administration, New Orleans District, 1971.
179. G.P. Markin, J.H. Ford, J.H. Spence, J. Davies and C.D. Loftis, *Environmental Monitoring for the Pesticide Mirex*, USDA APHIS 81-83, US Department of Agriculture, Washington, DC, November 1972.
180. L.H. Zakitis and E.M. McCray, *Bulletin of Environmental Contamination and Toxicology*, 1982, 28, 334.
181. M.E. Deutsch, W.E. Westlake and F.A. Gunther, *Agricultural and Food Chemistry*, 1970, 18, 178.
182. S.Y. Szeto, J. Yee, M.J. Brown and P.C. Oloffs, *Journal of Chromatography*, 1982, 240, 526.
183. E.M. Lores, J.C. Moore, J. Knight, J. Forester, J. Clark and P. Moody, *Journal of Chromatographic Science*, 1985, 23, 124.

184. S.D. West and W. Day, *Journal of the Association of Official Analytical Chemists*, 1986, 69, 856.
185. J.D. Brewster and A.R. Lightfield, *Analytical Chemistry*, 1993, 65, 2415.
186. D.A.J. Murray, *Fisheries Research Board, Canada*, 1975, 32, 457.
187. H. Steinwandter, *Fresenius Zeitschrift für Analytische Chemie*, 1987, 326, 139.
188. Y.K. Chau, P.T.S. Wong, G.A. Bengert and O. Kramer, *Analytical Chemistry*, 1979, 51, 186.
189. K. Lindstrom and R. Schubert, *Journal of High Resolution Chromatography and Chromatography Communications*, 1984, 7, 68.
190. D.W. Kuchl, H.L. Kopperman, G.D. Veith and G.E. Glass, *Bulletin of Environmental Contamination and Toxicology*, 1976, 16, 127.
191. H. Steinwandter and L. Zimmer, *Fresenius Zeitschrift für Analytische Chemie*, 1983, 316, 705.
192. T. Ramdahl, G.E. Carlberg and P. Kolsaker, *Science of the Total Environment*, 1986, 48, 147.
193. B. Bush and E.L. Bernard, *Analytical Letters*, 1982, 15, 1643.
194. C.S. Giam, H.S. Chan and G.S. Neff, *Analytical Chemistry*, 1975, 47, 2225.
195. U. Kiigemagi, J. Burnard and L.C. Terriere, *Journal of Agriculture and Food Chemistry*, 1975, 23, 717.
196. D.E. Wells and A.A. Cowan, *Analyst*, 1981, 106, 862.
197. J.L. Allen and J.B. Sills, *Journal of the Association of Official Analytical Chemists*, 1974, 57, 387.

4 Determination of Organometallic Compounds in Fish

4.1 ORGANOARSENIC COMPOUNDS

Fishman and Spencer [1] used an ultraviolet radiation or an acid persulphate digestion procedure to decompose organoarsenic compounds. The automated methods of Agemian and Cheam [2] use hydrogen peroxide and sulphuric acid for the destruction of organic matter, combined with permanganate-persulphate oxidation for the complete recovery of organoarsenic compounds from fish. An automated system based on sodium borohydride reduction with atomisation in a quartz tube is used for the determination of the inorganic arsenic thus produced.

Agemian and Cheam [2] found that, in the sodium borohydride reduction of inorganic arsenic to arsenic, concentrations from 0.5 to $1.5\,mol\,L^{-1}$ of hydrochloric acid gave the highest sensitivity; both As(III) and As(V) were equivalently detected. When the hydrochloric acid concentration was increased from 2 to $6\,mol\,L^{-1}$, the sensitivity for both species decreased, particularly for As(V). Replacement of the hydrochloric acid line with a sulphuric acid line reduced sensitivity for As(III) by about 30%, and As(V) gave a sensitivity for As(III) of about 50%. Replicate determination at a level of $0.10\,\mu g\ As\ g^{-1}$ in a fish sample gave a relative standard deviation of 15%.

Maher [3] has described a method for determining inorganic arsenic, monomethyl arsenic and dimethyl arsenic acids in fish. The procedure involves the use of solvent extraction to isolate the arsenic species, which are then separated by ion exchange chromatography and determined by arsine generation.

Beauchemin et al. [4] identified and determined organoarsenic species in a dogfish muscle reference sample using high-performance liquid chromatography (HPLC) coupled with inductively coupled plasma–mass spectrometry (ICP-MS), thin-layer chromatography and electron impact mass spectrometry–graphite furnace atomic absorption spectrometry. The major species (84% of total arsenic) was arsenobetaine, present at $16\,\mu g\,g^{-1}$ (as arsenic) in the dogfish sample.

Arsenobetaine has been determined in crab meat by inductively coupled plasma–atomic emission spectrometry.

Hanamura et al. [9] applied thermal vaporisation combined with plasma emission spectrometry to the determination of organoarsenic compounds in fish.

4.2 ORGANOLEAD COMPOUNDS

Chau et al. [5] have described a gas chromatographic method for the determination of organolead compounds. The extract was injected directly into the column injection port of the chromatograph. In this method, the frozen fish tissue was homogenised in a Hobart grinder and a Polytron homogeniser and 2 g of the fish homogenated with 5 ml of ethylenediaminetetraacetic acid (EDTA) reagent, and 5 ml of hexane was immediately placed in a 25 ml test tube with a Teflon-lined screw cap. The contents were shaken rigorously for 2 h in a reciprocating shaker and centrifuged to facilitate phase separation. A suitable aliquot, 5–10 μl of the hexane phase, was withdrawn and injected into the gas chromatographic–atomic absorption system. Chau et al. [6] pointed out that, as the authenticity of the compounds to be analysed must be preserved, any of the digestion methods with acids or alkalis is not suitable, and that extraction seemed to be the method of choice for removing these compounds from samples. For this extraction, they adopted hexane or benzene for the quantitative extraction for tetramethyl lead and tetraethyl lead from fish homogenates suspended in aqueous EDTA solution. Although ionic forms of lead, such as Pb(II), diethyl lead dichloride and trimethyl lead acetate, do not extract in the organic phase, any lead compounds that distribute into this phase as tetraalkyl lead will be determined.

Recovery experiments carried out by Chau et al. [6] showed that benzene, hexane and octanol gave the most satisfactory recovery of tetraalkyl lead compounds from fish tissue (78%–90%).

Chau et al. [6] found that tetraalkyl lead compounds have high vapour pressures and are not stable in water. It was observed that water containing $4.2 \,\mu g \, L^{-1}$ Me_4Pb decreased to 2.8 and $3.9 \,\mu g \, L^{-1}$ when stored, respectively, at room temperature and at 4°C overnight.

Results obtained from measurements of the accumulation of tetramethyl lead in rainbow trout indicated that the tetramethyl lead content of dead trout tissue increased from $0.43 \, mg \, kg^{-1}$ (1-day exposure) to $2.09 \, mg \, kg^{-1}$ (3-day exposure). The trout, after exposure for different periods of time to water containing $3.5 \,\mu g \, L^{-1}$ tetramethyl lead, were found to contain tetramethyl lead.

Butylation of di-, tri- and tetraalkyl compounds followed by gas chromatography has been used for the determination of alkyl lead compounds in fish samples [7].

In this method, fish samples were homogenised a minimum of five times in a commercial meat grinder. About 2 g of the homogenised paste was digested in 5 ml of tetramethylammonium hydroxide solution in a water bath at 60°C for 1–2 h until the tissue had completely dissolved to a pale yellow solution. After cooling, the solution was neutralised with 50% hydrochloric acid to pH 6–8 and the mixture was extracted with benzene for 2 h in a mechanical shaker after addition of 2 g of sodium chloride and sodium diethyldithiocarbamate. After centrifugation of the mixture, a measured amount (1 ml) of the benzene was transferred to a glass-stoppered vial and butylated with 0.2 ml of n-butyl magnesium chloride. The mixture was washed with sulphuric acid (1 N) to destroy the excess Grignard reagent. The organic layer was separated and dried with anhydrous sodium sulphate. Suitable aliquots were injected into the gas chromatography–atomic absorption spectrometry system for analysis.

The percentage recoveries of trialkyl lead and dialkyl lead species at 1–20 μg lead added levels were 72%–91% (trimethyl lead), 72%–86% (triethyl lead, dimethyl lead) and 92%–114% (diethyl lead). The relatively low recovery of dimethyl lead indicates that the dialkyl lead remained bound to the tissue, or existed in solution as a complex, unextractable with solvent. It was noticed that there was a large Pb(II) peak in the fish sample containing spiked dimethyl lead, but such was not found in the standard, which was run in parallel but without the sample. Such low recovery was attributed to the decomposition of dialkyl lead in the fish matrix.

This method was applied to a number of fish samples. For the first time, the occurrence of trimethyl and diethyl lead compounds was detected in fish.

Arsenobetaine has been determined in crab by a procedure of inductively coupled plasma–atomic emission spectrometry [8].

Sirota and Uthe [10] have described a fast, sensitive atomic absorption procedure for determining tetraalkyl lead compounds in biological materials such as fish tissue. Tissue homogenates were extracted by shaking with a benzene/aqueous EDTA solution, a measured portion of the benzene was removed and after digestion, the residue was defatted if necessary. The resultant Pb^{2+} was determined by flameless atomic absorption spectroscopy using a heated graphite atomiser. Using a sample weight of 5 g, 10 µg kg^{-1} of lead, PbR_r can be determined with a relative standard deviation of 5%. No other forms of lead that were tested (e.g. PbR_3X, PbR_2X_2) were found to partition into the benzene layer under these conditions. Recovery of tetramethyl lead was in the range of 103%–130%, and tetraethyl lead was in the range 70%–100%. A study of the distribution of inorganic lead and tetraalkyl lead in a range of fish, including cod and mackerel, showed that between 9% and 90% of the lead existed as tetraalkyl lead.

Chau et al. [11,12,14] and others [13] have described a simple and rapid extraction procedure involving extraction with benzene hexane or octanol of five tetraalkyl lead compounds (Me_4Pb, Me_3EtPb, Me_2Et_2Pb, $MeEt_3Pb$ and Et_4Pb) from fish samples. The extracted compounds are analysed in their authentic forms by a gas chromatographic–atomic absorption spectrometry system. Other forms of inorganic lead do not interfere. The detection limit for fish (2 g) was 0.025 mg kg^{-1}. The recovery of the five alkylated lead compounds from fish tissue averaged 74%, and the coefficient of variation 7.4%. While this method would be applicable to the determination of tetraalkyl lead compounds in fish, the main interest of Chau et al. was determining the organically bound lead produced by biological methylation of inorganic and organic lead compounds in the aquatic environment.

This method was applied to a number of fish samples, including carp, pike, white sucker and small-mouthed bass.

Tetramethyl lead and diethyl dimethyldene were found in carp at levels of 137 and 96 mg kg^{-1}, respectively. All the fish examined contained appreciable levels of methyl trimethyl lead (142–252 mg kg^{-1}), tetramethyl lead (780–7475 mg kg^{-1}) and trimethyl lead.

Birnie and Hodges [15] have given details of a procedure for the determination of down to 0.01 mg kg^{-1} of ionic species of alkyl lead in marine organisms by solvent extraction and differential pulse anodic stripping voltammetry. The sample is homogenised in the presence of a mixture of salts (lead nitrate, sodium benzoate, potassium iodide, sodium chloride and EDTA), which effectively releases the di- and trialkyl lead species present, and facilitates their transfer into toluene as a preliminary to back extraction into dilute nitric acid ready for differential pulse anodic stripping voltammetry. Recoveries were in the range 70%–90% ($Et Pb^{3+}$ and Me_3Pb^+) to 10%–40% (Me_2Pb^{2+}).

Chau et al. [16] applied gas chromatography atomic absorption to the determination of tetraalkyl lead compounds in fish samples in high lead areas. Of some 50 fish samples analysed, only one sample was found to contain detectable amounts (0.26 mg kg^{-1}) of tetramethyl lead in the fillet. The source of tetramethyl lead was not known. The possibility that it comes from *in vivo* lead methylation in the sediment or in the fish cannot be totally disregarded.

4.3 ORGANOMERCURY COMPOUNDS

Jones and Nickless [17] have described a dithizone spectrophotometric procedure for the measurement of trace concentrations of methylmercury salts in fish tissue. A simple equation was applied where absorbance measurements were taken at two wavelengths, cancelling out small differences in excess of dithizone arising between the blank and sample, thus ensuring good precision in the range of 0.1–4.0 µg ml^{-1}. Dithizone reacts with most organomercury salts of the type RHgX, where X is any anion. The extraction procedure used is based on that described by Westoo [18], in which 40 g of fish is homogenised with water, and then concentrated hydrochloric acid and sodium chloride are added. A benzene extract is centrifuged to provide a clear phase, which is then treated with aqueous 1% cysteine solution. After acidification, the aqueous phase is then again extracted with benzene, and to the extract is added a benzene solution of dithizone. This extract is evaluated spectrophotometrically at 628 and 475 nm. Extraction efficiencies were low (24%–32%) but reproducible. Methylmercury contents found in tuna fish flesh ranged from 0.15 to 0.69 mg kg^{-1}.

Various workers have applied cold vapour atomic absorption spectrometry to the determination of organomercury compounds in fish [19–41]. Various methods have been used to decompose organomercury compounds prior to atomic absorption spectrometry, including digestion with acidic potassium permanganate [40], sodium hydroxide [30], sulphuric acid–hydrogen peroxide [26–28,31] or sulphuric acid–nitric acid [32]; hydrochloric acid digestion–benzene extraction [37]; hydrochloric–acid digestion–toluene extraction [38]; and acidic copper bromide digestion–toluene extraction [37]. Steam distillation has also been used to isolate organomercury compounds from fish prior to determination by cold vapour atomic absorption spectrometry [39].

Stuart [32] used mercury-203-labelled methylmercury chloride for *in vivo* labelling of fish to study the efficacy of various wet ashing procedures, and obtained a 93% recovery of activity in digestion of fish with concentrated sulphuric acid and fuming nitric acid.

Magos [20] has described a simple method for the determination of total mercury in biological samples contaminated with inorganic mercury and methylmercury. The method is based on the rapid conversion of organomercurials, first into inorganic mercury and then into atomic mercury suitable for aspiration through the gas cell of a mercury vapour concentration meter, by a combined tin(II) chloride–cadmium chloride reagent. It was found that if 100 mg of tin(II) chloride alone was added instead of the tin(II) chloride–cadmium chloride reagent, then only the release of inorganic mercury influenced the peak deflection of the potentiometer, thus permitting the selective determination of inorganic mercury and then, after reacidification of the reaction mixture, methylmercury, by adding the tin(II) chloride–cadmium chloride reagent and sodium hydroxide. When total mercury and inorganic mercury were determined separately, the difference between results gave the methylmercury content of the sample.

In another method [37], fish sample is treated with boiling sulphuric acid, nitric acid and hydrogen peroxide, with excess peroxide being removed by boiling and the addition of potassium permanganate. A portion of the digest is reduced with hydroxylammonium chloride solution and stannous chloride, aerated through a cell in an atomic absorption spectrometer fitted with a cold cathode mercury lamp, and the absorption measured at 253.7 nm.

In a procedure involving digestion with hydrochloride [30], the fish sample is digested with 2 mol L^{-1} hydrochloric acid, and then organic mercury is extracted into benzene. Organic mercury is then extracted from the benzene phase with $3 \times 10^{-4}\%$ glutathione in 0.1 mol L^{-1} ammonia solution. The extract is digested with sodium hydroxide, 2 ml of 1000 ppm copper solution and 5 ml of 5% tin(II) chloride dihydrate solution. Nitrogen gas is passed to eliminate any mercury in the reagent solutions. Then the aqueous back extract from the sample is added. Mercury is concentrated on 1.5 g of gold granules by passing nitrogen gas. The gold granules are heated in a boat to 500°C in a furnace for 2 min, and the absorbance at 253.7 nm is measured by passing nitrogen gas at a flow rate of 1.2 L min^{-1}. Between 0.1 and 1.0 mg kg mercury in rockfish samples were determined by this method.

Capelli et al. [38] have described a procedure for the determination of down to 0.7 mg kg^{-1} mercury in fish.

A sample involving digestion with hydrobromic acid of 5.00–10.00 g of fish flesh is homogenised with 10 ml of water in a mortar and then transferred into a centrifuge tube with water; 47% m/m hydrobromic acid is then added (i.e. 2 N acid concentration in the total volume). Then 50 ml of toluene is added and the tube is shaken for 5 min. After centrifuging, 25 ml of the organic phase is transferred into another 100 ml test tube; 1% cysteine solution is added and the solution is shaken for 2 min and then centrifuged, and 5 ml of the aqueous phase is diluted with water to 50 ml. Mercury is determined in this solution by atomic absorption spectroscopy with the cold vapour technique, using the standard addition method. Then 5 ml of the solution is placed in the bubbler and 2 ml of 45 wt% sodium hydroxide solution and 1 ml of the reducing solution [tin(II) chloride–cadmium chloride] are added. A stream of air or argon at 0.3 L min^{-1} is used to strip the mercury vapour and convey it into the silica-walled cell across the spectrophotometric beam.

A comparison of determination of total mercury in tuna fish obtained by this method with that of methylmercury obtained by a gas chromatographic method [42] indicates that 77% of the total mercury occurs as methylmercury.

Shum et al. [37] have described a procedure for determining down to 0.8 mg kg^{-1} methylmercury in fish, in which an acetone extract of the sample is treated with acidic copper bromide to release methylmercury, which is then extracted into toluene. Addition of dithizone to the extract allows the determination of mercury by graphite furnace atomic absorption spectrometry. The average recovery of added methylmercuric chloride at 2.00 and 4.00 µg levels is 97.7% ± 5.5%. Determination of methylmercury in aqueous sodium thiosulfate, after partitioning from toluene, permits an autosampler to be used. A number of fish samples were analysed for methylmercury using this procedure. The total mercury contents of these samples were determined for comparison samples. The average recovery was 97.7% ± 5.5%. An average of 92.2% of the mercury in muscle samples was methylmercury. This showed that prewashing the samples with acetone did not remove appreciable amounts of tightly bound organic mercury. The swordfish liver samples, however, contained a much lower percentage of methylmercury than muscle. This was not surprising, because the liver had been suggested as one of the sites for demethylation of methylmercury.

In a gas chromatographic method [31], 5 g of the homogenised sample is placed in a centrifuge tube and 25 ml of sodium hydroxide added. The tube is heated at 100°C for 30 min, and then hydrochloric acid, freshly distilled toluene and copper(II) sulphate are added to the mixture, which is shaken and then centrifuged. The toluene layer is siphoned into a 125 ml separating funnel.

The combined toluene extracts are shaken successively with 3 and 2 ml portions of cysteine hydrochloride reagent solution and the extracts combined. Then 1 ml of hydrochloric acid is added prior to extraction with toluene. The cysteine solution is run into a second 25 ml separating funnel, retaining the toluene layer in a separate container. The toluene extracts are combined and made up to 25 ml. This solution is now ready for gas chromatography on a Carbosorb 20M–Chromosorb G column at 160°C using a nickel-63 electron capture detector and nitrogen as carrier gas.

Kamps and McMahon [43] determined methylmercury in fish by gas chromatography. The method involved the partitioning of methylmercury chloride in benzene, and analysis with electron capture detection. Down to 0.02 ppm of methylmercury chloride was detected in a 10 g sample. Longbottom et al. [44] used the Westoo clean-up procedure to detect down to 10 µg kg^{-1} of methylmercury. They improved the Westoo clean-up procedure by replacing cysteine with the more stable sodium thiosulphate when forming the methylmercury adduct. For the gas chromatography of methylmercury iodide, these workers recommended the use of a ^{63}Ni electron capture, as it does not form an amalgam at 280°C.

Cappon and Crispin Smith [45] have described a method for the extraction, clean-up and gas chromatographic determination of organic (alkyl and aryl) and inorganic mercury in fish. Methyl-, ethyl- and phenylmercury are first extracted as the chloride derivatives. Inorganic mercury is then isolated as methylmercury upon reaction with tetramethyltin. The initial extracts are subjected to thiosulphate clean-up, and the organomercury species are isolated as the bromide derivatives. Total mercury recovery ranges between 75% and 90% for both forms of mercury, and is assessed by using appropriate ^{203}Hg-labelled compounds for liquid scintillation spectrometric assay. The lower detection limit is 1 µg kg^{-1} for the detection of mercury. The mean deviation and relative accuracy averages are 3.2% and 2.2%, respectively.

Uthe et al. [42,46] have described a rapid semi-micromethod for determining methylmercury in fish. The procedure involves extracting the methylmercury into toluene as methylmercury(II) bromide, partitioning the bromide into aqueous ethanol as the thiosulphate complex and reextracting methylmercury(II) iodide into benzene, followed by gas chromatography on a glass column packed with 7% Carbowax 20M on Chromosorb W and operated at 170°C with nitrogen as carrier gas and electron capture detection. Down to 0.01 mg kg^{-1} of methylmercury in a 2 g sample could be detected.

Callum et al. [47] used the proteolytic enzyme subtilisin Carlsberg Type A for the breakdown of fish tissues prior to the release of methylmercury. The finely chopped tissue was homogenised with 1 mol L^{-1} tri(hydroxymethyl)amino methane–hydrochloric acid buffer (pH 8.5) and then incubated with the subtilisin for 1 h at 50°C. Then 40% w/w sodium hydroxide solution and 1% w/v cysteine hydrochloride solution were added and the samples stirred for 5 min at 50°C. When cool, 1 ml of 0.5 mol L^{-1} copper(II) bromide and 10 ml of acidic sodium bromide were added. The methylmercury(II) bromide was then extracted with two 5 ml portions of toluene. In each extraction, the mixtures were shaken for 2 min and then centrifuged at 600g for 10 min. The two toluene extracts were removed and combined, and the methylmercury was extracted twice with 1 ml of ethanolic thiosulphate solution (a 1:1 mixture of 95% ethanol and 0.005 mol L^{-1} sodium thiosulphate solution). During each extraction, the solutions were vortex mixed and centrifuged at 400g for 2 min. The lower aqueous layers were removed and combined, and then 0.5 ml of 3 mol L^{-1} potassium iodide was added to these combined aqueous extracts, followed by 0.5 ml of benzene (pesticide grade, distilled in glass) containing ethylmercury(II) iodide as an internal standard. These solutions were shaken and then centrifuged at 3000g for 1 min. Standard solutions of methylmercury(II) iodide were prepared in the benzene containing the internal standard. Samples were analysed by gas chromatography on a column comprising 5% w/w ethylene glycol adipate polyester on an 80- to 100-mesh Gas Chrom G at 155°C with electron capture detection.

Methylmercury recoveries of 96%–97% from tuna were obtained by this procedure. The coefficient of variation was in the range of 3.6%–5.5%.

Fischer et al. [48] determined methylmercury, diethylmercury and methylethylmercury in fish using gas chromatography–atomic absorption spectrometry, following derivatisation with sodium tetramethylborate. The sample was dissolved in alcoholic potassium hydroxide, and then reacted with sodium tetramethylborate and subjected to cryogenic trapping of reaction products passing through a chromatographic column. Down to 4 pg of CH$_3$Hg$^+$g^{-1} could be determined.

Hight [49] and Hight and Corkoran [50] extracted methylmercuric chloride from homogenised acetone-washed (hydrochloric acid) fish tissue using toluene. Toluene extracts were then analysed by gas chromatography with electron capture detection using a 5% DEGS-PS column pretreated with mercuric chloride. Samples of swordfish, shark, shrimp, oysters, clams and tuna were analysed for methylmercury. The detection limit for the method was 0.25 mg kg^{-1}.

Bye and Paus [51] determined alkylmercury compounds in fish tissue, using an atomic absorption spectrometer tuned in at the mercury wavelength as a specific gas chromatographic detector.

Bache and Lisk [52] used emission spectrometry in a helium plasma to detect organomercury compounds separated by gas chromatography of benzene extracts of hydrochloric acid digests of fish.

Holak [53] used a procedure based on HPLC with an atomic absorption spectrometric detector for the determination of methylmercury in fish. Methylmercury is isolated from the blended sample by chloroform elution from a diatomaceous earth–hydrochloric acid column. The organomercury compound is then extracted into a small volume of 0.01 N sodium thiosulphate solution. An aliquot of this solution is injected on a Zorbax ODS column and eluted with methanol–ammonium acetate solution (3:2) buffer, pH 5.5, containing merceptoethanol.

The reproducibility of multiple injections of 2.95 µg ml^{-1} of methylmercury(II) chloride was 3.2% in terms of relative standard deviation. The detection limit was 0.6 ng.

The recoveries obtained by this procedure ranged from 96% to 106% (i.e. 101% ± 5%). Precision was, in terms of relative standard deviation, 4.1%.

MacCrehan and Durst [54] also used an electrochemical detector in the HPLC of extracts of fish. Down to 2 µg kg^{-1} methylmercury was determined in tuna fish and shark meat. The samples were digested with sodium hydroxide, followed by acidification with hydrochloric acid to produce methylmercury chlorides. These organomercury cations are extracted with toluene as neutral chloride complexes. The separation of methyl-, ethyl- and phenylmercury was accomplished in a reversed-phase system by the formation of their neutral 2-mercaptoethanol complexes. Concentrations of 0.93 and 8.41 mg kg^{-1} methylmercury were found, respectively, in albacore tuna and Japanese shark by this method, agreeing well with the 0.95 and 7.4 mg kg^{-1} total mercury contents. Ethyl- and phenylmercury were both absent.

Pillay et al. [55] have used neutron activation analysis to determine down to 0.01 mg kg^{-1} organomercury as total mercury in frozen homogenised fish samples. The samples were then wet ashed with concentrated sulphuric–70% perchloric acid at 120°C–160°C with mercury carrier. A preliminary precipitation as sulphide in acidic medium was followed by a further precipitation as sulphide, and electrode electrodeposition or precipitation was used to isolate mercury. The radioactivities of mercury-197 and mercury-197m were measured by scintillation spectrometry using a thin sodium iodide detector. These techniques were used to carry out a survey of mercury levels of edible fish taken in Lake Erie. In general, the fish from the western basin of Lake Erie had elevated levels of mercury in their edible tissues, 0.2–0.79 mg kg^{-1}, when compared with similar species caught from the central and eastern basins.

Montero-Alvarez et al. [56] have described a sensitive and accurate quantitative method for the speciation of inorganic Hg^{2+} and methylmercury by ICP-HPLC–ICP-MS implemented for marine fish samples. Quantitative extraction of mercury species was achieved using a 0.1% (v/v) 2-mercaptoethanol, 0.05% (w/v) L-cysteine and 0.10% (v/v) hydrochloric acid solution by sonication for 30 min. Chromatographic separation of mercury species was carried out on a C8 reversed-phase column with 0.05% (v/v) 2-mercaptoethanol, 0.75% (w/v) L-cysteine and 0.06 mol L^{-1} ammonium acetate as the mobile phase. A species-specific isotope dilution analysis approach, using ^{201}CH$_3$Hg$^+$ and ^{200}Hg^{2+}, was employed for the quantification of both species. Two biological Certified Reference Materials (DOLT-2 and DORM-2) were analysed to assess the analytical performance. No significant differences were found between the obtained concentrations and the certified reference values for total Hg and CH$_3$Hg$^+$. The results indicate that no species interconversion reactions occurred during the extraction and chromatographic separation procedures. The detection limits for CH_3Hg^+ and Hg^{2+} species were 7.7 and 5.2 ng g^{-1}, respectively. The method recovery (expressed as the sum of both species contents in relation to total mercury concentration analysed by ICP-MS) was about 97% ± 5%. The procedure was applied to speciation of mercury in 12 species of the most commonly consumed commercial fish in Cuba.

4.4   ORGANOTIN COMPOUNDS

Smith [57] discussed the determination of tin in fish. McKie [60] determined total tin and tributyltin in fish by graphite furnace atomic absorption spectrometry following extraction by digestion with nitric acid for total tin, and by *n*-hexane after treatment with hydrochloric acid for tributyltin.

Short [58] has compared two methods for the determination of tributyltin in salmon. One method was a simple screening procedure, determining tin by flameless atomic absorption spectrometry, while the other method was specific for tributyltin and involved separation of tributyltin by gas chromatography, its reduction to metallic tin and determination by atomic absorption spectrometry. The screening method tended to overestimate tributyltin in fish flesh, but could be useful for identifying samples requiring more detailed examination.

Sasaki Fishizaka et al. [59] determined tri-*n*-butyltin and di-*n*-butyltin in fish by gas chromatography with flame photometric detection. The method involved extraction with acidified solvent, gel permeation chromatography clean-up, methyl derivatisation with Grignard reagent and gas chromatographic analysis using flame photometric detection. Recoveries from fish samples spiked at 0.2 and 1.0 mg kg^{-1} were 80%–104% (i.e. 92% ± 12%) for tri-*n*-butyltin and 92%–105% (i.e. 98 ± 6%) for di-*n*-butyltin. The detection limit for both compounds was 0.2 mg kg^{-1}. The levels determined in reared yellowtail fish were similar for both flame photometric detection gas chromatography and gas chromatography–mass spectrometry.

Navarro et al. [86] described a method for simultaneous determination of mercury and butyltin species using a multiple-series specific isotope dilatation methodology. The method was applied to the European *Anguilla anguilla* glass eel and yellow eel.

4.5   ORGANOSILICON COMPOUNDS

Wanatabe et al. [61] have described a method for the separation and determination of siloxanes in water, sediment and samples of fish tissues, using inductively coupled plasma–emission spectrometry. Organosilicone extracted with petroleum ether is evaporated to dryness. The damp residue is dissolved in methyl isobutyl ketone, aspirated into the plasma. The detection limit is 0.01 mg kg^{-1}. Recoveries are about 50% with a coefficient of variation of about 13%.

4.6   DETECTION LIMITS

Detection limits achievable for the analysis of organometallic compounds in fish are reviewed in Table 4.1.

TABLE 4.1

Methods for the Determination of Organometallic Compounds in Fish

Determined	Technique	Detection Limit (mgkg⁻¹ Unless Otherwise Stated)	Reference
Methylmercury compounds	HPLC with absolute atomic absorption spectrometric detector	0.0037 µg	[85]
	HPLC with electrochemical detector	0.002	[62]
Alkylmercury compounds	Sediment extraction, gas chromatography	0.01	[63]
		0.02	[64]
		0.001	[65]
Methylmercuric chloride in swordfish, shark, shrimp, oyster, clam, tuna	Gas chromatography with electron capture detector	0.25	[66–68,70]
Organomercury compounds	Gas chromatography with helium plasma detector, ethylation with sodium borohydride, cryogenic trapping of ethylation produces and then gas chromatography	—	[69]
		4 ng kg⁻¹ as Hg (CH₃Hg⁺)	
		75 ng kg⁻¹ as Hg (labile Hg)	
Methylmercury compounds	Neutron activation analysis enzymic breakdown–gas chromatography	0.01	[71]
Organolead Compounds			
Tetramethyl lead, methyl lead, trimethyl dimethyl lead, diethyl trimethyl lead, ethyl tetraethyl lead	Gas chromatography	0.025	[72,73]
	Butylation of organolead – lead compounds–gas chromatography	—	[74]
Alkyl lead	Differential pulse analysis anodic scanning voltammetry	0.01	[75]
Organotin Compounds			
Tributyltin	Flameless atomic absorption spectrometry	—	[76]
Tributyltin	Gas chromatography with atomic absorption detector	—	[76]
Organotin compounds	Gas chromatography with flame photometric detector	—	[77]

(Continued)

TABLE 4.1 (*Continued*)
Methods for the Determination of Organometallic Compounds in Fish

Determined	Technique	Detection Limit ($mg\,kg^{-1}$ Unless Otherwise Stated)	Reference
Tributyltin compounds	Graphite furnace atomic absorption spectrometry	—	[78]
	Organoselenium Compounds		
Organoselenium compound metabolites	Anion exchange chromatography analysis	—	[79]
	Organoarsenic Compounds		
Mono- and dimethyl arsenic compounds	Sodium borohydride reduction, then atomic absorption spectrometry	<1	[80,81]
Organoarsenic compounds in dogfish muscle	HPLC with absolute ICP-MS detector; also, thin-layer chromatography, electron impact mass spectrometry and graphite furnace atomic absorption spectrometry	$0.3\,\mu g$ As	[82]
	Organoboron Compounds		
$(Et_3NH)_2B_{12}H_{12}CS_2B_{12}H_{11}$ $SHH_2OC_{15}H_{32}B_{10}O_5$	Direct current (DC) plasma atomic emission	0.1	[83]
	Organosilicon Compounds		
Siloxanes	Inductively coupled plasma–atomic emission spectrometry	0.01	[84]

REFERENCES

1. M. Fishman and R. Spencer, *Analytical Chemistry*, 1977, 49, 1599.
2. H. Agemian and V. Cheam, *Analytica Chimica Acta*, 1978, 101, 193.
3. W.A. Maher, *Analytica Chimica Acta*, 1981, 126, 157.
4. D. Beauchemin, M.E. Bednas, S.S. Berman, J.W. McLaren, K.W.M. Siu and R.E. Sturgeon, *Analytical Chemistry*, 1988, 60, 2209.
5. Y.K. Chau, P.T.S. Wong and P.D. Goulden, *Analytica Chimica Acta*, 1976, 421, 85.
6. Y.K. Chau, P.T.S. Wong, G.A. Bengert and O. Kramer, *Analytical Chemistry*, 1979, 51, 186.
7. Y.K. Chau, P.T.S. Wong, G.A. Bengert and J.L. Dunn, *Analytical Chemistry*, 1984, 56, 271.
8. K.A. Francskoni, P. Hicks, R.A. Stockton and K.J. Irgolic, *Chemosphere*, 1985, 14, 144.
9. S. Hanamura, B.W. Smith and J.D. Winefordner, *Analytical Chemistry*, 1983, 55, 2026.
10. C.R. Sirota and J.F. Uthe, *Analytical Chemistry*, 1977, 49, 823.
11. Y.K. Chau, P.T.S. Wong, G.A. Bengert and O. Kramer, *Analytical Chemistry*, 1979, 51, 186.
12. Y.K. Chau, P.T.S. Wong and P.D. Goulden, *Analytica Chimica Acta*, 1981, 421, 85.
13. J.R. Rice and H.J. Dishbergen, *Agriculture and Food Chemistry*, 1968, 16, 867.

14. Y.K. Chau, P.T.S. Wong, G.A. Bengert and J.L. Dunn, *Analytical Chemistry*, 1984, 56, 271.
15. S.E. Birnie and D.J. Hodges, *Environmental Technology Letters*, 1981, 2, 433.
16. Y.K. Chau, P.T.S. Wong and H.J. Suitoh, *Journal of Chromatographic Science*, 1976, 162, 14.
17. P. Jones and G. Nickless, *Analyst*, 1978, 103, 1121.
18. G. Westoo, *Analytion Chimica Scandinavia*, 1967, 21, 1790.
19. M.P. Stainton *Analytical Chemistry*, 1971, 43, 625.
20. L. Magos, *Analyst*, 1971, 96, 847.
21. J.F. Koff, M.C. Longbottom and L.B. Labring, *Journal of the American Water Works Association*, 1972, 64, 20.
22. Environmental Protection Agency, Mercury in water – Provisional method; Mercury in fish – Provisional method, Analytical Quality Control Laboratory, Environmental Protection Agency, Cincinnati, OH, 1972.
23. S. Yamanaka and K. Ueda, *Bulletin of Environmental Contamination and Toxicology*, 1975, 14, 409.
24. M.R. Hendzel and D.M. Jamieson, *Analytical Chemistry*, 1976, 48, 926.
25. C.D. Schultz, D. Clear, J.E. Pearson and J.B. Rivers, *Bulletin of Environmental Contamination and Toxicology*, 1976, 15, 230.
26. Analytical Methods Committee, *Analyst*, 1976, 92, 403.
27. Analytical Methods Committee, *Analyst*, 1976, 101, 62.
28. M.T. Friend, C.A. Smith and D. Wishart, *Atomic Absorption Newsletter*, 1977, 16, 46.
29. H. Agemian and A.S.Y. Chau, *Analytica Chimica Acta*, 1975, 75, 297.
30. K. Matsunaga and S. Takahashi, *Analytica Chimica Acta*, 1976, 87, 487.
31. Analytical Methods Committee, Chemical Society London, *Analyst*, 1977, 102, 769.
32. D.C. Stuart, *Analytical Chemistry*, 1978, 96, 83.
33. H. Agemian and V. Cheam, *Analytica Chimica Acta*, 1978, 101, 193.
34. I.M. Davies, *Analytical Chemistry*, 1978, 102, 189.
35. P. Jones and G. Nickless, *Analyst*, 1978, 103, 1121.
36. K.I. Aspila and J.M. Carron, *Interlaboratory Quality Control Study No 1: Total Mercury in Sediments*, Report series, Inland Waters Directorate Water Quality Branch Services Section, Department of Fisheries and Environment, Burlington, ON, 1970.
37. G.T.C. Shum, H.C. Freeman and J.F. Uthe, *Analytical Chemistry*, 1979, 51, 414.
38. R. Capelli, C. Fezia and A. Franchi, *Analyst*, 1979, 104, 1197.
39. D.L. Collett, D.E. Fleming and G.E. Taylor, *Analyst*, 1980, 105, 897.
40. M.D.K. Abo-Rady, *Fresenius Zeitschrift für Analytische Chemie*, 1979, 299, 187.
41. A.V. Holden, *Pesticides Science*, 1973, 4, 399.
42. J.F. Uthe, J. Solomon and B. Grift, *Journal of the Association of Official Analytical Chemists*, 1972, 55, 583.
43. L.R. Kamps and J. McMahon, *Journal of the Association of Official Analytical Chemists*, 1970, 18, 351.
44. J.E. Longbottom, R.C. Dressman and J.J. Lichtenberg, *Journal of the Association of Official Analytical Chemists*, 1973, 56, 1297.
45. C.J. Cappon and V. Crispin-Smith, *Analytical Chemistry*, 1977, 49, 365.
46. J.F. Uthe, J. Solomon and B. Grift, *Journal of the Association of Official Analytical Chemists*, 1972, 55, 583.
47. G.I. Callum, M.M. Ferguson and J.M.A. Lenihan, *Analyst*, 1981, 106, 1009.
48. R. Fischer, S. Rapsomanikis and M.O. Andreae, *Analytical Chemistry*, 1993, 65, 763.
49. S.C. Hight, *Journal of the Association of Official Analytical Chemists*, 1987, 70, 667.
50. S.C. Hight and M.T. Corcoran, *Journal of the Association of Official Analytical Chemists*, 1987, 70, 24.
51. R. Bye and P.E. Paus, *Analytica Chimica Acta*, 1979, 107, 169.
52. C.A. Bache and D.J. Lisk, *Analytical Chemistry*, 1971, 43, 950.

53. W. Holak, *Analyst*, 1982, 107, 1457.
54. W.A. MacCrehan and R.A. Durst, *Analytical Chemistry*, 1978, 50, 2108.
55. K.K.S. Pillay, C.C. Thomas, J.A. Sondel and C.M. Hyche, *Analytical Chemistry*, 1971, 43, 1419.
56. A. Montero-Alvarez, M.R. Fernandez de la Campa and A. Sanz-Medel, *International Journal of Environmental Analytical Chemistry*, 2014, 94, 36.
57. J.D. Smith, *Nature*, 1970, 225, 103.
58. J.W. Short, *Bulletin of Environmental Contamination and Toxicology*, 1987, 39, 412.
59. K. Sasaki, T. Ishizaka, T. Suzuki and Y. Saito, *Journal of the Association of Official Analytical Chemists*, 1988, 71, 360.
60. J.C. McKie, *Analytica Chimica Acta*, 1987, 197, 303.
61. N. Wanatabe, Y. Yasuda, K. Kato, T. Nakamura, R. Funasaka, K. Shimokawa, E. Sato and Y. Ose, *Science of the Total Environment*, 1984, 34, 169.
62. W.A. MacCrehan and R.A. Durst, *Analytical Chemistry*, 1978, 50, 2108.
63. J.F. Uthe, J. Solomon and B. Grift, *Journal of the Association of Official Analytical Chemists*, 1972, 55, 583.
64. L.R. Kamps and I. McMahon, *Journal of the Association of Official Analytical Chemists*, 1970, 18, 351.
65. C.J. Cappon and V. Crispin-Smith, *Analytical Chemistry*, 1977, 49, 365.
66. Analytical Methods Committee, Chemical Society London, *Analyst*, 1977, 102, 769.
67. S.C. Hight and M.T. Corcoran, *Journal of the Association of Official Analytical Chemists*, 1987, 70, 24.
68. S.C. Hight, *Journal of the Association of Official Analytical Chemists*, 1987, 70, 667.
69. C.A. Bache and D.J. Lisk, *Analytical Chemistry*, 1971, 43, 950.
70. R. Fischer, S. Rapsomanikis and M.O. Andreae, *Analytical Chemistry*, 1993, 65, 763.
71. K.K.S. Pillay, C.C. Thomas, J.A. Sondel and C.M. Hyche, *Analytical Chemistry*, 1971, 43, 1419.
72. Y.K. Chau, P.T.S. Wong, G.A. Bengert and O. Kramer, *Analytical Chemistry*, 1979, 51, 186.
73. Y.K. Chau, P.T.S. Wong and P.D. Goulden, *Analytica Chimica Acta*, 1976, 421, 85.
74. Y.K. Chau, P.T.S. Wong, G.A. Bengert and J.L. Dunn, *Analytical Chemistry*, 1984, 56, 271.
75. S.E. Birnie and D.J. Hodges, *Environmental Technology Letters*, 1981, 2, 433.
76. J.W. Short, *Bulletin of Environmental Contamination Toxicology*, 1987, 39, 412.
77. K. Sasaki, T. Ishizaka, T. Suzuki and Y. Saito, *Journal of the Association of Official Analytical Chemists*, 1988, 71, 360.
78. J.D. Smith, *Nature (London)*, 1970, 225, 103.
79. A.J. Blotcky, A. Ebrahim and E.P. Rack, *Analytical Chemistry*, 1988, 60, 2734.
80. W.A. Maher, *Analytica Chimica Acta*, 1981, 126, 157.
81. H. Agemian and V. Cheam, *Analytica Chimica Acta*, 1978, 101, 193.
82. D. Beauchemin, M.E. Bednas, S.S. Berman, J.W. McLaren, K.W.M. Siu and R.E. Sturgeon, *Analytical Chemistry*, 1988, 60, 2209.
83. R.J. Barth, D.M. Adams, A.H. Soloway, E.B. Mechetner, F. Alam and A.K.M. Anisuzzamen, *Analytical Chemistry*, 1991, 63, 890.
84. N. Wanatabe, Y. Yasuda, K. Kato, T. Nakamura, R. Funasaka, K. Shimokawa, E. Sato and Y. Ose, *Science of the Total Environment*, 1988, 34, 109.
85. W. Holak, *Analyst*, 1982, 107, 1457.
86. A. Navarro, S. Clemens, V. Perrot, V. Bolliet, H. Tabonret, J. Guerin, M. Montperrus and D. Amouroux, *International Journal of Environmental Chemistry*, 2013, 93, 166.

5 Determination of Metals in Crustacea and Other Non-Fish Creatures

5.1 ARSENIC

Spectrophotometry [1] has been used to determine arsenic in crustacea with a recovery of 98%–100%. Hydride generation atomic absorption spectrometry has been used to determine arsenic in oyster tissue [2,3].

Uthus et al. [3] and Brook and Evans [2] described a method in which arsine generated from dry combusted samples was measured by an atomic absorption spectrometer equipped with a graphite furnace.

With this procedure, the arsenic found in National Bureau of Standards (NBS) Standard Reference Material (SRM) 1566, the standard for oyster tissue (13.7 ± 0.34 mg kg^{-1}), agreed well with the certified value of 13.4 ± 1.9 mg kg^{-1}. The sensitivity and absolute detection limits of the method were 0.11 and 0.14 ng, respectively.

Siu et al. [4] determined arsenic in standard reference materials, including lobster hepatopancreas and oyster tissue. Biological samples were analysed after digestion with concentrated acid and derivatisation with 2,3-dimercaptopropanol. Gas chromatography was used to determine organoarsenic compounds in crustacea with electron capture detection. Results of analysis compared favourably with certified values. A detection limit of 10 pg was reported with an analytical precision of 10%.

Brzezinska-Paudyn et al. [5] compared detection limits for arsenic in various standard reference samples (oyster, lobster, scallop) determined by five different analytical techniques. The results obtained by graphite furnace atomic absorption, combined furnace–flame atomic absorption, non-destructive neutron activation analysis, conventional inductively coupled plasma–atomic emission spectroscopy (ICP-AES), and flow injection/hydride generation inductively coupled atomic emission spectrometry showed all these methods were appropriate for arsenic determination at concentrations higher than 5 mg kg^{-1}. Graphite furnace atomic absorption, with a L'vov platform and nickel matrix modifier, was the most suitable method for analysis of arsenic in biological materials. This method had a detection limit of 0.5–1.0 mg kg^{-1}.

5.2 CADMIUM

Atomic absorption spectrometry has been applied to the determination of cadmium in mussels [6,7] and clams [8].

Molybdenum and lanthanum-coated graphite furnace atomic absorption spectrometry [8] has been used for the determination of cadmium in clam tissue. An average value of 1.3 mg kg^{-1} was found. The determined cadmium content

$(0.31 \pm 0.05 \, \text{mg kg}^{-1})$ of NBS SRM 1577, the bovine liver standard, is in good agreement with the nominal value $(0.34 \pm 0.04 \, \text{mg kg}^{-1})$. The average analytical recovery of cadmium in the clam sample is $104 + 10\%$.

Cadmium determination in clam tissue digests obtained by this above procedure agreed with those obtained by anodic scanning voltammetry in the range $1.0–2.3 \, \text{mg kg}^{-1}$.

Ashworth and Farthing [7] have described a procedure for extracting cadmium from common mussels prior to analysis by atomic absorption spectrometry. The individual whole mussels were dehydrated to constant weight at 50°C, digested under reflux in nitric acid, and the solution buffered to pH 5 with sodium hydroxide and sodium citrate. The cadmium was extracted into dithizone in methyl isobutyl ketone and the organic layer analysed by atomic absorption spectrometry.

Mussels collected from the same region of Port Phillip Bay were found to have a cadmium concentration of approximately $0.5 \, \text{mg kg}^{-1}$ dry weight. A surprisingly high variability of $\pm 0.4 \, \mu\text{g/g}$ was found in a group of 100 individuals, ranging in size from 2 to 5 g dry weight, with no correlation of cadmium concentration with size.

The wavelength modulation inductively coupled plasma–echelle spectrometric technique [9] has been applied to the determination of cadmium in crab tissue. Freeze-dried crab tissue was digested in open tubes with nitric and perchloric acids. Spectrometric evaluation was carried out using the cadmium 226.502 nm line, which is not subject to arsenic interference as is the cadmium 228.803 nm line, but does need a two-point background correction. Very good agreement was obtained in determination of cadmium in crabs by three different methods of analysis: graphite furnace atomic emission spectrometry, $0.76 \pm 0.6 \, \text{mg kg}$; inductively coupled plasma–mass spectrometry (ICP-MS), $0.83 \pm 0.08 \, \text{mg kg}$; and graphite furnace atomic absorption spectrometry, $0.71 \pm 0.8 \, \text{mg kg}^{-1}$.

Mazzucotelli et al. [10] point out that interference by inorganic elements frequently occurs in the determination of cadmium in mussels by methods based on electrothermal atomic absorption spectrometry and ICP-AES. Electrothermal atomic absorption spectrometry of cadmium in solutions containing $50 \, \mu\text{g kg}^{-1}$ plus increasing amounts $(0.5–500 \, \text{mg kg}^{-1})$ of interfering elements showed that sodium, potassium and calcium acted as enhancing agents, whereas iron and magnesium did not. In similar experiments using ICP-AES (wavelengths: 228.802 and 14.438 nm), calcium and iron acted as enhancing agents at both wavelengths, whereas sodium and potassium acted as enhancing agents at 228.802 nm but depressive agents at 214.438 nm. Liquid anion exchange extraction was suggested as a way of overcoming metal interaction (only applicable to electrothermal atomic absorption spectrometry), but separation was necessary when an absolute cadmium value was required.

Greenberg [11] has developed a radiochemical neutron activation procedure for the determination of cadmium in oysters. The procedure is based on irradiation of the sample in a quartz tube with neutrons. Then following a 3-day decay period, the sample is digested with concentrated nitric and sulphuric acids in a sealed polytetrafluoroethylene (PTFE)-lined bomb at 140°C for 2 h, followed by treatment with hydrofluoric acid to remove silica and hydrogen peroxide to destroy nitrogen oxides.

Zinc nitrate is added as a holdback carrier, and a chloroform solution of nickel diethyldithiocarbamate is added to extract mercury into the organic phase (which can also be determined by this procedure). The remaining aqueous fraction is extracted with a chloroform solution of zinc diethyldithiocarbamate. Back extraction of this organic phase with aqueous hydrochloric acid provides an extract containing cadmium. The hydrochloric acid solution is allowed to decay for 24 h to establish the equilibrium between cadmium-115 and its daughter, indium-115m. The 336 keV line from indium-115m and the 528 keV line from cadmium-115 were both used to evaluate the cadmium content of the sample.

Some 2.49 mg kg^{-1} cadmium was found by this method by a certified sample of oyster containing 2.3 mg kg^{-1} cadmium.

5.3 COBALT

Van Raaphorst et al. [12] have investigated the loss of cobalt-60 during the dry ashing of marine mussels. They observed no loss by volatilisation in porcelain crucibles when ashing was carried out at temperatures up to 1000°C. After ashing at 450–550°C, the cobalt could be removed from the crucible with hydrochloric acid prior to counting with a thallium-activated sodium iodide crystal corrected to a single-channel analyser.

5.4 LEAD

The atomic absorption spectrometry method described under cadmium above [8] has also been applied to the determination of lead in clams. The results on clam tissues obtained in a spiking recovery experiment carried out on an authenticated reference sample (NBS 1577, bovine liver) gave a lead content of 0.33 ± 0.08 mg kg^{-1}. The average analytical recovery of lead in the clam sample is 100 ± 6%. A value of 0.83 mg kg^{-1} lead was obtained on a clam sample.

Lead determinations in clam tissue digests obtained by the above procedures are in excellent agreement with those obtained by anodic scanning voltammetry in the concentration range 0.9–2.4 mg kg^{-1}. The relative standard deviation obtained by flame atomic absorption spectrometry in this concentration range is between 18% and 42%.

The wavelength modulation inductively coupled plasma–echelle spectrometric technique [13] has been applied to the determination of lead in crab tissue. Freeze-dried crab tissue was digested in open tubes with nitric and perchloric acids. Spectrometric evaluation was carried out using the lead 220.353 nm line. Very good agreement was obtained in determination of lead in crab tissue by three different methods of analysis, namely ICP-AES, 3.0 ± 0.5; ICP-MS, 2.9 ± 0.1; and graphite furnace atomic emission spectrometry, 2.4 ± 0.3.

5.5 MERCURY

Nondispersive atomic fluorescence spectrometry has been applied to the determination of mercury in shrimps [14].

Various digestion procedures for mussels and oysters, including wet oxidation with nitric–sulphuric acids [15], digestion with concentrated nitric acid in a PTFE-lined bomb [16], Wickbold combustion and digestion with concentrated sulphuric and nitric acids, have been used to digest these materials prior to the determination of mercury by cold vapour atomic absorption spectrometry. Recoveries of between 90% and 105% (i.e. 91.5% ± 1.5%) are claimed with a detection limit of 0.01 mg kg^{-1}.

The cold vapour atomic absorption spectrometric method [17] has been applied to the determination of mercury in oysters. A 97% recovery of mercury was obtained in spiking experiments with oysters following open tube digestion with hydrochloric, nitric and sulphuric acids at 70°C–95°C.

The radiochemical neutron activation analysis procedure [11] has been applied to the determination of mercury in oysters. The final hydrochloric acid extract containing nickel diethyldithiocarbamate and mercury obtained in this procedure was counted immediately after separation for 67.0 keV Au X-ray and the 77.5 keV combination gamma ray and Au X-ray produced by the decay of mercury-197 or, after decay for several weeks, for the 279 keV gamma ray from mercury-203.

Determined mercury contents obtained on an International Atomic Energy Agency oyster homogenisate sample MAM-1 of 0.15 ± 0.012 µg kg^{-1} obtained by this procedure are in reasonably good agreement with the certified values of 0.20 ± 0.02 µg kg^{-1}.

Lo et al. [18] preconcentrated mercury with lead diethyldithiocarbamate prior to its determination by neutron activation analysis in oysters.

5.6   SELENIUM

Magee [19] determined selenium(IV) in oysters by cathodic stripping voltammetry; arsenic(III), copper(II), lead(II), iron(III) and zinc(II) did not interfere. The oyster sample was digested at 50°C with aqueous Mumatron (Hans Kurner, Nurnberg, Germany), then methanol was added and then the solution was acidified with hydrochloric acid prior to polarography. Pre-electrolysis was carried out at −0.0.5 V for 120 s and the solution was then cathodically polarised and quantified by standard addition and measurement of the peak heights of selenium at $Ep = 0.47$ V versus standard electrode (SCE):

$$Se(IV) + 4e + Hg \rightarrow HgSe \ (-0.05\,V)$$

$$HgSe + 2H^+ + 2e \rightarrow HgSe + H_2 \ (-0.41\,V)$$

Good agreement was obtained by this procedure in determination of selenium in NBS SRM 1577 (oyster tissue), for which a value of 2.26 ± 0.24 mg kg^{-1} was obtained against a certified value of 2.1 ± 0.5 mg kg^{-1}.

Maher [20] has reported on selenium levels in prawns (4.01 mg kg^{-1}) and scallops (1.24 mg kg^{-1}), and has pointed out that the selenium is predominantly associated with soluble protein and is not present in inorganic selenium.

5.7   TIN

Thin-film anodic stripping voltammetry has been applied to the determination of tin in distillates of sulphuric acid–hydrobromic acid digests of marine organisms [21].

5.8   VANADIUM

Blocky et al. [22] have described a pre-irradiation chemistry neutron activation analysis procedure employing cation exchange chromatography for the determination of trace-level vanadium in marine biological specimens, including shrimps, crabs and oysters. The procedure, utilising a low-power nuclear reactor ($1 \times 10^{11}\,n\,cm^{-2}\,s^{-1}$), consists of wet digestion of the sample with concentrated nitric acid at 65°C, cation exchange chromatography employing a nitric acid wash to remove the major radioactivatable contaminants (sodium and chloride ions), ammonium hydroxide elution to remove vanadium from the resin and neutron irradiation and radioassay for ^{52}V. The limit of detection of the method is $30\,\mu g\,kg^{-1}$.

The vanadium content obtained in industrial areas is distinctly higher ($3\,mg\,kg^{-1}$) than that in nonlancustrine areas ($0.53\,mg\,kg^{-1}$).

5.9   ZINC

It has been reported [17] that no loss of zinc occurs when mussels are dry ashed at 450°C–550°C in porcelain crucibles and the zinc subsequently dissolved in hydrochloric acid.

5.10   MULTIMETALS

The nitric acid digestion–atomic absorption spectrometric procedure [23] has been applied to the determination of these elements in shrimps, crabs, oysters and mussels.

Four arbitrarily chosen sizes of Mediterranean mussels, collected near Gemlik on the Sea of Marmara in western Turkey, were analysed for mercury, cadmium, copper and lead.

Depending on shell size, the concentrations of mercury, cadmium, copper and lead in Mediterranean mussels were, respectively, 0.03–0.05, 0.07–0.40, 0.75–2.65 and 0.48–0.61 $mg\,kg^{-1}$. There is a distinct variation in concentrations of these metals with mussel shell size, and consequently of shell age.

Topping [24] has reported on interlaboratory comparison studies on atomic absorption spectrometric and other methods for the determination of lead and cadmium in crabs and lobsters. He demonstrated that the majority of the participants in this study can produce comparable (i.e. interlaboratory coefficient of variation [CV] of 17%) and accurate data for cadmium at a tissue concentration of around $10\,\mu mol\,kg^{-1}$, which is typically encountered in shellfish monitoring programmes. Unfortunately, difficulties were encountered in the analyses of lead.

Schlemmer and Welz [25] investigated the determination of arsenic, lead, cadmium, copper and selenium in lobsters and mussels by acid extraction–Zeeman atomic absorption spectrometry. Factors investigated were the importance of Zeeman background compensation and the relative merits of two different extraction methods, one an extraction under pressure with sulphuric acid and the other the complete combustion of the sample in a stream of oxygen. The sulphuric acid extraction under pressure appeared to give more acceptable results, except for arsenic and very low levels of cadmium, for which the Trace-O-Mat combustion process was preferable. The advantages of the Zeeman background correction were confirmed.

Welz and Melcher [26] investigated three different decomposition procedures for lobsters and scallops prior to the determination of arsenic, selenium and mercury by using hydride generation and cold vapour atomic absorption spectrometry. These procedures involved the following:

1. Decomposition with nitric acid under pressure in a PTFE-lined bomb. This resulted in low values for arsenic and selenium but was adequate for the subsequent determination of mercury.
2. Decomposition with nitric, sulphuric and perchloric acids. This method gave the highest values for arsenic and selenium, whereas mercury was partly lost under these conditions.
3. Combustion in a stream of oxygen which could be applied for all three elements, and gave results that were in good agreement with the mean values of an intercalibration.

Pressure decomposition with nitric acid was recommended for mercury, followed by a sulphuric and perchloric acid treatment for the subsequent determination of arsenic and selenium. Detection limits under routine conditions are 0.3 mg kg^{-1} for arsenic, 0.3 mg kg^{-1} for selenium and 0.005 mg kg^{-1} for mercury.

Solchaga and De La Guardia [27] have proposed a method for the nitric acid pressure extraction, using nitric acid, of cadmium, copper, iron, lead and zinc in stoppered borosilicate glass, followed by flame atomic absorption spectroscopy. The five metals were determined in a 300 mg single sample of mussel meat. Detection limits were 0.3 mg kg^{-1} for cadmium, 0.7 mg kg^{-1} for copper, 33.0 mg kg^{-1} for iron, 0.7 mg kg^{-1} for lead and 6.0 mg kg^{-1} for zinc. Precision was estimated from the CV for 20 independent analyses and was 7%, 7%, 6%, 14% and 8% for the five metals. Recoveries were between 90% and 107%, (i.e. 98.5 ≤ 8.5%).

Amiard et al. [28] applied Zeeman atomic absorption spectrometry to the determination of silver, cadmium, chromium, copper, manganese, nickel, lead and selenium in oysters and lobsters. Aliquots (100 mg powdered sample) were digested in 1 ml of concentrated nitric acid at 95°C for 1 h, the volume adjusted to 4 ml with deionised water, and analysed using a graphite furnace coated with tantalum carbide. Detection levels were less than 1 μg kg^{-1} for silver, cadmium and manganese, about 1 μg kg^{-1} for chromium and lead, 5 μg kg^{-1} for copper and nickel, and 15 μg kg^{-1} for selenium. Coefficients of variation were 5%–10% for two series of six determinations and experimental values agreed with certified values.

De Oliviera et al. [29] have described a technique based on continuous hydride generation coupled to an inductively coupled plasma–echelle spectrometer for the determination of arsenic, antimony and selenium in oysters in amounts down to 1 μg L^{-1}.

Ridout et al. [30] used ICP-AES to determine various elements in nitric acid digests of lobster hepatopancreas.

Rancevic et al. [31] used ICP-AES to determine low levels of aluminium, barium, cadmium, copper, iron, magnesium, manganese, sodium, nickel, lead, strontium and zinc in mussel shells.

Shell samples taken near harbours, waste or sewage outlets usually had enhanced levels of these metals, proving that analysis of mussel shells provides useful information on marine environmental status.

Vitoulova et al. [32] characterised the metal in binding biomolecules in *Chamelea gallina* by biodimensional liquid chromatography with in-series ultraviolet and ICP-MS detection containing iron, zinc, copper and manganese. Chisela et al. [33] used epithermal and thermal neutron activation analysis to determine arsenic, bromine, cadmium, iron, manganese, molybdenum, nickel, rubidium, selenium, strontium and zinc in lobsters. Figure 5.1 shows a typical X-ray spectrum of an oyster tissue sample fluorescence.

Anodic scanning voltammetry has been applied to the determination of trace elements in crustacea.

Zeisler et al. [34] determined 44 elements in the digest of marine bivalve tissue using X-ray fluorescence spectrometry. He also used neutron capture, gamma activation analysis and neutron analysis.

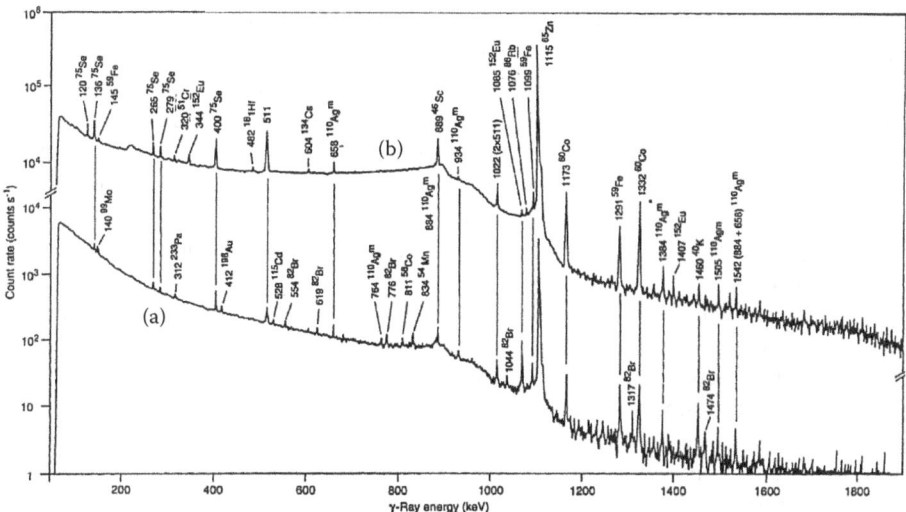

FIGURE 5.1   (a) Gamma-ray spectrum of NBS SRM 1566 (oyster tissue) for brief irradiation. Upper spectrum: Activation with total neutron spectrum; $t_i = 1$ h, $t_w = 6$-day, $t_e = 3600$ s. Lower spectrum: Activation with epithermal neutrons; $t_i = 1$ h, $t_w = 2$ h, $t_e = 3600$ s. ^{24}Na-pp = ^{24}Na gamma-ray pair peak. (b) Gamma-ray spectrum of NBS SRM 1566 (oyster tissue) for prolonged irradiation. Upper spectrum: Activation with total neutron spectrum; $t_i = 48$ h, $t_w = 50$-day, $t_e = 7200$ s. Lower spectrum: Activation with epithermal neutrons; $t_i = 48$ h, $t_w = 15$-day, $t_e = 7200$ s. (t_i = irradiation time, t_w = waiting time, t_e = counting time.)

The photon activation analysis technique [35] has been applied to the determination of manganese, nickel, copper, zinc, arsenic, strontium, cadmium, lead, sodium, manganese and calcium in reference oyster hepatopancreas samples (NBS SRM 1566 and TORT-1). The reference samples were homogenised, spray dried, acetone extracted, vacuum dried, screened, blended, bottled and, finally, irradiation sterilised. All samples were dried at 105°C to remove residual water, and photon activation analysis was performed on the samples in aluminium vials using the National Research Council of Canada electron linear accelerator, using the bremsstrahlung produced by the impact of a focussed electron beam on a tungsten converter. After irradiation, the samples were allowed to decay for a specified time and the gamma-ray spectrum was prepared. Results attained were in fair agreement with those for the standard oyster hepatopancreas sample. Detection limits ranged from 0.004 mg kg^{-1} for calcium to 34 mg kg^{-1} for copper.

Gamma-ray spectrometry has been used to determine trace elements in mussels and oysters [36]. Fourie and Peisach [37] studied the loss of traces of chromium, manganese, iron, cobalt, zinc, arsenic, selenium, cadmium, antimony and lead from oyster tissue during dehydration of wet samples. They used radioactive tracers in these studies, and following an oven drying at 50°C–120°C or freeze-drying dehydration procedure, they estimated unvolatilised residual elements by gamma-ray spectrometry. Live oysters were fed with various radioactive elements and then subjected to various dehydration procedures to establish whether element loss occurred.

The results indicate the following: (1) for the elements chromium, manganese, iron, cobalt and zinc, where possibly no losses occurred during drying, the existing techniques are applicable and reliable; (2) for the elements selenium, cadmium and lead, where appreciable losses were detected, the application of existing techniques without additional precautions or corrections would probably lead to inaccurate results; and (3) antimony and arsenic were not studied owing to 100% mortality among the oysters. This study points to the need to reinvestigate the analytical validity of dehydration processes in other biological systems.

Perhaps the most extensive set of data relating metal concentrations in mussels and oysters in the environment have come from the Mussel Watch Program [38]. Both geographical and temporal trends in seawater concentrations are sought through soft tissue analyses. In the US programme [39], bivalves (oysters and mussels) were collected at more than 100 stations along the East, West, and Gulf Coasts of the United States. Animals of uniform size were sought, where possible, approximately 5–8 cm long, although oysters were slightly larger. Elements determined in this programme included lead, cadmium, silver, zinc, copper, nickel, plutonium and uranium.

Two laboratories carried out the metal analyses, usually by atomic absorption spectrometry – the Scripps Institution of Oceanography and the Loss Landing Marine Laboratories, both in California.

The general picture that emerges for most of the metals is a distributional pattern that repeats itself year after year.

Such a situation can result from long biological half-lives of metals in the organisms (half-lives of the order of a year or more), or from uniform levels of the metals in the seawater, or a combination of the two.

The high lead concentrations in seawaters adjacent to the urban areas result from the combustion of lead alkyls as antiknock agents in gasolines [40]; also, atmospheric input and sewage, storm and river runoff contribute to the anthropogenic lead burden of surface waters. It has been estimated that the annual lead inputs to Southern California coastal waters were 310 metric tons in the early 1970s from the atmosphere, 200 tons from sewage, 190 tons from storm runoff and 40 tons from natural sources. As lead alkyls are phased out from use in gasolines, it is expected that lead concentrations in the waters, and in the bivalves, will decrease with time.

Mussel analyses led to the identification of 'hot spots', where the lead concentrations in the mussels, and presumably in their environmental waters, are raised over adjacent areas as a consequence of fluxes from highly populated industrial areas.

5.10.1 SILVER

In comparison to lead, the sources of silver in the coastal waters of the United States have not been clearly identified. The photographic industry is, perhaps, the largest consumer of silver. Inputs from the plating industry to sewage is a secondary source.

There appears to be no significant atmospheric input of silver into the marine environment. Thus, elevated levels of lead in mussels living adjacent to urban areas may or may not be accompanied by complementary increased concentrations of silver. Such appears to be the case in two West Coast stations, the Farallon Islands and the San Pedro Harbour. Both have high lead concentrations in mussels and low silver values for the 3-year sampling period. Neither of these stations receives sewage. The San Pedro Harbour area does not receive outflow from the Los Angeles River, which undoubtedly carried anthropogenic lead, washed into it from storm runoff.

In contrast, mussels from stations having exposure to sewage show elevated amounts of both lead and silver.

In addition to lead, plutonium serves as an example of the usefulness of isotopic composition for the identification of sources. The Pu-238/Pu239 $\leq$ 240 ratio resulting from the entry of nuclear weapon debris ranges between 0.03 and 0.88 in the byssal threads of mussels taken from waters where there are no localised nuclear point sources. On the other hand, the Pu-238/Pu-239 $\leq$ 240 ratio in the byssal threads of mussels taken near the site of nuclear reactors in San Onofre, California, had values of 0.21 and 0.16. The Pu-238 is used as a fuel and probably leaked into the marine environment in the cooling water discharges. Thus, the plutonium burden of these coastal waters has been increased by a factor of three or four over that of the background fallout plutonium on this basis. Monitoring of byssal threads, in which the plutonium is enriched, is a far simpler task than the monitoring of the waters themselves. Plutonium content in waters off the California coast, for example, ranged from values of $37 \, \text{mg} \, \text{kg}^{-1}$ in 1978 to $2.1 \, \text{mg} \, \text{kg}^{-1}$ in 1981.

The silver, lead, cadmium and plutonium data quoted above illustrate the importance of bivalve monitoring programmes. First, they provide evidence of metal pollution along parts of the conterminous United States. Clean environments can be defined without actual measurements, within the water column. The US Mussel Watch Program suggests a lead baseline of $1.0 \, \text{mg} \, \text{kg}^{-1}$, a West Coast silver baseline

(*Mytilus californianus*) of $0.1 \, \text{mg kg}^{-1}$ and an East Coast silver baseline (*Mytilus edulis*) of $0.05 \, \text{mg kg}^{-1}$ for organisms inhabiting a clean environment.

Second, without expensive and time-consuming water analyses, natural phenomena influencing metal concentrations in seawater can be identified. Clearly, there is a crucial importance for a confirmation of such hypotheses through actual water studies. Without systematic surveys, elevated lead and cadmium might have been interpreted as the result of a localised anthropogenic input rather than a natural physical phenomenon such as upwelling.

Ramelow et al. [41] used atomic absorption spectrometry to determine several metals in marine organisms. Samples were digested with nitric acid in Teflon decomposition vessels.

Vitoulova et al. [42] characterised metal binding biomolecularly in the clam *Chamelea gallina* using a combination of liquid. Clams are very popular for human consumption in the Atlantic southwest coast of Spain. This area is affected by metal pollution from mining activities, which can modify biomolecule expression in this bivalve. The total content of elements was determined by ICP-MS and revealed Fe, Zn, Cu, As and Mn. A metallomics approach has been optimised for this mollusc using size exclusion chromatography on a column of Superdex 30 pg HiLoad 26/60 within series UV and ICP-MS detection. At least four fractions with molecular weight in the range of 425–1540 Da were observed with UV detection, but the ICP-MS chromatogram showed the presence of metals of interest only in the first two fractions. The apparent molecular weights of these metal-containing fractions were from 1325 to 1764 Da. The fractions containing metal compounds were collected and lyophilised for further purification of reconstituted extracts with a second orthogonal chromatographic separation using reversed-phase high-performance liquid chromatography with ICP-MS detection. Several peaks were obtained in this second dimension separation, which allows the isolation of As-, Cu- and ZN-containing biomolecules.

The monitoring of heavy metals in the marine environment is often carried out by using bioindicator organisms. In most cases, metal concentrations in an area are evaluated using pooled samples from a single sampling site. Thus, a large number of individuals are analysed per site, but neither the statistical distribution of the data nor the intrasite variability is known. In order to optimise the monitoring of heavy metals using this kind of sample, some authors have suggested that the variability among individuals should be studied at least in one site.

Saavedra et al. [43] carried out work to determine the frequency distribution and the interindividual variability of Hg, Cd, Pb, Cr, Ni, As, Ag, Cu and Zn in four bioindicator organisms (the blue mussel *Mytilus galloprovincialis*, the clam *Venerupis pullastra*, the king scallop *Pecten maximus* and the cockle *Cerastoderma edule*). In most cases, metals in one-individual samples were shown to follow a log-normal distribution. As the pooled samples included more individuals, they approached the normal distribution but were still closer to the log-normal one, suggesting that, in all cases, a logarithmic transformation should be used to normalise the data. The interindividual variability observed indicated that at least two pooled samples of 30 individuals (a hundred in a few cases) must be analysed to detect differences of 25% (both between sites and with time).

TABLE 5.1

Methods of Determination of Metals and Non-Metals in Water Creatures Other than Fish

Determined	Type of Sample	Technique	Detection Limit (mg kg^{-1} Unless Otherwise Stated)
Arsenic	Oyster		0.11 ng absolute
	Oyster	Graphite furnace atomic absorption spectrometry	0.5
	Oyster	ICP-AES	5
	Oyster	Flow injection hydride generation ICP-AES	5
	Lobster heptopancreas	Spectrophotometric derivativisation with 2,3–dimercaptopropanol–gas chromatography	10 pg absolute
	Oyster	Neutron activation analysis	5
Cadmium	Clam	Atomic absorption spectrometry	—
	Crab	ICP-AES	—
	Oyster	Neutron activation analysis	—
Gadolinium	Crab	Solvent extraction–graphite furnace atomic absorption spectrometry	—
Iodine	Oyster	Microwave digestion and preconcentration, neutron activation analysis	5 ng absolute
Iodine	Oyster	Isotope dilution laser resonance ionisation mass spectrometry	mg kg^{-1} level
Lead	Crab, lobster		—
	Crab	ICP-AES	—
Magnesium	Oyster		—
Manganese	Oyster		—
Mercury	Mussel, oyster		0.01
Selenium	Oyster	Cathodic stripping voltammetry	—
Tin	Marine organism	Anodic stripping voltammetry	—
Vanadium	Shrimp, crab, oyster	Neutron activation analysis	0.03

(Continued)

TABLE 5.1 (Continued)

Methods of Determination of Metals and Non-Metals in Water Creatures Other than Fish

Determined	Type of Sample	Technique	Detection Limit (mg kg⁻¹ Unless Otherwise Stated)
Ni, Pu, U	Mussels		—
Cd, Cu, Fe, Pb, Zn	Mussel	Atomic absorption spectrometry	Cd 0.36 Cu 0.7 Fe 3.3 Pb 0.7 Zn 6.0
As, Se, Hg	Lobster, scallop	Cold vapour atomic absorption spectrometry	As 0.3 Se 0.2 Hg 0.005
Miscellaneous elements	Oyster	Magnetron relating direct current (DC) arc plasma with graphite furnace sample introduction	—
As, Se, Hg	Lobster, scallop	Hydride generation, atomic absorption spectroscopy	As 0.03 Se 0.2 Hg 0.005
As, Pb, Cd Ag, Cd, Cr, Cu, Mn PB, Se, Ni	Lobster, mussel	Zeeman atomic absorption spectrometry	Ag 1 µg absolute Cd 1 Cr 1 Cu 5 Mn 1 Ni 5 Pb 1 Se 15

(Continued)

TABLE 5.1 (*Continued*)

Methods of Determination of Metals and Non-Metals in Water Creatures Other than Fish

Determined	Type of Sample	Technique	Detection Limit (mg·kg^{-1} Unless Otherwise Stated)
Transition metals, rare earths	Oyster	Chelation–ion chromatography	—
As, Br, Cd, Fe, Mn, Mo, Ni, Rb, Se, Cr	Lobster	Neutron activation analysis	As 0.16
			Br 0.26
			Fe 2.8
			Mn 0.16
			Mo 0.16
			Ni 0.1
			Rb 0.07
			Se 0.02
			Sr 1.5
			Zn 0.18
44 elements	Marine bivalves	X-ray fluorescence spectrometry, neutron activation analysis, prompt γ-activation analysis	—
Cr, Mn, Fe, Co, Zn, As, Se, Cd, Sb, Pb	Mussel, oyster	γ-Ray spectrometry	—
Mn, Ni, Cu, Zn, As, Cd, Pb, Na, Mg, Cl, Ca, Sr	Oyster heptopancreas	Photon activation analysis	Mn 4
			Ni 0.6
			Cu 3.4
			Zn 2.0
			As 0.3
			Sr 12

(*Continued*)

TABLE 5.1 (*Continued*)
Methods of Determination of Metals and Non-Metals in Water Creatures Other than Fish

Determined	Type of Sample	Technique	Detection Limit (mg kg^{-1} Unless Otherwise Stated)
		Metals in Biological Materials	
Indium	Blood, liver, kidney, urine	Ion pair extraction–electrothermal absorption spectrometry	Cd 2 Pb 3 Na 200 Mg 0.4 Cl 100 Ca 3 µg L^{-1}
Selenium	Blood, liver	ICP-MS using hydride generation sample introduction system	6.4 absolute
Strontium	Biological materials	Preconcentration–extraction chromatography using crown ether (bis(tertbutyl-cyclohexano)-18-crown-6) in 1-octanol	—
Al, Ba, Ca, Cd, Co, Fe, K, Mg, Mn, Na, P, S, Si, Se, Ti, V, Zn		Parr bomb digestion–ICP-AES	—

Metal Content of Creatures

Element	Lobster	Mussel	Oyster	Crab	Scallop	Clam	Shrimp	Prawn	Whelk
Pb	1	0.7–1		0.43–0.61		0.48–0.6	16.6–1.5		
Mo	0.1–0.2						0.1		
Sr	1.4–1.5	12			0.1				
Na			200						
Mg			0.4						
Ca			3						
Az	0.86–0.93		0.4						
Sb	0.02–0.09	—	0.3						
Bi		0.0025							
Co	0.34–0.44	—	0.75–2.65						
As	0.5–5.0		3–132.4						
Cd		0.3–25.6	0.0025–25.6	0.03–7.7		13			3.2
Hg	0.05	0.01	0.01	1.5	0.04	0.003		14	
V			0.03–1.43	0.03–0.04					
Fe	2.8–219	3.3	0.48–0.61	0.03–1.8			0.03–3.05		
Zn	0.18–350	6.0	2.0						
Se	0.2–15		17–23		0.2–2.0			0.03–13	
Cr	0.75–1.0	0.75–26					0.75–2.69		
Cu	5	5	3.4						
Mn	0.16–19	0.16	4						
Mi	0.1–5	5	0.6						

Locatelli and Melucci [44] have reported a method for the sequential voltammetric determination in mussels and clams of Hg(II) and Cu(II) at gold electrode, and of Cu(II), Pb(II), Cd(II) and Zn(II) at hanging mercury drop electrode by square wave anodic stripping voltammetry environmental biomonitor. The analytical procedure was verified by the analysis of standard reference materials: oyster tissue (NBS SRM 1566), mussel tissue (BCR-CRM 278) and cod muscle (BCR-CRM 422). Precision and accuracy, expressed as relative standard deviation and relative error, respectively, were always less than 6%. The analytical procedure was transferred and applied to mussel and clam samples in two lagoon ecosystems connected with the Adriatic Sea (Italy): the Goro Bay, located in the Po River mouth area, and the lagoon ecosystem located in proximity to Ravenna. A critical comparison with spectroscopic measurements is also discussed.

Some idea of the concentrations of metals that can occur in sea reactions can be gained by examination of Table 5.1.

Metal concentrations vary over a wide range and certainly cover the region where adverse effects or mortalities in the creatures would occur and where the suitability of the creature as an item of human diet would be queried.

REFERENCES

1. W.A. Maher, *Analyst*, 1983, 108, 939.
2. P.J. Brook and W.H. Evans, *Analyst*, 1981, 106, 574.
3. E.V. Uthus, E. Collins, W.E. Cormatzor and F.H. Nelson, *Analytical Chemistry*, 1981, 55, 2221.
4. K.W. Siu, S.Y. Roberts and S.S. Berman, *Chromatographia*, 1984, 19, 398.
5. A. Brzezinska-Paudyn, J. Vanham and R. Hancock, *Atomic Spectroscopy*, 1986, 7, 72.
6. R.L. Gabrielli, J.P. Marletta and L. Favrelo, *Atomic Spectroscopy*, 1980, 1, 35.
7. M.J. Ashworth and R.H. Farthing, *International Journal of Analytical Chemistry*, 1981, 10, 35.
8. J.E. Poldoski, *Analytical Chemistry*, 1980, 52, 1147.
9. J.W. McLaren and S.S. Berman, *Applied Spectroscopy*, 1981, 35, 403.
10. A. Mazzucotelli, A. Vinrengo, G. Martino and R. Frache, *Environmental Research*, 1988, 24, 129.
11. R.R. Greenberg, *Analytical Chemistry*, 1980, 52, 676.
12. J.G. Van Raaphorst, A.W. van Weers and H.M. Haremaken, *Analyst*, 1974, 99, 523.
13. J.W. McLaren and S.S. Berman, *Applied Spectroscopy*, 1981, 35, 403.
14. R.C. Hatton and B. Preston, *Analyst*, 1980, 105, 981.
15. W. Zhe-Ming, Xia Chun and H. Heng Bin, *Analytica Chimica Acta*, 1986, 186, 147.
16. W. Holak, B. Frnitz and J.C. Williams, *Journal of the Association of Official Analytical Chemists*, 1972, 55, 741.
17. J.G. Van Raaphorst, A.W. van Weers and H.M. Haremaken, *Analyst*, 1974, 99, 523.
18. J.M. Lo, J.C. Wei, M.H. Yang and S.J. Yen, *Journal of Radioanalytical Chemistry*, 1982, 72, 371.
19. R.J. Magee, *Analyst*, 1983, 108, 835.
20. W.A. Maher, *Marine Pollution Bulletin*, 1985, 16, 33.
21. T.M. Florence and Y.I. Farren, *Electroanalytical Chemistry*, 1974, 51, 191.
22. A.J. Blocky, C. Falone, V.A. Medina and E.P Rack. *Analytical Chemistry*, 1979, 51, 178.
23. G. Ramelow, M.A. Ozkan, G. Tuncel, C. Saydam and T.I. Balkas, *International Journal of Environmental Analytical Chemistry*, 1978, 5, 125.

24. G. Topping, Report of the 6th ICES trace metal intercomparison exercise for cadmium and lead in biological tissue, ICES Liverpool Research Report No. 111, International Council for the Exploration of the Sea, Copenhagen, 1982.
25. G. Schlemmer and B. Welz, *Fresenius Zeitschrift für Analytische Chemie*, 1985, 320, 648.
26. B. Welz and M. Melcher, *Analytical Chemistry*, 1985, 57, 427.
27. M. Solchaga and M. De La Guardia, *Journal of the Association of Official Analytical Chemists*, 1986, 69, 874.
28. J.C. Amiard, A. Pineau, H.I. Boiteau, C. Metayer and C. Amiard-Trigneb, *Water Research*, 1987, 21, 693.
29. E. De Oliviera, J.W. McLaren and S.S. Berman, *Analytical Chemistry*, 1987, 59, 610.
30. P.S. Ridout, R. Jones and J.G. Williams, *Analyst*, 1988, 113, 1383.
31. S. Roncevic, L.P. Svedruzic, J. Smetisko and D. Medakovic, *International Journal of Environmental Analytical Chemistry*, 2010, 90, 620.
32. E. Vitoulova, T. Garcia-Barrera, J.L. Gomez-Ariza and M. Fisera, *International Journal of Environmental Analytical Chemistry*, 2011, 91, 1282.
33. F. Chisela, D. Gavlik and P. Bratten, 1986, *Analyst*, 1986, 111, 405.
34. R. Zeisler, S.F. Stone and R.W. Sanders, *Analytical Chemistry*, 1988, 60, 2760.
35. S. Hansberger and W.F. Davidson, *Analytical Chemistry*, 1985, 57, 197.
36. J.W.R. Dutton, Technical report, Friu Fisheria Radiological Laboratory, Hamilton Lowestoft, Suffolk, UK, 1969.
37. H.O. Fourie and M. Peisach, *Analyst*, 1977, 102, 193.
38. NAS, International Mussel Watch, US National Academy of Sciences XVI, Washington, DC, 1980.
39. E.D. Goldberg, V.T. Bowen, J.W. Farrington, G. Harvey, J.H. Martin, P.L. Parker, R.W. Risebrough, W. Robertson, E. Schneider and E. Gamble, *Environment Conservation*, 1978, 5, 101.
40. D. Patterson, D. Settle, B. Schaule and M. Burnett, Transport of pollution lead for the oceans and within ocean ecosystems, in H.L. Wisdom and R.H. Duce (eds), *Marine Pollutant Transfer*, Lexington Books, Lanham, MD, 1976, p. 23.
41. R. Ramelow, M.A. Ozkan, G. Tuncel, C. Saydam and T.I. Balkas, *International Journal of Environmental Analytical Chemistry*, 1978, 5, 125.
42. E. Vitoulova, T. Garcia-Barreri, J.L. Gomez-Ariza and M. Fisera, *International Journal of Environmental Analytical Chemistry*, 2011, 91, 1282.
43. Y. Saavedra, A. Gonzalez and J. Blanco, *International Journal of Environmental Analytical Chemistry*, 2009, 89, 559.
44. C. Locatelli and D. Melucci, *International Journal of Environmental Analytical Chemistry*, 2010, 90, 49.

6 Determination of Organic Compounds in Crustacea and Other Non-Fish Creatures

A variety of substances can occur in sea creatures other than fish. It is seen in Table 6.1 that, in many instances, the concentrations are of a similar order of magnitude.

Rice and White [1] determined the concentrations of polychlorobiphenyls in fathead minnows (*Pimephales promelas*) and fingernail clams (*Sphaerium striatinum*) before, during and up to 6 months after completion of dredging polychlorobiphenyl-contaminated sediments in the Shiawassee River, Owosso, Michigan, 1 km downstream of the pollution outfall. The concentration of polychlorobiphenyls found in fathead minnow tissue was 32.1–6.1 mg kg^{-1} dry weight, and the concentration in clams was 13.2–15.3 mg kg^{-1} dry weight. It was concluded that dredging had increased the bioavailability of polychlorobiphenyls to these organisms.

Exposure of *Asellus aquaticus* crustacea to water containing 5 µg L^{-1} lindane for 5 days led to a pickup by the organism of 0.2 mg kg^{-1} lindane in the tissue [2]. Depuration was rapid, with more than 40% of accumulated lindane being eliminated within 1 day.

Methods used for the determination of a wide range of organic substances in creatures other than fish are shown in Table 6.2. Detection limits are usually well below 0.01 mg kg^{-1}. Gas chromatography combined with mass spectrometry or various spectrometry methods is the most commonly employed technique, as shown in Table 6.3.

6.1 ALIPHATIC AND AROMATIC HYDROCARBONS

Morgan [44] has described a gas chromatographic method for the determination of Bunker C fuel oil in marine organisms at the 0.5 mg kg^{-1} level. Pentane-methanol extraction of tissues, using a blender, is followed by adsorption chromatography.

Meyers [18] investigated the occurrence of non-biogenic hydrocarbons in shrimp occurring in the vicinity of offshore drilling and petroleum exploration in the Gulf of Mexico. Samples were stored in glass at –20°C prior to analysis. The crushed sample was treated with 0.5 N potassium hydroxide in 1:1 v/v benzene–methanol, and following the addition of water, non-saponifiable lipids were extracted with petroleum ether. Column chromatography using alumina oxide silica gel 50/50 separated saturated from unsaturated plus aromatic hydrocarbons. Gas–liquid chromatography

TABLE 6.1

Occurrence of Organic Substances in Creatures Other than Fish

Substance Determined	Creatures Other than Fish, Type of Creature	Concentration Found in Creatures Other than Fish (mg kg⁻¹)	Reference	Concentration Found in Fish for Comparison (mg kg⁻¹)	Reference
		Polyaromatic Hydrocarbons			
Naphthalene	Mussels	0.003–0.1	[3]	0.001	[5]
Phenanthrene and anthracene		0.008–0.032	[3]	0.008	[4]
Fluoranthene		0.042–0.080	[3]	0.004–1.8	[6]
Pyrene		0.034–0.092	[3]	0.004–1.5	[6]
Benz[*a*]anthracene and chrysene		0.029–0.059	[3]	0.004–0.022	[6]
Benzo[*a*]pyrene		0.055	[7]	0.007–0.083	[6]
Polychlorobiphenyls		0.41–0.9	[3]	0.48–5.0	[3]
		0.011–0.56	[8]		
	Oyster	0.002	[9]		
		0.00015	[9]		
Dibenzothiophane	Mussel	0.0001–0.8	[10]	—	
Diethyl hexyl phthalate	Shrimp, crab	0.003–0.02	[11]	—	
Aliphatic hydrocarbons	Mussel	0.54	[12]		
	Oyster	0.65	[12]		
	Clam	0.49–1.41	[12]		
Ascorbic acid–2-sulphate	Brine				
	Shrimp	0.03–0.1	[4]		

with a flame ionisation detector was employed to resolve and quantify the various components of each hydrocarbon fraction. Both a non-polar column and a polar column were used.

Although the organisms are from an order other than Crustacea and were collected during different sampling periods, their traces are very similar. Few normal alkanes are found in any of the samples, and the saturated hydrocarbon compositions of these animals appear to be composed mostly of branched compounds. The unsaturated hydrocarbon composition of these organisms also displays a fairly simple pattern. Usually, four to six peaks dominate the chromatograms obtained from both non-polar and polar columns. Hydrocarbon compositions are dominated by only a few peaks, and some peaks having the same Kovats indices are common to all five crustaceans examined. The largest peak from the non-polar chromatograms of the saturated fraction has an index of 2506–2510 in all five samples. However, a peak having an index around 2500 is not a major contributor to polar chromatographs of

TABLE 6.2

Methods for the Determination of Organic Compounds in Water Creatures Other than Fish

Determined	Type of Sample	Technique	Detection Limit (mg kg⁻¹ Unless Otherwise Stated)	Reference
Aliphatic hydrocarbons	Oyster	Gas chromatography	—	[13]
Hydrocarbons	Mussel, oyster, clam	Headspace gas chromatography–mass spectrometry	—	[14,15]
Aromatic hydrocarbons	Mussel	Atomic fluorescence spectrometry	—	[16]
Hydrocarbon oils	Marine organisms	Gas chromatography	0.5	[17]
Non-biogenic hydrocarbons	Shrimp	Gas chromatography	—	[18]
	Shrimp	Gas chromatography–mass spectrometry	—	[18]
Polyaromatic hydrocarbons	Mussel	Capillary gas chromatography	—	[7,16]
Naphthalene, phenanthrene + anthracene, fluoranthene, pyrene, benz[a]anthracene + chrysene	Marine organisms	Gas chromatography	—	
Benzopyrene, hexachlorobenzene pentachlorophenol	Marine organisms	Gas chromatography	0.0002	[19]
Polyaromatic hydrocarbons	Lobster	Gas chromatography with ultraviolet fluorescence detector	—	[20,21]
	Mussel	Atomic fluorescence spectroscopy	—	[22]
Phthalate esters	Shrimp, crab	Gas chromatography	0.001	[11]
Trichloroethylene, tetrachloroethylene chloroform, carbon tetrachloride	Marine organisms	Gas chromatography	—	[23,24]
Polychlorobiphenyls	Crab, shrimp	Gas chromatography	0.01	[25–28]

(*Continued*)

TABLE 6.2 (*Continued*)
Methods for the Determination of Organic Compounds in Water Creatures Other than Fish

Determined	Type of Sample	Technique	Detection Limit (mg kg⁻¹ Unless Otherwise Stated)	Reference
Polychlorobiphenyls	Oyster, clam	Electron capture ion chemical ionisation mass spectrometry	0.0065	[29,30]
Polychloroterphenyl	Oyster	Gas chromatography–mass spectrometry	—	[16,31, 32]
Chlorinated insecticides	Oyster	Gas chromatography	0.01	[33]
	Mollusc	Gas chromatography	—	[34]
	Crustacea	Gas chromatography	—	[25, 31–37]
	Oyster	Gas chromatography–mass spectrometry	0.00004 (lindane)	[29]
DDT, DDE, DDD	Scallops	Gas chromatography–mass spectrometry	—	[37,38]
DDT	Crab, shrimp	Gas chromatography	0.001	[11, 20–23]
Mirex		Gas chromatography	0.005	[25–28]
Dursban	Crustacea	Gas chromatography	0.01	[38]
Polychloro-benzothiophenes	Crab, lobster	Gas chromatography–mass spectrometry	—	[39]
Sulphur-containing hydrocarbons, e.g. dibenzothiophene	Mussel	Gas chromatography–mass spectrometry and gas chromatography with flame photometric detector	—	[10,40]
Ascorbic acid–2-sulphate	Oyster, brine shrimp	Solid-phase extraction	—	[41]
Coprostanol	Crustacea	Electron ionisation mass spectrometry	75 ng absolute	[42]
Miscellaneous compounds	Oyster	Magnetron rotating direct current arc plasma	—	[43]

this fraction. Instead, the most common major peak in these latter distributions has a Kovats index of 2140–3144. Major peaks, comprising the unsaturated fractions of the samples, are grouped between indices of 1900 and 2500 on the non-polar chromatograms. A shift to indices between 2200 and 2850 on polar chromatograms is indicative of the relatively polar nature of these unsaturated hydrocarbons.

TABLE 6.3
Review of Methods and Detection Limits Applicable to the Determination of Organic Compounds in Creatures Other than Fish

Technique	Compounds	Detection Limit (mg kg^{-1} Unless Otherwise Stated)
Gas chromatography	Aliphatic hydrocarbons	0.5
	Hydrocarbon oils	
	Non-biogenic hydrocarbons	
	Polyaromatic hydrocarbons	
	Benzopyrene	0.0002
	Phthalate esters	0.001
	Chloroaliphatic compounds	
	Haloforms	
	Polychlorobiphenyls	0.0065–0.01
	Chlorinated insecticides	≤0.01
	DDT	0.001
	Mirex	0.005
	Dursban	0.01
With ultraviolet fluorescence detector	Polyaromatic hydrocarbons	
	Non-biogenic hydrocarbons	
Gas chromatography–mass spectrometry	Chlorinated insecticides	0.00004 (lindane)
	Polychloroterphenyls	
	Polychlorobiphenyls	
	Polychlorobenzothiophenes	
Electron capture negative ion chemical ionisation mass spectrometry	Polychlorobiphenyls	
	Coprostanol	75 ng absolute
High-resolution mass spectrometry	Polychlorodibenzo-*p*-dioxins, dibenzofurans	
Atomic fluorescence spectroscopy	Aromatic hydrocarbons	
	Polyaromatic hydrocarbons	

Clearly, a combination of gas chromatography and mass spectrometry would provide more useful information in studies of hydrocarbons in crustacea and would facilitate a clearer distinction to be made between contaminant hydrocarbons and non-biogenic naturally occurring hydrocarbons.

Chesler et al. [12] have described a headspace sampling–gas chromatographic method for the determination of petroleum hydrocarbons in mussels, oysters and clams. This procedure utilises dynamic headspace sampling of an aqueous caustic tissue homogenate to extract and collect volatile organic components. Interfering polar biogenic (non-anthropogenic) components are removed by normal-phase high-performance liquid chromatography. Quantitation and identification of the individual compounds are accomplished using gas chromatography and gas chromatography–mass spectrometry (GC-MS). The non-volatile polynuclear aromatic hydrocarbons which remain in the homogenate after headspace sampling are solvent extracted, and then analysed by reversed-phase liquid chromatography. The crustacea samples were kept at a low temperature (–10°C) between sampling and analysis. Approximately 30 g of tissue, 500 ml of hydrocarbon-free water, and 50 g of sodium hydroxide were combined in a flask, together with aliphatic or aromatic hydrocarbon internal standards, and the mixture homogenised. The tissue homogenate was heated to 70°C, and the headspace sampled for 18 h at a nitrogen flow rate of 150 ml min^{-1}. The headspace vapours were passed into a Tenax gas chromatography packed stainless steel column. The homogenate solution remaining in the flask after headspace sampling was extracted with pentane to remove non-volatile polyaromatic hydrocarbons (PAHs). This extract was concentrated by nitrogen purge, and devolved in 1 ml of acetonitrile for high-performance liquid chromatography.

Following headspace sampling and drying, the Tenax gas chromatography was connected as part of the injection loop of a liquid chromatograph and the organic compounds eluted with pentane onto a μ Bondepak NH$_2$ clean-up column, the first 15 ml of eluate containing the hydrocarbons. This fraction was reduced to 300 μl by nitrogen purge and the residue was washed onto a Tenax gas chromatography column, the contents of which were thermally purged onto a gas chromatographic column for analysis.

Using an 18 h headspace sampling period, recoveries from water for the higher-molecular-weight aromatic and aliphatic components (i.e. trimethyl naphthalene, MeC$_{16}$ and MeC$_{18}$) were nearly 100%. Exposed live *Mytilus* (mussels) gave a recovery of 78% ± 12%. These were exposed to ^{14}C-naphthalene and then analysed using the 4 h headspace sampling procedure and no high-performance liquid chromatography clean-up. Aliphatic hydrocarbon recoveries were found to be much lower than aromatic hydrocarbon recoveries in the headspace sampling of the tissue homogenate. Using caustic digestion, recoveries from mussel tissue homogenate approached 100% for the higher aromatics but were only 30% for the aliphatic components. It is assumed that the aliphatic hydrocarbons were being retained in the lipophilic portion of the tissue homogenate and that the partition coefficient for these hydrocarbons between the headspace sampling gas and the lipophilic fraction was quite unfavourable. Recovery data for the complete analytical scheme indicate that some losses of the internal standard also occur during the liquid chromatographic clean-up and concentration step. The losses that occur during the concentration step amount to 25% for mesitylene, 30% for 2-methylundecane, 40% for naphthalene, 11% for 5-methyl tetradecane, 5% for trimethyl naphthalene and less than 1% for 7-methyl hexadecane, 2-methyl naphthalene, phenanthrene and hydrocarbons of higher molecular weight.

The headspace sampling procedure for the analysis of hydrocarbons in marine biota offers several advantages over solvent extraction procedures. The headspace sampling technique requires minimal sample handing, few sample transfers, and only a minimal amount of organic solvent, thereby reducing the risk of contamination. A system blank for the headspace sampling method results in a value of only ~5 µg kg^{-1} based on a sample of 600 ml of water [45,46]. In addition, only one solvent concentration step is involved, thereby reducing the losses of more volatile components.

Berthou et al. [47] used gas chromatography to determine weathered aliphatic and aromatic hydrocarbons in oyster samples.

Mason [16] studied the feasibility of using fluorescence spectroscopy to determine aromatic compounds in mussel tissues and compared the results with those obtained by gas chromatography. There were significant correlations between the concentrations of aromatic hydrocarbons found by fluorescence spectroscopy and both aliphatic and aromatic hydrocarbon concentrations obtained by gas chromatography. Analysis of the aliphatic fraction by gas chromatography and of the aromatic fraction by fluorescence spectroscopy would give a reasonable estimation of the relative degree of contamination of mussel by petroleum hydrocarbons.

Warner [67] has described a semitic procedure for the determination of aliphatic and aromatic compounds in marine organisms. The procedure involves digestion with sodium hydroxide, followed by gas chromatography of an ether extract. The detection limit is 0.1–1.0 µg g^{-1}. The method is applicable to all organisms.

6.2 POLYAROMATIC HYDROCARBONS

Bjorseth et al. [20] described a capillary gas chromatographic method for determining PAHs in mussels. Up to 34 PAHs were identified. Dunn and Stich [7] have described a monitoring procedure for PAHs, particularly benzo[a]pyrene in marine organisms in coastal waters. This procedure involves extraction and purification of hydrocarbon fractions from the organism, and determination of compound by thin-layer chromatography and fluorimetry, or gas chromatography.

To avoid possible photodecomposition of PAHs, all extraction and purification procedures were carried out under subdued yellow tungsten light. Between 20 and 40 g of tissue was placed in a 300 ml flask, and 150 ml of ethanol, 7 g of potassium hydroxide, boiling chips, and an aliquot of radioactive benzo[a]pyrene (either 1000 dpm ^{14}C-benzo[a]pyrene, around 5 µg, or 25,000 dpm ^{3}H-benzo[a]pyrene, around 0.1 ng) were added. The tissue was digested by refluxing gently for 1.5 h with occasional swirling. The digest was added while hot to 150 ml of water in a 2 L separatory funnel, and the digestion flask rinsed out with an additional 50 ml of ethanol. The water–ethanol mix was extracted three times with 200 ml of isooctane, combined and washed with water. This extract was then passed down a Florisil clean-up column. Polycyclic aromatic hydrocarbons were eluted from the column with 3 × 100 ml of benzene. The combined eluate was reduced to 5 ml by rotary evaporation, 50 ml of isooctane was added, and the volume again reduced to 5 ml to remove the benzene.

PAHs were extracted from the isooctane with 3 × 5 ml of dimethyl sulphoxide. The dimethyl sulphoxide extracts were combined with 30 ml of water, and the PAHs extracted into 2 × 20 ml of isooctane. The isooctane extracts were combined, washed with water and dried by passage through 10 g of sodium sulphate in a 15 ml coarse fritted glass Buchner funnel. This extract was used for thin-layer chromatography benzo[a]pyrene being detected under long-wavelength ultraviolet light.

The absorbent at the position of the benzo[a]pyrene band was scraped off the plate while still damp, and placed in a fine fritted Buchner funnel. The benzo[a]pyrene was removed from the cellulose acetate by washing with 4×4 ml of hot (65°C) methanol, using gentle suction. The methanol was added to 10 ml of a solution of 20% hexadecane in isooctane, and methanol and isooctane were removed by rotary evaporation to leave the benzo[a]pyrene in 2 ml of hexadecane, ready for fluorimetry.

Benzo[a]pyrene was measured fluorometrically in hexadecane using the baseline technique of Kunte [21]. Samples and standards of 10–200 ng ml^{-1} benzo[a]pyrene in hexadecane were excited at 3–5 nm in an Aminco-Bowman spectrophotofluorimeter, and the emission spectrum was recorded from 375 to 500 nm. An artificial baseline was drawn between minima in the fluorescence spectrum occurring at 418 and 448 nm, and the height of the true peak at 430 nm above this baseline was measured. Where necessary, highly fluorescent samples were diluted with hexadecane to bring their fluorescence within the range of the standard used.

Uthe and Musal [49] carried out an intercomparison study on the determination of polynuclear aromatic hydrocarbons in lobster. Intercomparative kits comprising lobster digestive gland powder and lobster digestive gland were sent to participants in Europe, the United States and Canada. The participants were requested to measure a suite of non-alkylated PAHs and to analyse each material. The methods used were either liquid chromatography with UV absorption–fluorescence detection or GC-MS on cleaned-up extracts. Intralaboratory relative standard deviations for PAH concentrations in oil ranged from 4.3% to 24.1%. Interlaboratory relative standard deviations ranged from 39% to 96%. Laboratories using GC-MS reported a greater number of compounds, whereas those using liquid chromatography–ultraviolet spectroscopy reported higher concentrations.

Giam et al. [19] have reported on the uptake and depuration of benzopyrene, hexachlorobenzene and pentachlorophenol in marine organisms. Methods capable of determining down to 0.2 µg kg^{-1} of these substances are discussed.

Dunn and Stich [7] applied a method based on fluorimetry to determine benzo[a]pyrene in mussels (*Mytilus edulis*) taken from the outer Vancouver harbour and showed lower benzo[a]pyrene levels in the summer than in the winter, perhaps a result of seasonal discharges of sewage and storm drain water into the harbour. Elevated levels of benzo[a]pyrene in mussels, growing near creosoted timbers or piling, suggested that creosote may be a significant source of this substance in the marine environment. Direct evidence for this suggestion was obtained by comparison of gas chromatography profiles of polycyclic aromatic hydrocarbons isolated from mussels and creosoted wood.

6.3   CHLORINATED ALIPHATIC COMPOUNDS

Murray and Riley [23,48] described gas chromatography methods for the determination of trichloroethylene, tetrachloroethylene, chloroform and carbon tetrachloride in marine organisms. These substances were separated and determined on a glass column packed with 3% of SE-52 on Chromosorb W (AW DMCS) (80–100 mesh) and operated at 35°C, with argon (30 ml min^{-1}) as carrier gas. An electron capture detector was used, with argon-methane (9:1) as quench gas. A limitation of this procedure is that compounds which boil considerably above 100°C could not be determined.

6.4   CHLORINATED INSECTICIDES

The determination of chlorinated insecticides in crustacea has been discussed by several workers [25,31–36,50]. Mills et al. [33] dehydrated oyster samples by mixing them with a 9:1 mixture of anhydrous sodium sulphate and Quso (a microfine silica). They could be held at room temperature for up to 15 days without loss or degradation of chlorinated insecticides. The tissues of oysters were homogenised. Approximately 30 g of the homogenate was added to a second Mason jar and blended with a 9:1 mixture of sodium sulphate and Quso. By alternately chilling and blending, a free-flowing powder was obtained. The blended sample was wrapped in aluminium foil and shipped to the laboratory. Upon receipt of the sample, it was weighed and extracted in a Soxhlet apparatus for 4 h with petroleum ether. The extracts were then purified by concentrating and transferring the extract to separatory funnels. The extracts were diluted to 25 ml with petroleum ether and partitioned with two 50 ml portions of acetonitrile previously saturated with petroleum ether. The acetonitrile was evaporated to dryness and the residue eluted from a Florisil column [33]. In this technique, increasing proportions of ethylether to petroleum ether were used to elute fractions containing increasingly polar insecticides. The extracts were analysed by gas chromatography. Recoveries of DDE, DDD and dichlorodiphenyltrichloroethane (DDT) were between 79% and 96%. The detection limit for a 30 g oyster sample was 10 µg kg^{-1}.

Arias et al. [34] have described a method for the determination of organochlorine insecticide residues in molluscs. The method involves extraction, Florisil column clean-up and analysis of the extract by thin-layer chromatography on silica gel G or alumina with hexane or hexane-acetone (49:1) as solvent, or gas chromatography on a polar column of 10% DC-200 on Chromosorb W HMDS and on a semipolar column of 5% DC-200 plus 7.5% QF-1 on Chromosorb W, with electron capture detection.

Ernst et al. [31,32] and Williams and Holden [50] have determined, by GC-MS, residues of DDT, DDE, DDD and polychlorinated biphenyls (PCBs) in scallops from the English Channel.

Neudorf and Khan [51] investigated the uptake of ^{14}C-labelled DDT, dieldrin and photodieldrin by *Ankistrodesmus amalloides*. The results of liquid scintillation spectrometric analyses show that the total pickup of DDT during a 1–3 h period was 2–5 times higher than that of dieldrin, and 10 times higher than that of photodieldrin. The algae metabolised 3%–5% of DDT to DDE, and 0.8% to DDD. The metabolism of DDT by *Daphnia pulex* was also monitored by exposing 100 organisms to 0.31 ppm of the labelled pesticide for 24 h without feeding. The metabolites were then extracted and separated by thin-layer chromatography, and the *R* values of radioactive spots were compared to the *R* values for non-radioactive DDD and radioactive DDE. The results show a conversion of DDT to DDE of about 13.6%.

6.5   POLYCHLORINATED BIPHENYLS

Markin et al. [25] have also discussed the possible confusion between mirex and PCBs in the analyses of crabs and shrimps. In their method, the samples were ground whole and mixed in a Waring blender to make a composite sample. A 20 g subsample of the composite was removed and analysed as follows. The homogenised sample was extracted with a mixture of hexane and isopropanol, and the extract subjected to a concentrated sulphuric acid clean-up. The sulphuric acid destroys dieldrin, endrin and organophosphorus insecticides. The final extract was cleaned up on a Florisil column and concentrated to the desired level of analysis. If PCBs were suspected in the first analysis, their presence usually being indicated by a series of characteristic peaks, the sample was reprocessed to separate the PCBs from the insecticides, as described by Armour and Burke [26], Gaul and Cruz-La Grange [27] and Markin et al. [28]. After concentrating to the appropriate volume, the extract from both methods of clean-up was chromatographed on a dual-column gas chromatograph equipped with dual electron capture detection. Each sample was analysed on the two different columns; the first column was a mixture of 1.5% OV-17 and 1.95% QF-1 on Gas Chrom Q. The temperatures of the injector, oven and detector were 250°C, 200°C and 210°C, respectively. The second column was 2% DC-200 on Gas Chrom Q, and the injector, oven and detector temperatures were 245°C, 175°C and 205°C, respectively. Argon methane at 80 ml min^{-1} was the carrier gas. The level of detection was 0.001 ppm for DDT and its metabolites, 0.005 ppm for mirex and 0.01 ppm for Aroclor 1260.

Markin et al. [25] commented that they found mirex in only a minority of the samples they analysed, contrary to results obtained by earlier workers [29,52,53].

Teichman et al. [29] separated PCBs from chlorinated insecticides in oyster samples using gas chromatography coupled to mass spectrometry. PCBs were separated from DDT and its analogues and from the other common chlorinated insecticides by adsorption chromatography on columns of alumina and charcoal. Elution from alumina columns with increasing fractional amounts of hexane first isolated dieldrin and heptachlor from a mixture of chlorinated insecticides and PCBs. The remaining fraction, when added to a charcoal column, could be separated into two fractions, one containing the chlorinated insecticides, and the other containing the PCBs by eluting with acetone-diethylether (25:75) and benzene, respectively. The PCBs and the insecticides were then determined by gas chromatography on the separate column eluates without cross-interference.

Teichman et al. [29] and others [52–54] used an electron capture gas chromatograph (Areograph 1200) containing a glass column (6 in. × 0.125 in.) packed with 4% SE-30 and 6% SP-4201 on Chromosorb W (100 to 120 mesh). Heptachlor epoxide and dieldrin were removed from the column by extending the solvent beyond the 30 ml volume with an additional, but separate, elution volume of 30 ml. The PCBs remained an integral part of the mixture containing the insecticides in the first 30 ml of eluate. The elution pattern of alumina column fraction on the charcoal column shows that the insecticides were subsequently removed from the charcoal column with benzene. Recoveries from oyster samples were between 68% (PCBs) and 102% (heptachlorepoxide). Limits of detection were between 0.04 µg kg^{-1} (lindane) and 6.5 µg kg^{-1} (PCBs).

Tanabe et al. [8] used mussels as bioindicators of PCB pollution. When uncontaminated lipped mussels were transplanted in severely contaminated Hong Kong Bay water, total PCB concentrations increased from $11 \mu g \, kg^{-1}$ wet weight to $560 \mu g \, kg^{-1}$ wet weight in 17 days. When the remaining mussels were returned to clean waters after 17 days, total PCB concentrations decreased from $630 \mu g \, kg^{-1}$ wet weight to $12 \mu g \, kg^{-1}$ wet weight in 32 days. Lower chlorinated PCBs (isomers and congeners containing two to four chlorine atoms) were taken up and depurated more rapidly than the more mophilic higher-chlorinated PCBs (hexa-, hepta- and octachlorobiphenyls). It was suggested that time bulking (combining samples collected at frequent intervals from a zinc location) in PCB monitoring studies involving bivalves would provide a more accurate picture of average contamination conditions.

Ya Ma Bayne [55] discriminated polychlorobiphenyls in clam tissue using electron capture negative ion chemical ionisation mass spectrometry.

Careo et al. [65] used microwave-assisted extraction as an initial extraction step and membrane-assisted solvent extraction as a clean-up extraction–concentration step for the determination of PCBs in mussel samples. The purified extract was analysed using gas chromatography–electron capture detection. The results show that extraction time was statistically significant for polychlorobiphenyls 31, 28, 118 and 180 and shaking speed for polychlorobiphenyls 153, 138 and 156; both factors had positive estimated effects. The extraction efficiency of the whole method was between 74% and 100%, and the relative standard deviation ranged from 2% to 15%. The detection limits were about $0.1–0.9 \mu g \, kg^{-1}$.

6.6   POLYCHLOROTERPHENYLS

The presence of polychloroterphenyls has been reported in oyster tissue [9]. To determine polychloroterphenyls in oyster tissue, hexane extract was cleaned up on an alumina or Florisil column and then the analyses performed using a combination of mass spectrometry and gas chromatography m/e values of 436 and 470. Approximately $0.15 \mu g \, kg^{-1}$ of polychloroterphenyls could be detected.

A concentration of $6.2 \mu g \, kg^{-1}$ PCB was found in oyster tissue.

6.7   POLYCHLORINATED DIBENZO-p-DIOXINS AND DIBENZOFURANS

Taguchi et al. [56] used high-resolution mass spectrometry to determine these substances in a crustacea tissue.

Buser [57] determined polychlorobenzothrophenes, the sulphur analogues of polychlorobenzofurans, in crabs, lobsters and worms using various gas chromatographic–mass spectrometric techniques.

6.8   PHTHALATE ESTERS

A gas chromatographic procedure [11] has been used in the determination of phthalate esters in shrimps and crabs. Between 3 and $20 \mu g \, kg^{-1}$ of diethyl hexyl phthalate was found in crabs taken from the Gulf of Mexico, while dimethyl, diethyl and dibutyl phthalate occurred at concentrations less than the detection limit of the method ($1\,\mu g\,kg^{-1}$).

6.9   ORGANOPHOSPHORUS INSECTICIDES

Deutsch et al. [38] determined Dursban in crustacea. After a preliminary clean-up, the extract was chromatographed on a column packed with 3% Carbowax 20 M on Gas Chrom Q (60–80 mesh), which gives excellent separation of Dursban from other organophosphorus insecticides. Both thermionic and flame photometric detectors (FPDs) are satisfactory. Recoveries range from 75% to 105% depending on the nature of the sample. This procedure will detect as little as 0.5 ng of Dursban, corresponding to a level of $0.01\,mg\,kg^{-1}$ in a 10 g sample.

6.10   ORGANOSULPHUR COMPOUNDS

Organosulphur compounds are minor components (0.002%–3%) of crude oil and of some fuel oils.

In a field study, these compounds were found in benthic organisms after an oil spill [58]. Researchers have presented several papers on the accumulation of these compounds in eels and short-necked clams [59] and have also identified dibenzothiophene using GC-MS [59] and capillary GC-MS [38] in biota samples after experimental exposure to crude oil suspension. Moreover, mussels are a well-known biological monitor of marine pollutants in 'the mussel watch' [60]. Many investigators have reported the susceptibility of this organism to petroleum hydrocarbons [61] and polynuclear aromatic hydrocarbons [62]. Kira et al. [10] and Otaga et al. [40] have identified several organosulphur compounds using GC-MS and measured the levels of dibenzothiophene using a GC-FPD, in both mussels and water. The calculated concentration ratio of dibenzothiophene in mussels to that in water ranged up to 500 in the field sample and 800 or higher after an experimental exposure. The estimated biological half-life of dibenzothiophene from field mussel samples was about 9 days in clean seawater.

Dibenzothiophene levels were measured by gas chromatography using an FPD. In field samples, the levels of dibenzothiophene ranged from less than 0.1 to more than $800\,\mu g\,kg^{-1}$.

Dibenzothiophene was separated from other organosulphur compounds, even at parts per billion levels. The presence of dibenzothiophene was indicated by simultaneous detection of M + 184 and 186 on the GC-MS single ion monitor. Accumulation studies of the compound dibutyl phthalate in mussels showed levels that were approximately 600 and 800 times higher than those in water after 4- and 8-day exposures, respectively. A concentration ratio of 800 obtained after an 8-day exposure was close to that of petroleum hydrocarbons.

6.11   COPROSTANOL

Matusik et al. [42] used capillary column gas chromatography to separate coprostanol. Mass spectrometry in the electron ionisation mode was used to confirm its identity at the 75 ng level in sewage-contaminated crustacea.

6.12   ARSENOBETAINE

Francesconi et al. [63] gave details of equipment and a procedure for the identification and determination of arsenic compounds in species in various types of Alaskan crab (*Paralithodes camtschaticus*) and (*Chemisette bardii*) using high-performance liquid chromatography and inductively coupled plasma–atomic emission spectrometry. The only water solubility arsenic compound found in crabs was arsenobetaine.

6.13   NEUTRAL PRIORITY POLLUTANTS

The extraction gas chromatographic procedure [64] has been used for determining neutral priority pollutants in mussel homogenates. The tissue was macerated with distilled water in a blender, anhydrous sodium sulphate was added, and the mixture ground until dry and powdery. The powder was sonified with acetonitrile and the clear phase recovered.

Polychlorinated diphenyl, naphthalene, anthracene, fluoranthene, pyrene, benzo[a] anthracene and chrysene were all found in mussel homogenates in amounts between 0.003 and 0.47 mg kg^{-1}; these compounds were all determined in mussel homogenates at between 0.003 and 0.47 mg kg^{-1}.

6.14   METALLOTHIONENE

Ivankovic et al. [66] found this compound in *Mytilus galloprovincialis* digestive gland homogenate extracts. Polarographic and spectrometric methods were compared in combination with samples subjected to heat treatment (at 70°C and 85°C) as well as to solvent precipitation. With spectrophotometric measurements ranging within 0.081–0.144 mg g^{-1} *Mytilus galloprovincialis* were at least one order of magnitude lower in comparison with the polarographic ones (2.21–2.88 mg g^{-1} *Mytilus galloprovincialis*) depending on the particular sample treatment as well as the specific reagents and references applied. Sephadex G-75 column chromatography of the differently treated samples and subsequent polarographic analysis in the pooled elution fractions indicated that a level of non-metallothionene-interfering thiolic component in homogenate extract may account for more than 50% of the total polarographic signal, irrespective of the method of sample treatment applied. Consequently, the large discrepancy between methods could not be attributed only to the sample treatment procedures, but more likely to certain not entirely explained analytical problems, particularly calibration with an appropriate reference standard.

REFERENCES

1. C.P. Rice and D.S. White, *Science of the Total Environment*, 1987, 6, 259.
2. E. Thyband and S. Le Braz, *Bulletin of Environmental Contamination and Toxicology*, 1988, 40, 731.
3. L. Maltby, J.O.H. Smart and P. Calow, *Environmental Bulletin*, 1987, 43, 271.
4. H.J. Nelis, G. Merchie, P. Lavens, P. Sorgeloos and A.P. De Leenheer, *Analytical Chemistry*, 1994, 66, 1330.

5. J.A. Zilchke and J.W. Arthur, *Archives of Environmental Contamination and Toxicology*, 1987, 16, 225.
6. R.J. Klonda, R.E. Palmer and M.J. Kenkevich, *Estuaries*, 1987, 10, 44.
7. D.P. Dunn and H.F.J. Stich, *Journal of the Fisheries Research Board of Canada*, 1976, 33, 2040.
8. S. Tanabe, R. Tatsukawa and D.J.H. Phillips, *Environmental Pollution*, 1987, 47, 41.
9. J. Freudenthal and P.A. Greve, *Bulletin of Environmental Contamination and Toxicology*, 1973, 10, 108.
10. S. Kira, T. Izumi and M. Ogata, *Bulletin of Environmental Contamination and Toxicology*, 1983, 31, 518.
11. C.S. Giam, H.S. Chan and G.S. Nett, *Analytical Chemistry*, 1975, 47, 2225.
12. S.N. Chesler, B.H. Gump, H.S. Hertz, W.E. May and S.E. Wise, *Analytical Chemistry*, 1978, 50, 865.
13. F. Berthou, Y. Gourmelun and Y. Dreano and M.P. Friocourt, *Journal of Chromatography*, 1981, 203, 279.
14. W.A. May, S.N. Chesler, S.P. Cram, B.H. Gump and D.S. Hertz, *Analytical Chemistry*, 1978, 50, 867.
15. S.N. Chesler, B.H. Gump, H.S. Hertz, W.E. May, S.M. Dyszel and D.P. Enagonio, Technical Note No. 889, National Bureau of Standards, Washington, DC, 1976.
16. P.R. Mason, *Marine Pollution Bulletin*, 1987, 18, 528.
17. N.L. Morgan, *Bulletin of Environmental Contamination and Toxicology*, 1975, 14, 36.
18. P.A. Meyers, *Chromosphere*, 1978, 7, 385.
19. Y. Giam, D. Trujillo, S. Kira and Y. Hrung, *Bulletin of Environmental Contamination and Toxicology*, 1986, 25, 824.
20. A. Bjorseth, J. Krutsen and J. Skei, *Science of the Total Environment*, 1979, 13, 71.
21. H. Kunte, *Archiv für Hygiene und Bakteriologie,* 1967, 151, 193.
22. G.S. Nett and C.S. Musal, *Journal of the Association of Official Analytical Chemists*, 1988, 63.
23. A.J. Murray and J.P. Riley, *Analytica Chimica Acta*, 1973, 65, 261.
24. A.J. Murray and J.P. Riley, *Analytica Chimica Acta*, 1973, 242, 37.
25. G.R. Markin, B. Hawthorne, H.L. Collins and J.H. Ford, *Pesticide Monitoring Journal*, 1974, 17, 139.
26. J.A. Armour and J. Burke, *Journal of the Association of Official Analytical Chemists*, 1970, 3, 761.
27. J. Gaul and P. Cruz-la Grange, Separation of mirex and PCBs in fish, *Laboratory Information Bulletin*, Food and Drug Administration, New Orleans District, 1983.
28. J.P. Markin, J.H. Ford, J.C. Hawthorne, J.H. Spence, J. Davies and C. Loftus, *Environmental Monitoring for the Insecticide Mirex*, APHS 81–83, US Department of Agriculture, Washington, DC, November 1972.
29. J. Teichman, J. Benenue and J.W. Hylin, *Journal of Chromatography*, 1978, 151, 155.
30. J. Freudenthal and P.A. Greve, *Bulletin of Environmental Contamination and Toxicology*, 1973, 10, 108.
31. W. Ernst, H. Goerke, G. Der and R.C. Schaefer, *Bulletin of Environmental Contamination and Toxicology*, 1976, 15, 55.
32. W. Ernst, H. Schaefer, H. Goerke and G.Z. Eder, *Analytical Chemistry*, 1974, 227, 378.
33. P.A. Mills, F.J.F. Caley and R.A. Grithen, *Journal of the Association of Official Analytical Chemists*, 1963, 46, 106.
34. C. Arias, A. Vidal and J. Maria, *Analytical Bromat (Spain)*, 1980, 22, 273.
35. H.H. Kouyomjiam and R.F. Uglow, *Environmental Pollution*, 1984, 7, 103.
36. US Environmental Protection Agency, EPA-600–4–74 1974T, Washington, DC, 1974, p. 108.

37. A.J. Wilson, J. Forester and J. Knight, US Wildlife Service, Circular, Centre for Estuaries and Research, Gulf Breeze, FL, 1969.
38. M.E. Deutsch, W.E. Westlake and F.A. Gunter, *Agricultural and Food Chemistry*, 1970, 18, 178.
39. R.H. Buzer and C. Rappe, *Analytical Chemistry*, 1991, 63, 1210.
40. M. Otaga, Y. Mujake and Y. Yamazaki, *Water Research*, 1979, 13, 1179.
41. H.J. Nelis, G. Merchie, P. Lavens, P. Sorgeloos and A.P. De Lenheer, *Analytical Chemistry*, 1994, 66, 1330.
42. J.E. Matusik, G.P. Haskin and J.A. Splon, *Journal of the Association of Official Analytical Chemists*, 1988, 71, 994.
43. D. Slinkman and R. Sacks, *Analytical Chemistry*, 1991, 63, 343.
44. N. Morgan, *Bulletin of Environmental Contamination and Toxicology*, 1975, 14, 309.
45. W.E. May, S.N. Chesler, S.P. Cram, B.H. Gump, D.S. Hertz, D.P. Enagonio and S.M. Dyszel, *Journal of Chromatographic Science*, 1975, 13, 535.
46. S.N. Chesler, B.H. Gump, H.S. Hertz, W.E. May, S.M. Dyszel and D.P. Enagonio, Technical Note No. 889, National Bureau of Standards, Washington, DC, 1976.
47. F. Berthou, Y. Gourmelun, Y. Dreano and M.P. Friocourt, *Journal of Chromatography*, 1981, 203, 279.
48. A.J. Murray and J.P. Riley, *Nature*, 1973, 242, 37.
49. J.F. Uthe and C.J. Musal, *Journal of the Association of Official Analytical Chemists*, 1988, 71, 363.
50. R. Williams and A.V. Holden, Private communications, National Institute of Oceanography, Wormley, Surrey, UK.
51. S. Neudorf and M.A.Q. Khan, *Bulletin of Environmental Contamination and Toxicology*, 1975, 13, 443.
52. P.A. Butler, *Biological Science*, 1969, 19, 889.
53. M.D. McKenzie, *Fluctuations in Abundance of the Blue Crab and Factors Affecting Mortalities*, South Carolina Wildlife Resources Division Technical Rep. No. 1, South Carolina Department of Natural Resources, Columbia, 1970.
54. R.K. Mahood, M.D. McKenzie, D.P. Middough, S.J. Bellar, J.R. Davies and D. Spitsbergen, *A Report on the Cooperative Blue Crab Study in South Atlantic States*, Projects No. 2–79-R-1 and 2–82-R-1, US Department of the Interior Bureau of Commercial Fisheries, Washington, DC.
55. L.K. Ya Ma Bayne, *Analytical Chemistry*, 1993, 65, 772.
56. V.W. Taguchi, E.J. Reiner, D.T. Wong, O. Meresz and B. Hallam, *Analytical Chemistry*, 1988, 60, 1429.
57. H.R. Buser, *Analytical Chemistry*, 1991, 63, 12 10.
58. O. Grahl-Nielsen, J.T. Steveland and S. Wilhelmsen, *Journal of the Fisheries Research Board of Canada*, 1978, 35, 615.
59. M. Ogata, Y. Miyake, K. Fujisawa, S. Kira and Y. Yoshida, *Bulletin of Environmental Contamination and Toxicology*, 1980, 25, 130.
60. E.D. Goldberg, *Marine Pollution Bulletin*, 1975, 6, 111.
61. R.F. Lee, R. Sanerheber and A.A. Benson, *Science*, 1972, 177, 344.
62. B.P. Dunn and D.R. Young, *Marine Pollution Bulletin*, 1976, 7, 231.
63. K.A. Francesconi, P. Hicks, R.A. Stockton and K.J. Irgolic, *Chemosphere*, 1985, 14, 1443.
64. R.J. Ozretich and W.P. Schroeder, *Analytical Chemistry*, 1986, 58, 2041.
65. N. Careo, I. Garcia, M. Ignacio and A. Mouteira, *International Journal of Environmental Analytical Chemistry*, 2009, 89, 759.
66. D. Ivankovic, J. Pavicic, B. Raspor, I. Falnoga and M. Tusec-Znidaric, *International Journal of Environmental Analytical Chemistry*, 2003, 83, 219.
67. J.S. Warner, *Analytical Chemistry*, 1976, 48, 578.

7 Determination of Organometallic Compounds in Crustacea and Other Non-Fish Creatures

7.1 ORGANOMERCURY COMPOUNDS

Uthe et al. [1] have described a rapid semi-micromethod for determining methylmercury in crustacea. The procedure involves extracting the methylmercury into toluene as methylmercury(II) bromide, partitioning the bromide into aqueous ethanol as the thiosulphate complex and reextracting methylmercury(II) iodide into benzene, followed by gas chromatography on a glass column (4 ft $\times$ 0.25 in.) packed with 7% Carbowax 20M on Chromosorb W and operated at 170°C with nitrogen as the carrier gas (60 ml min^{-1}) and electron capture detection. Down to 0.01 µg kg^{-1} of methylmercury in a 2 g sample could be detected. A comparison for the results with those obtained by atomic absorption (total Hg content) indicated that all the fish samples examined contained more than 41% of the mercury as methylmercury.

7.2 ORGANOLEAD COMPOUNDS

Alkyl lead compounds have been encountered in crustacea and fish in the concentration ranges of 0.01–0.05 mg kg^{-1} and 5–20 mg kg^{-1}, respectively. An atomic absorption spectrometric method [2,3] with detection limits of 0.01–0.02 mg L^{-1} is perfectly suitable for analysing alkyl lead compounds at the lowest levels encountered in these types of sample.

Birnie and Hodges [4] have described the combination of solvent extraction and differential pulse anodic scanning voltammetric techniques for the isolation and determination of trialkyl lead (Et$_3$Pb$^+$ and Me$_3$Pb$^+$) and dialkyl lead (Et$_2$Pb^{2+} Me$_2$Pb^{2+}) species in oysters and macuma.

In this method, the sample is homogenised in the presence of a mixture of salts (lead nitrate, sodium benzoate, potassium iodide and sodium chloride), which effectively releases the di- and trialkyl lead present and facilitates their transfer to toluene as a preliminary to back extraction into dilute nitric acid. The differentiation and determination of the alkyl lead species are achieved by differential pulse anodic stripping voltammetry. The efficiency of the extraction procedure was examined at alkyl lead concentrations up to 2 mg kg^{-1} as lead, and a detection limit of 0.01 mg kg^{-1}

was established. The recovery of ionic diethyl lead and the two trialkyl lead species from various marine vertebrates and molluscs was in the range of 80%–90%, while that of dimethyl alkyl lead was appreciably lower (30%–40%) using this method.

Concentrations of tetramethyl, triethyl and diethyl lead found in macuma were generally between 0.02 and 0.05 mg kg^{-1}. The concentration of organolead found in creatures other than fish, for example, the concentration of organolead found in periwinkle (*Littorina irrorata*) samples taken from Maryland and Virginia, was 0.02 mg kg^{-1}, whereas the level found in fish was in the range 0.92–7.9 mg kg^{-1}.

7.3   ORGANOTIN COMPOUNDS

Alkyltin compounds have been encountered in invertebrates (oysters and mussels) in the range 0.02–0.7 mg kg^{-1}. Methods based on the reduction of organotin to hydride with sodium borohydride, followed by atomic absorption spectrometry, have detection limits as low as 0.023 mg kg^{-1}. These compounds have been the subject of environmental studies for two obvious reasons. First is the increasing worldwide use of inorganic and organotin compounds in many industrial, chemical and agricultural areas, with very little being known about their environmental fate. Second, there is a great difference in toxicity of various organotin compounds according to the variation of the organic moiety in the molecules.

Organotin compounds occur extensively in the ecosystem, being found in natural and marine waters, sea creatures and sediments. This originates from the use of these compounds in industry (polyvinyl chloride [PVC] stabilisers and agricultural fungicides, biocides and bacteriocides) and in the marine environment (algicides and molluscicides) [5–9]. Compounds such as triphenyltin acetate and triphenyltin hydroxide are used as molluscicides in antifoulant paint compositions for ships and harbour works.

Alkyltin compounds have been encountered in invertebrates (oysters and mussels) in the range of 0.02–0.7 mg kg^{-1}. Methods based on reduction of organotin to hydrides with sodium borohydride, followed by atomic absorption spectrometry, have detection limits as low as 0.023 mg kg^{-1}.

Han and Weber [10] studied the speciation of methyl- and butyltin compounds and inorganic tin in oysters using hydride generation atomic absorption spectrometry. Recoveries from spiked samples of oyster tissue were about 100%, and no organotin decomposition products were observed. Detection limits of methyltin, butyltin and inorganotin were respectively 0.023, 0.025 and 0.011 µg kg^{-1} oyster sample (wet weight). A comparative study of monomethyltin levels in shellfish from the Great Bay Estuary, New Hampshire, and the Mediterranean Sea (Turkish coast) suggested that monomethyltin in Great Bay oysters was a result of biological methylation of inorganic tin, whereas in the Mediterranean Sea, mono- and dimethyltin compounds resulted from degradation of anthropogenic trimethyltin. Comparisons were also made of butyltin levels in oysters from the Great Bay Estuary and English shellfish samples.

Stephenson and Smith [11,12] applied graphite furnace atomic absorption spectrometry to the determination of down to 11–2.5 ng butyltin (absolute) in oysters.

A variety of seashells have been analysed for organotin compounds using the gas chromatographic procedure described by Braman and Tomkins [13]. The average total tin content of seashell and egg samples was between 0.001 and 0.002 mg kg^{-1}, and methylated tin compounds were detected (Me_4Sn^{3+} and Me_2Sn^{2+}). Less than $0.01 \, \mu g \, kg^{-1}$ Me_3Sn^+ was present.

Higher concentration of tin in the seashells relative to the water in which they were found would indicate the presence of a bioaccumulation process.

Gas chromatographic methods have been described for the determination of tetraalkyl and trialkyltin [14] in amounts down to $10^{-12} \, g$ (absolute) as trialkyltin chlorides in biological materials. Unfortunately, these methods were not easily applicable to the determination of the dialkyl homologues because of their absorption and decomposition during chromatography.

Gas chromatography with flame photometric detection using quartz surface-induced luminescence has been used [15] to determine butyltin compounds in mussels with an absolute detection limit of 0.3 pg tin for tetrapropyltin or 2–3 pg tin for trimethylpentyltin, dimethyldipentyltin or methyltripentyltin. Between 0.04 and $0.09 \, \mu g \, g^{-1}$ organic tin was found in mussel samples.

A high-performance liquid chromatography–hydride generation direct current plasma emission spectrometric technique for the determination of organotin compounds in water has been applied to the analysis of clams and tuna fish [16].

Pinel and Mediac [21] determined heavy organic pollution in shellfish using a modified hydride atomic absorption procedure.

Tin speciation in an aquatic environment is very complex. To the natural Sn^{IV} and methylated compounds, human activities add mainly butylated, octylated, phenylated or even methylated derivatives. The most environmentally significant, due to their high toxicity and direct introduction in water through biocidal use, are the trisubstituted ones. Several sophisticated speciation procedures have been proposed for common use.

Pinel and Mediac [21] proposed a simple and fast procedure which makes it possible to determine 'heavy' tin species that are most susceptible to exerting harmful effects. This atomic desorption spectrometric method uses the differences in volatility of stannanes generally by reduction with sodium hydride. Sn^{IV} and the methylated species have very close response coefficients, whereas heavy compounds respond very slightly at room temperature and are eliminated in a $-40°C$ cold trap.

7.4 ORGANOARSENIC COMPOUNDS

Maher [17] used ion exchange chromatography to separate inorganic arsenic and methylated arsenic species in marine organisms and sediments. The method determines nanomethyl arsenic and dimethyl arsenic. The procedure involves the use of solvent extraction to isolate the arsenic species, which are then separated by ion exchange chromatography and then determined by arsine generation.

Organoarsenic levels encountered range from $0.00001–23 \, mg \, kg^{-1}$ in molluscs to $0.4–0.001 \, mg \, kg^{-1}$ in prawns. The most sensitive method listed is high-performance chromatography with an inductively coupled plasma–mass spectrometric detector. This method has a detection limit of $0.002 \, mg \, kg^{-1}$. It was not sensitive enough to determine organoarsenic compounds at the $0.00001 \, mg \, kg^{-1}$ level but will be adequate for many types of samples.

A review of further methods for the determination of organometallic compounds in creatures other than fish is given in Table 7.1.

TABLE 7.1

Methods for the Determination of Organometallic Compounds in Water Creatures Other than Fish

Determined	Type of Sample	Technique	Detection Limit (mg kg^{-1} Unless Otherwise Stated)	Reference
		Organotin Compounds		
Butyltin	Oyster	Hydride generation atomic absorption absolute spectrometry	1.1–2.5 ng	[18]
Butyltin	Mussels	Gas chromatography with flame photometric detector	Pr$_4$Dn 0.3 pg as Sn absolute	[15]
Me$_3$SnC$_5$H$_{11}$			Me$_3$SnC$_5$H$_{11}$	
Me$_2$Sn(C$_5$H$_{11}$)$_2$			Me$_2$Sn(C$_5$H$_{11}$)$_2$	
MeSn(C$_5$H$_{11}$)$_3$			MeSn(C$_5$H$_{11}$)$_3$	
Pr$_4$Sn			2–3 pg as Sn absolute	
Methyltin and butyltin	Oyster	Hydride generation atomic absorption	MeSn	[10]
			0.023	
			BuSn	
			0.01–0.25	
		Organolead Compounds		
Et$_3$Pb$^+$	Oyster macuma	Differential pulse anodic scanning voltammetry	—	[4]
Me$_3$Pb$^+$				
Et$_2$Pb^{2+}				
Me$_2$Pb^{2+}				
		Organomercury Compounds		
Methylmercury	Crustacea	Extraction gas chromatography	0.000 01	[1]
		Organoarsenic Compounds		
Arsenobetaine	Crab	Inductively coupled plasma–atomic emission spectrometry	—	[19]

TABLE 7.2
Detection Limits for Various Types of Organometallic Compounds

Techniques	Compounds	Detection Limit (mg kg^{-1} Unless Otherwise Stated)
Hydride generation atomic absorption spectrometry	Butyltin compounds	1–2.5 ng as Sn absolute or 0.000 025
	Methyltin compounds	0.000 023
Gas chromatography	Methylmercury compounds	0.000 01
With flame photometric detector	Butyltin compounds	
	Propyltin compounds	3 pg as Sn absolute
	Me$_3$Sn(C$_5$H$_{11}$)	
	Me$_2$Sn(C$_5$H$_{11}$)$_2$	2–3 pg as Sn absolute
	MeSn(C$_5$H$_{11}$)$_3$	
Inductively coupled plasma–atomic emission spectrometry	Organoarsenic compounds	
Differential pulse anodic scanning voltammetry	Et$_3$Pb$^+$	
	Me$_3$Pb$^+$	
	Et$_2$Pb^{2+}	
	Me$_2$Pb^{2+}	

The best available detection limits for various types of organometallic compounds in crustacea are tested in Table 7.2.

Navaro et al. [20] have described a simultaneous determination of mercury and butyltin compounds in eel species, including glass eel and yellow eel. These workers used a multiple-species-specific isotype dilution technique. Low detection limits were achieved, namely 0.007–0.13 mg kg^{-1} for mercury and 0.42–0.71 mg kg^{-1} for tin species.

REFERENCES

1. J.F. Uthe, J. Solomon and B. Grist, *Journal of the Association of Official Analytical Chemists*, 1972, 55, 583.
2. Y.K. Chau, P.T.S. Wong, G.A. Bengert and O. Kramer, *Analytical Chemistry*, 1979, 51, 186.
3. Y.K. Chau, P.T.S. Wong and H. Saitoh, *Journal of Chromatographic Science*, 1976, 162, 14.
4. S.E. Birnie and D.E. Hodges, *Environmental Technology Letters*, 1981, 2, 433.
5. WHO Task Group, *Environmental Health Criteria*, 1980, 51, 1.
6. J.J. Zuckerman, P.R. Residorf, H.V. Ellis and R.R. Wilkinson, in F.E. Brinkman and J.M. Bellama (eds), *Organometals and Organometalloids, Occurrence and Fate in the Environment*, ACS Symposium Series No. 28, American Chemical Society, Washington, DC, 1978, p. 388.
7. R. Bock, *Residue Review*, 1981, 79, 216.
8. R. Gächter and H. Muller, *Handbuch der Kunstoff–Additive Hanser*, Springer Verlag, Munich, 1979.
9. L. Fishbein, *Science of the Total Environment*, 1974, 2, 341.

10. J.S. Han and J.H. Weber, *Analytical Chemistry*, 1988, 60, 311.
11. J.D. Smith, *Nature*, 1970, 225, 103.
12. M.D. Stephenson and D.R. Smith, *Analytical Chemistry*, 1988, 60, 696.
13. R.S. Braman and M.A. Tompkins, *Analytical Chemistry*, 1979, 51, 12.
14. Y. Arakawa, T.E. Wade and H. Iwai, *Journal of Chromatography*, 1981, 216, 209.
15. G.B. Giang, B. Maxwell, K.W.M. Siu, V.T. Luong and S.S. Berman, *Analytical Chemistry*, 1991, 63, 1506.
16. I.S. Krull and R.W. Panaro, *Applied Spectroscopy*, 1985, 39, 960.
17. W.A. Maher, *Analytica Chimica Acta*, 1981, 126, 157.
18. A. Jones, *Analytical Chemistry*, 1988, 60, 316.
19. K.A. Francesconi, P. Hicks, R.A. Stockton and K.J. Irgolic, *Chemosphere*, 1985, 14, 1443.
20. P. Navaro, S. Clemens, V. Perrot, V. Bolliet, H. Talouret, T. Guerin, M. Monperrus and D. Amourkus, *International Journal of Environmental Analytical Chemistry*, 2012, 12, 166.
21. R. Pinel and M. Mediac, *International Journal of Environmental Analytical Chemistry*, 1986, 27, 265.

8 Toxic Effects of Metals on Fish and Invertebrates

This chapter reviews the available information on the toxicity towards fish and creatures other than fish of various types of metallic pollutants. Information has also been reported on the concentrations of toxicants found in tissues of creatures that are known to have been killed by pollutants, that is, a study of acute exposure resulting in fish killed. Although there is a vast amount of literature available concerning the presence of metal in fish, it deals almost exclusively with levels in muscle tissue or in whole fish after chromic exposure. Van Hoof and Van Son [1] pointed out that investigation of the cause of fishkills by water analysis alone has serious drawbacks since, in many cases, the causative agent may have been diluted, biodegraded or volatilised to a level which does not allow an unambiguous interpretation at the time of sampling. Eventually, it may have been displaced from the site where killed fish are localised. The work of Van Hoof and Van Son [1] on copper, cadmium, zinc and chromium is discussed where relevant in this chapter.

Mount and Stephen [2] developed an autopsy technique for zinc-caused fish mortalities and found that the ratio of opercle to gill zinc concentrations gave valuable information for discrimination between acute and chronic exposure.

Mount and Stephen [2] found that cadmium intoxications in the bluegill sunfish (*Lepomis macrochirus*) and catfish (*Ictalurus nebulosus*) could be demonstrated through analysis of gill tissue. Martin et al. [3] and Kariya et al. [4] found similar results for copper in five different fish species. These findings were not confirmed by the work of Brings et al. [5], which found no significant differences between copper tissue levels in *I. nebulosus* exposed to acute lethal and subacute nonlethal concentrations.

8.1 TOXICITY OF METALS IN WATER ON FISH AND INVERTEBRATES

The toxicity of metals towards fish and invertebrates is now discussed in order of increasing toxicity. For comparison, each element is identified by the mean (S_x) and 95% percentile (S_{95}) concentrations. The lethal concentration values to kill 50% (LC_{50}) of fish are given in Table 8.1 as a function of the type of water (nonsaline or saline) and type of creature, and the exposure time values obtained for some invertebrates are included for comparison.

The natural concentrations of trace metals in relatively unpolluted open seawater where the effect of coastal discharges is minimal are very low and the accurate determination of these concentrations has presented a great challenge to the analytical chemist. As such, low-level contamination of the sample by sampling equipment and neighbouring ships is an important factor affecting the accuracy of results, and it

TABLE 8.1
Concentration of Elements in Water Causing Mortalities (LC$_{50}$) for Salmonid (s) and Non-Salmonid (n/s) Fish and Invertebrates

Element	S$_x$ (µg L^{-1})	S$_{95}$ (µg L^{-1})	Fish		Invertebrates	
			Exposure Time (days)	LC$_{50}$ (µg L^{-1})	Exposure Time (days)	LC$_{50}$ (µg L^{-1}) [23–26]
Ni	220	900 (n/s, s)	Long term	500 (s, n/s)	—	—
			100	2200 (s, n/s)		
			15	8000 (s, n/s)		
			4	35,000 (s, n/s)		
S	200	1300 (n/s)	4 (8)	2900–3060	2	1100 (as SeO3)
			10	300	2	5300 (as SEO4)
					2	680 (adult)
						750 (juvenile)
						(*Daphnia magna*)
V	100	1000–1600 (n/s)	7	2400–3000 (s) (salt waters)	—	—
				2900–5000 (s) (hard waters)		
Cr	100	800 (s)	100	1150 (s)	3	30–80 (as CrVI)
			10	18,300 (s)		(Crustaceans)
			4	3300–65,000 (s)		
As	80	600 (n/s, s)	60	200 (n/s)	3–5	1000
			4	25,000–169,000 (n/s)		
			4	14,400		
Ag	70	850 (n/s)	Short term	10–10,000 (as Ag No$_3$)	—	—

(*Continued*)

TABLE 8.1 (Continued)

Concentration of Elements in Water Causing Mortalities (LC$_{50}$) for Salmonid (s) and Non-Salmonid (n/s) Fish and Invertebrates

Element	S$_x$ (µg L^{-1})	S$_{95}$ (µg L^{-1})	Fish		Invertebrates	
			Exposure Time (days)	LC$_{50}$ (µg L^{-1})	Exposure Time (days)	LC$_{50}$ (µg L^{-1}) [23–26]
Zn	23	200 (n/s)	500–1000	260 (soft waters)	4	70 (*Daphnia magna*)
			500–1000	1050 (hard waters)	4	10,000 (annelids, insect larvae)
				2600 (juvenile)	60	200–600 (snail, *Ancyclus flaviatus*)
			4	2400 (adult) rainbow trout (*Salmo gairdnera*)		
			4	13,300–33,000 (*Tilapiazilli*)	60	200 (amphipod, *Allorchestes compressa*)
			4	2600–52,000 (*Clarias lazena*)		
Pb	20	100 (n/s)	90	5500		Similar to fish
			40	900		
			4	1500		
Cu	4	17 (s)	72	80 (s)	3	24 (crustaceans)
			30	200 (s)	4	400–2000 (Mollusc adult)
			6	250–400 (s)	1	500 (clam)
			3	—	40	140 (junior clam)
			3	—		
					4	480 (amphipod, *Allorchestes compressa*)

(Continued)

TABLE 8.1 (*Continued*)
Concentration of Elements in Water Causing Mortalities (LC$_{50}$) for Salmonid (s) and Non-Salmonid (n/s) Fish and Invertebrates

Element	S$_x$ (µg L^{-1})	S$_{95}$ (µg L^{-1})	Fish		Invertebrates	
			LC$_{50}$ (µg L^{-1})	Exposure Time (days)	Exposure Time (days)	LC$_{50}$ (µg L^{-1}) [23–26]
Cd	4	16 (n/s)	180 (n/s)	100	4	680 (crustaceans)
			4000 (n/s)	10	4	
	2	6 (n/s)	2 (s)	700		
			<10,000 (*Notropis lutrensis*)	4	4	780 (amphipod, *Allorchestes compressa*)
			<10,000 (fathead minnow, *Pimephales promelas*)	4	25	10 (*Daphnia magna*)
			12,600 (*Punctius conchonus*)	4		
			350,000 (male)	4		
			371,000 (female) (*Herbistes reticulates*)	4		
Hg	2	22 (n/s)	2 (as MeHg [21])	30	3	0.2 (crayfish)
					2	110 (slipper limpet, *Cripidula fornicate*)
Al	—	—	3800 (rainbow trout, *Salmo gairdneri*)	4		
Fe	—	—	—	—	4	25–43 (isopod, *Asselus aquaticus*)
NH$_4$	—	—	2170 (fathered minnow, *Pimephales promelas*)	4	—	—

TABLE 8.2
Metal Determination (µg L⁻¹) in Seawater (1974–2012)

	Year		
Element	1974	1977	1978–2012
Copper	0.5–6	0.1	0.1–0.2
Cadmium	—	0.17	15–70
Zinc	—	15	0.03–0.35
Nickel	—	2.0	0.25–0.39

has only been in recent years, in fact, that reliable techniques have evolved. For this reason, in the discussion which follows, only results obtained since 1975 are quoted, as these represent the most accurate available. In general, lower values for metals in seawater have been obtained in recent years, compared with earlier, due to the control of contamination (Table 8.2).

Information on the effects of metals on marine creatures is summarised in Table 8.3A–D. The durations of the toxicity tests are not included in this table. They are, however, generally short-term tests and the concentrations quoted are toxic effect concentrations, that is, concentrations above which mortalities can occur. Table 8.3A–D is a useful guide in that it highlights those species at risk when concentrations of the stated elements exceed the levels quoted in short-term exposures. If the analytical composition of a marine water is known, then reference to Table 8.3A–D shows the adult and larval species at risk. It can be seen, for example, that when concentrations of copper or mercury exceed $1 \mu g L^{-1}$, adult and larval bivalve molluscs are at risk, and when concentrations of cadmium exceed $1 \mu g L^{-1}$, adult bivalve molluscs are at risk. When concentrations of mercury exceed $10 \mu g L^{-1}$, the following species are at risk: molluscs and crustacea (adult and larval) and adult fish. When concentrations of copper exceed $10 \mu g L^{-1}$, then adult larval annelids, bivalve molluscs, crustacea and fish are at risk, as are adult echinoderms and hydrozoans. With cadmium, hydrozoans are at risk. Zinc at this concentration causes mortalities in bivalve molluscs and larval bivalves.

At-risk creatures in open sea and coastal waters are calculated in Table 8.4.

In Tables 8.5 and 8.6, respectively, are compared the short-term 4-day LC_{50} values obtained for various metals in nonsaline and saline waters with typical concentrations of these elements which have been found in natural waters.

When the concentration of metal in an environmental water is lower than the 4-day LC_{50} value, then less than 50% mortalities occur in this period.

The higher the 4-day LC_{50} value relative to the observed concentration in the environmental water, the fewer the mortalities. Thus, if 4-day LC_{50} is $3000 \mu g L^{-1}$ and $5 \mu g L^{-1}$ is present in environmental water, then few or zero mortalities will occur. If $1000–2000 \mu g L^{-1}$ of the metal is present in the environmental water, then some mortalities (<50%) and adverse effects will take place.

Fewer types of creatures will undergo mortalities in saline waters due to the lower environmental concentrations of metals that occur, compared with concentrations present in non-saline inland waters. Thus, the only observed case where the

TABLE 8.3A
Toxicity of Metals to Marine Fish and Invertebrate Species at Risk during 4- to 14-day Exposure to Stated Concentrations

Adult Species

Annelids   Bivalve Molluscs   Crustaceans

Concentration above Which Mortalities Can Occur

$\mu g\,L^{-1}\,mg\,L^{-1}$

Hg Cu Cd Zn Pb Cr As Ni

0.1–1
0.0001–0.001
1–10
0.001–0.01
10–100
0.01–0.1
100–1000
0.1–1
1–10
10–100

TABLE 8.3B

Toxicity of Metals to Marine Fish and Invertebrates Species at Risk during 4- to 14-day Exposure to Stated Concentrations

Adult Species

Concentrations above Which Mortalities Can Occur	Echinoderms								Fish								Gastropods								Hydrozoans							
µg L⁻¹ mg L⁻¹	Hg	Cu	Cd	Zn	Pb	Cr	As	Ni	Hg	Cu	Cd	Zn	Pb	Cr	As	Ni	Hg	Cu	Cd	Zn	Pb	Cr	As	Ni	Hg	Cu	Cd	Zn	Pb	Cr	As	Ni
0.1–1																																
0.0001–0.001																																
1–10																																
0.001–0.01																																
10–100																																
0.0–0.1																																
100–1000																																
0.1–1																																
1–10																																
0–100																																

TABLE 8.3C
Toxicity of Metals to Marine Fish and Invertebrates Species at Risk during 4- to 14-day Exposure to Stated Concentrations

Larval Species

Concentration above Which Mortalities Can Occur	Larval Annelids								Larval Bivalves							
µg L^{-1} mg L^{-1}	Hg	Cu	Cd	Zn	Pb	Cr	As	Ni	Hg	Cu	Cd	Zn	Pb	Cr	As	Ni
0.1–1																
0.0001–0.001																
1–10																
0.001–0.01									→→	→→						
10–100											→→	→→	→→		→→	→→
0.01–0.1														→→		
100–1000		→→		→→												
0.01–1																
1–10																
10–100																

TABLE 8.3D
Toxicity of Metals to Marine Fish and Invertebrates Species at Risk during 4- to 14-day Exposure to Stated Concentrations

Larval Species

Concentration above Which Mortalities Can Occur	Larval Crustacea								Larval Echinoderms								Larval Fish							
$\mu g\,L^{-1}$ $mg\,L^{-1}$	Hg	Cu	Cd	Zn	Pb	Cr	As	Ni	Hg	Cu	Cd	Zn	Pb	Cr	As	Ni	Hg	Cu	Cd	Zn	Pb	Cr	As	Ni
0.01–1																								
0.0001–0.001																								
1–10																								
0.001–0.01																								
10–100																								
0.01–0.1																								
100–1000																								
0.1–1															→				→					
1–10		→		→	→	→	→	→									→							
10–100	→		→																→	→				

TABLE 8.4

At-Risk Creatures in Typical Environmental Open Waters and Coastal Waters, Short-Term (4- to 14-day) Exposure to Metallic Contaminants

Creature	Element Concentration ($\mu g\,L^{-1}$) at Which Creatures Do Not Survive						
	Hg	Cu	Cd	Zn	Pb	Cr	Ni
Annelids (adult)	>100	>10	>1000	>1000	>100	>00	>10,000
Annelid (larval)		>10		>100			
Bivalve mollusc (adult)	>0.1	>1	>1	>10	>1000	>1000	>10,000
Bivalve mollusc	>1	>1	>100	>10	>100	>1000	>100
Crustacea (adult)	>10	>10	>10	>100	>1000	>1000	>100
Crustacea (larval)	>1	>10	>100	>100	>100	>1000	>1000
Echinoderm (adult)		>10		>1000	>1000	>10,000	>10
Gastropods		>10			>100		>10,000
Hydrozoans		>10	>1				
Fish (adult)	>10	>10	>100	>1000	>1000	>10,000	>100
Fish (larval)		>10		>1000			
Range of metal concentrations found in open seawaters ($\mu g\,L^{-1}$)	0.002–0.078	0.0063–2.8	0.01–0.126	0.05–10.9	0.000041–9.0	0.005–1.26	0.15–0.93

(Continued)

TABLE 8.4 (*Continued*)
At-Risk Creatures in Typical Environmental Open Waters and Coastal Waters, Short-Term (4- to 14-day) Exposure to Metallic Contaminants

Creature	Element Concentration ($\mu g\,L^{-1}$) at Which Creatures Do Not Survive						
	Hg	Cu	Cd	Zn	Pb	Cr	Ni
Creatures at risk during short-term exposure in open seawater	None	Bivalve molluscs (adult and larval) at high end of concentration range	None	Bivalve molluscs (adult and larval) at high end of concentration range	None	None	None
Range of concentrations found in bay, coastal and estuary waters ($\mu g\,L^{-1}$)	0.00002–15.1	0.069–20.0	0.015–5.0	0.007–200	0.038–7.44	0.095–3.3	0.2–5.33
Creatures at risk during short-term exposure in above waters when metal concentrations are at higher range quoted	Bivalve molluscs (adult and larval), crustacea (adult and larval), fish	Annelids (adult and larval), crustacea (adult and larval), echinoderms, hydrozoans, fish (adult and larval)	Bivalve molluscs (adult), hydrozoans	Bivalve molluscs (adult and larval), annelids (larval), crustacea (adult and larval)	None	None	None

TABLE 8.5

Effect of Short-Term (4-day) Exposure of Creatures to Typical Concentrations of Metals Found in Freshwater (Nonsaline)

Element	Creature	4-day LC$_{50}$	Typical Concentration of Metal (µg L^{-1}) in Freshwater Max	Min
Nickel	Fish	3060 max	40	1.5
		2900 min	40	1.5
Vanadium	Fish	5000 max	24	0.1
		2900 min	24	0.1
Chromium	Fish	65,000 max	23	0.05
		3300 min	23	0.05
	Crustacea	80 max	23	0.05
		30 min	23	0.05
Arsenic	Fish	14,400	490	0.42
Silver	Fish	6700 max	32	0.3
		7 min	32	0.3
Zinc	Fish	52,000 max	630	0.86
		2400 min	630	0.86
	Daphnia magna	70	630	0.86
	Annelid	10,000	630	0.86
Lead	Fish	1500	60	0.13
	Invertebrates	1500	60	0.13
Copper	Fish	400 max	200	0.48
		250 min	200	0.48
	Crustacea	24	200	0.48
	Mollusc	2000 max	200	0.48
		400 min	200	0.48
	Amphipod	480	200	0.48
Cadmium	Fish	371,000 max	5	0.013
		<10,000 min	5	0.013
	Amphipod	780	5	0.013
Mercury	Crayfish	0.2	1.3	0.009
	Slipper limpet	1100	1.3	0.009
Aluminium	Fish	3800	3600	14
Iron	Isopod	43,100 max 28,610	5000	1

environmental concentration exceeds the 4-day LC$_{50}$ value is that of crabs exposed to 15.1 µg L^{-1} of mercury in seawater. Other cases where low percentage mortalities or adverse effects might occur (i.e. the 4-day LC$_{50}$/environmental concentration ratio is low) include bivalve molluscs and plankton, as well as planktonic crustacea in the presence of 5.3 µg L^{-1} nickel and young crustacea and molluscs in the presence of 20 µg L^{-1} copper.

TABLE 8.6
Effect of Short-Term (4-day) Exposure of Creatures to Typical Concentrations of Metals Found in Saline Waters

| Element | Creature | 4-day LC$_{50}$ | Typical Concentration of Metal (μg L^{-1}) in Saline Water | | | |
| | | | Open Seawater | | Coastal Bay and Estuary Waters | |
			Max	Min	Max	Min
Nickel	Fish	35,000	1.58	0.099	5.3	0.2
	Marine organisms	10,000	1.58	0.099	5.3	0.2
	Planktonic	600 max	1.58	0.099	5.3	0.2
	Crustacea	50 min	1.58	0.099	5.3	0.2
	Bivalve	600 max	1.58	0.099	5.3	0.2
	mollusc	50 min	1.58	0.099	5.3	0.2
Selenium	Invertebrates	>10,000 max	0.029	0.001	0.4	0.4
		2900 min	0.029	0.001	0.4	0.4
Vanadium	Fish	>10,000	2.0	0.45	5.1	<0.01
	Invertebrates	>10,000	2.0	0.45	5.1	>0.01
Chromium	Annelid	8000 max	1.26	0.005	3.3	0.15
		2200 min	1.26	0.005	3.3	0.15
	Mollusc	105,000 max	1.26	0.005	3.3	0.15
		14,000 min	1.26	0.005	3.3	0.15
	Crustacea	640,000 max	1.26	0.005	3.3	0.15
		2000 min	1.26	0.005	3.3	0.15
	Fish	190,000 max	1.26	0.005	3.3	0.15
		12,400 min	1.26	0.005	3.3	0.15
Arsenic	Fish	28,000 max	—	—	1.04	1.0
		15,000 min	—	—	1.04	1.0
	Invertebrates	4000	—	—	1.04	1.0
Zinc	Fiddler crab	76,950 max	10.9	0.05	250	0.007
		66,420 min	10.9	0.05	250	0.007
Copper	Crustacea (young)	300	8.6	0.006	20	0.065
	Mollusc (young)	300	8.6	0.006	20	0.065
	Crustacea (adult)	30,000	8.6	0.006	20	0.065
	Mollusc (adult)	30,000	8.6	0.006	20	0.065

TABLE 8.7
Effect of Exposure Time on Mortality

Environmental Concentration of Cadmium ($\mu g\,L^{-1}$)	Exposure Time (days)	
	4% Mortality	365% Mortality
0.005	Very low	<50
0.05	Very low	50
0.5	<50	>50
0.7	50	>50
5	>50	>50

It will be noted that in all the above considerations, only the 4-day LC_{50} test is discussed. This parameter gives the concentration of the test metal in the test water that will kill 50% of the creatures under test in 4 days. Obviously, if the duration of the LC_{50} test is increased, then the particular creature under discussion will only tolerate a lower concentration of the test metal over the extended period for 50% fatalities to occur. Thus, the LC_{50} value of $0.7\,\mu g\,L^{-1}$ obtained for salmonid fish in nonsaline waters when exposed in the short term to cadmium for 4 days is approximately 15 times greater than the value of $0.05\,\mu g\,L^{-1}$ obtained in a long-term 365-day exposure.

The effect of exposure of the fish to various concentrations of cadmium in nonsaline waters is shown in Table 8.7.

Thus, long-term exposure is more likely to produce mortalities at lower exposure concentrations. This is illustrated in Table 8.8, where it is seen, for example, that exposure to 780 and $52\,\mu g\,L^{-1}$ cadmium for 4 and 365 days, respectively, would kill 50% of amphipods. In the case of environmental waters containing $5\,\mu g\,L^{-1}$ cadmium, adverse effects on the amphipods and possibly a small number of fatalities are more likely to occur during the long-term exposure. In the case of zinc, short-term exposure under stated conditions would lead to more than 50% mortalities; in the case of *Daphnia magna*, long-term exposure would, in addition, have a similar effect on certain types of fish.

The concentration of substances picked up from water by sea creatures such as fish and invertebrates and by algal and plant matter is dependent on the concentration of the substances in the water and, to some extent, their concentration in sedimentary matter. Many creatures bioaccumulate toxicants from the water, and as a consequence, their concentration in the organism is many times higher than that present in the water. Once the concentration of a toxicant in the organism exceeds a certain level, then harmful effects or mortalities occur.

Analysis of creatures is therefore a very useful means of ascertaining the cause of adverse effects or death in creatures, and analysis of plant and algal material is valuable for obtaining an early warning that excessive levels of toxicants may be present. Much work has been carried out in the determination of toxicants in creatures and plant life, and this is discussed below.

TABLE 8.8
Effect of Duration of Exposure to Metals in Water on Mortalities

		Short-Term Exposure (4 days)			Long-Term Exposure (1 year)		
		Metal Concentration in Freshwater ($\mu g\,L^{-1}$)			Metal Concentration in Freshwater ($\mu g\,L^{-1}$)		
		4-day LC_{50}			365-day LC_{50}		
Element	Creature	($\mu g\,L^{-1}$)	Max	Min	($\mu g\,L^{-1}$)	Max	Min
Cadmium	Fish (max)	37,100	5	0.013	24,733	5	0.013
	(min)	<10,000	5	0.013	<666	5	0.013
	Amphipod	780	5	0.013	52	5	0.013
Zinc	Fish (max)	52,000	630	0.86	3466	630	0.86
	(min)	2400	630	0.86	160	630	0.86
	Daphnia magna	70	630	0.86	4.6	630	0.86
	Annelid	10,000	630	0.86	666	630	0.86
Mercury	Crayfish	0.2	1.3	0.009	0.013	1.3	0.009
	Slipper limpet	1100	1.3	0.009	73	1.3	0.009

8.2 TOXICITY OF METALS IN FISH AND INVERTEBRATE ORGANS

The concentrations of metals in fish organs are a useful indicator of the cause of mortalities, while the metal content, of gill, muscle and skin, does not provide any such indicator. Thus, the data in Table 1.4 (Chapter 1) show the maximum concentrations of chromium, zinc, copper and cadmium found in opercle, liver and kidney organs taken from environmental fish samples that would lead to fish mortalities, while fish would survive at the lowest concentrations encountered in environmental fish samples.

Cadmium content of muscle taken from juvenile *Tilapia aurea* fish exposed to water containing $6.8–52\,\mu g\,L^{-1}$ cadmium ranged from $0.12\,mg\,kg^{-1}$ at the $6.8\,\mu g\,L^{-1}$ level to $0.92\,mg\,kg^{-1}$ at the $52\,\mu g\,L^{-1}$ level in water. Few national or international authorities have set limits for the cadmium in foodstuff. Norway and the Netherlands are reported to have set a limit of $0.5\,mg\,kg^{-1}$ in fish [6].

Starved or fed yearling roach (*Rutilus rutilus*) exposed to sublethal copper contamination ($80\,\mu g\,L^{-1}$ copper) for 7 days accumulated $19\,\mu g\,g^{-1}$ copper dry weight in gill tissue, but only starved fish accumulated significant quantities of copper in water ($95\,mg\,kg^{-1}$ copper, dry weight) [7].

Exposure of juvenile rainbow trout (*Salmo gairdneri*) to $55\,\mu g\,L^{-1}$ copper for 28 days led to uptakes of copper in the whole body from $1.2\,\mu g\,g^{-1}$ copper on day 1 to $6.6\,mg\,kg^{-1}$ copper on day 28. Liver copper increased from $25\,\mu g\,kg^{-1}$ copper dry weight on day 1 to $69\,mg\,kg^{-1}$ on day 2 and $113\,mg\,kg^{-1}$ copper on day 28, both dry weight [8].

Mosquito fish (*Gambusia affinis*) in a reservoir at San Joaquin Valley, California, were found to contain $30\,\mu g\,kg^{-1}$ selenium (as selenate) originating in drainage waters. All other species of fish had died [9].

Concentrations of zinc found in *Tilapia zilli* gills, liver and muscle after a 4-day exposure to zinc were, respectively, 38,000, 23,000 and $2000\,mg\,kg^{-1}$ zinc dry weight. The corresponding figures for *Clarius lazera* were 49,000, 34,000 and $5000\,mg\,kg^{-1}$ dry weight [10].

8.3 DETAILED TOXICITY OF METALS

Listed below are comments on various metals affecting fish and invertebrates.

8.3.1 ARSENIC

The chemical form of arsenic in marine environmental samples is of interest from several standpoints. Marine organisms show widely varying concentrations of arsenic, and knowledge of the chemical forms in which the element occurs in tissues is relevant to the interpretation of these variable degrees of bioaccumulation and to an understanding of the biochemical mechanisms involved. Different arsenic species have different levels of toxicity and bioavailability, and this is important in food chain processes, while physiochemical behaviour in processes such as adsorption onto sediments also varies with the species involved. It has been shown that inorganic arsenic(III and V), monomethyl arsenic and dimethyl arsenic acids are present in natural waters, biological materials and sediments.

The US Arsenic in Food Regulations 1959 states that foodstuffs must not contain more than $1\,mg\,kg^{-1}$ of total arsenic. Certain exceptions are listed which include fish and edible seaweed and their products, where arsenic contents of above $1\,mg\,kg^{-1}$ are present naturally. The UK total diet survey suggests that at least 75% of total arsenic ingested as food originates from fish and shellfish. It is accepted that the arsenic in fish and shellfish is mainly organically bound; hence, if any of the more toxic inorganic arsenic is present, this is of greater interest. If the levels of the total inorganic arsenic approach $1\,mg\,kg^{-1}$, the proportion of arsenic(III) relative to arsenic(V) also assumes importance, as the latter is considered to be more toxic than the former.

Crab are susceptible to arsenic at the larval life stage. Crustacea are the most sensitive to arsenic and annelids are the least sensitive. For instance, toxicity decreased in the order As(V) > organic arsenic > As(III). Fish are less susceptible (4-day $LC_{50} = 1500\text{--}28,000\,\mu g\,L^{-1}$) than invertebrates (4-day $LC_{50} = 40\,\mu g\,L^{-1}$).

8.3.2 CAESIUM

Madigan et al. [11] have measured the toxicity of radio caesium in Pacific bluefin tuna (*Thunnus orientalis*) using a tracer technique.

8.3.3 COPPER

The importance of complexing agents in the mineral nutrition of phytoplankton and other marine organisms has been recognised for more than 20 years. Complexing agents have been held responsible for the solubilisation of iron, and therefore its greater biological availability. In contrast, complexing agents are assumed to reduce the biological availability of copper and minimise its toxic effects. Experiments with pure cultures of phytoplankton in chemically defined media have demonstrated that copper toxicity is directly correlated to copper ion activity and independent of the total copper concentration. In these experiments, copper ion (Cu^{2+}) concentrations can be varied in media containing a wide range of total concentrations through the use of artificial complexing agents. When Cu^{2+} concentration was calculated for earlier experiments with phytoplankton in defined media, it appeared that Cu^{2+} was toxic to a number of phytoplankton species in concentrations as low as $10^{-6}\,\mu mol\,L^{-1}$. Since copper concentrations in the world oceans typically range from 10^{-4} to $10^{-1}\,\mu mol\,L^{-1}$, complexing agents and other materials affecting the solution chemistry of copper must contain the Cu^{2+} activity at sublethal levels.

Copper may exist in particulate, colloidal and dissolved forms in seawater. In the absence of organic ligands or particulate and colloidal species, carbonate and hydroxide complexes account for more than 98% of the inorganic copper in seawater [11–13].

The young life stage of crustacea and molluscs is more sensitive to copper (2-day $LC_{50} = 300\,\mu g\,L^{-1}$) than the adult stage (2-day $LC_{50} = 30{,}000\,\mu g\,L^{-1}$). This also applies to fish. Low salinity increases the toxicity of copper. Bivalve molluscs are the most copper-sensitive species yet examined, undergoing reduced growth in the presence of $3–10\,\mu g\,L^{-1}$ copper for prolonged periods. Beyond a 500-days exposure, mortalities occurred [14]. Fiddler crabs *Uca annulipes* and *Uca triangularis*, when exposed to copper, gave 4-day LC_{50} values of $12{,}820–14{,}810\,\mu g\,L^{-1}$.

Further information on the toxicity of copper is presented in Table 8.9. Rivera-Duarte et al. [59] have discussed larval copper toxicity of marine invertebrates.

Temporal and spatial measurements of the toxicity (EC_{50}), chemical speciation and complexation capacity (Cu-CC) of copper in waters from San Diego Bay suggest control of the Cu-CC over copper bioavailability. While spatial distributions of total copper (Cu_T) indicate an increase in concentration from the mouth towards the head of San Diego Bay, the distribution of aqueous free copper ion ($Cu(II)_{aq}$) shows the opposite trend. This suggests that the bioavailability of copper to organisms decreases toward the head of the bay, and is corroborated by the increase in the amount of copper needed to reach an EC_{50}, observed for the larval states of three marine invertebrates (Mediterranean mussel, *Mytilus galloprovincialis*; sand dollar, *Dendraster excentricus*; and purple sea urchin, *Strongylocentrotus purpuratus*), and by the increase in Cu-CC heading into the head of the bay. The amount of $Cu(II)_{aq}$ required to produce a 50% reduction in normal larval development (referred to here as $_pCu_{Tox}$) of the mussel, the most sensitive of the three marine invertebrates, was generally at or above $\sim 1 \times 10^{-11}\,mol\,L^{-1}$ equivalents of Cu (i.e. $_pCu_{Tox} = 11 = -(\log\,[Cu(II)_{aq}])$). These results suggest that the copper complexation

TABLE 8.9
Toxicity of Copper to Saltwater Life

Species	Life Stage	Static (S) or Flowing (F)	Salinity (g kg⁻¹)	pH	Temperature (°C)	Chemical Form	Nominal (N) or Observed (O) LC_{50}	Concentration (mg L⁻¹)	Duration (days)	Effect	Reference
Companularia flexuosa	Mature	S	Seawater			Cu^{2+}	—		11	Growth reduced	[32]
Eirene viridula	Mature	S	Seawater			Cu^{2+}		0.03–0.06	14–21	Growth reduced	[33]
Ctenophora											
Mnemlopis macrodyi	—		Seawater			Cu^{2+}		0.017	1 LC_{50}	—	[34]
Pleurobrachia pileus	—		Seawater			Cu^{2+}		0.029 0.033	1 LC_{50}	—	
Chaetognatha											
Sagitta hispida	Adult	—	Seawater	—	—	Cu^{2+}		0.300	4 LC_{50}	—	[35]
Rotifera											
Brachronus pileatilis	Adult	—	Seawater	—	—	Cu^{2+}		0.077 0.200 0.044	4 LC_{50} Mortality 28 LC_{50}	50%	[36]
Ltenodrilus serratus	Adult	S	Seawater (5–34, 14%)	6–8		—		0.250	37 LC_{50}	—	[37]
Neanthes arenaceodentata	Adult		Seawater					0.20–0.48	4 h LC_{50}	—	[38]
Ncreoc	Adult	S	Seawater			CW Na atrate	N	0.05	3 h LC_{50}	—	[39]

(Continued)

TABLE 8.9 (Continued)
Toxicity of Copper to Saltwater Life

Species	Life Stage	Static (S) or Flowing (F)	Salinity (g kg⁻¹)	pH	Temperature (°C)	Chemical Form	Nominal (N) or Observed (O) LC₅₀	Concentration (mg L⁻¹)	Duration (days)	Effect	Reference
Ophryotrocha labronica	Larva	S	Seawater		21–22	Cu²⁺	N	1.0, 0.25, 0.1	5 h / 12	100% mortality / 100% mortality / No mortality	[40]
Phyllodoce maculate	Adult	S	Seawater				—	0.12	4 LC₅₀		[41]
Mollusca											
Crassostrea virginica	Adult					Cu²⁺	—	0.08		9.50% mortality	[42]
	Embryo / Adult	S / F	Seawater		26	Cu²⁺	O	0.022 / 0.0–12		50%–70% / Mortality / 100% mortality	[43]
Crassostrea gigas	Adult	F	33%	—		CuSo⁴	O	0.56 / 0.01	4 LC₅₀	14%–20% mortality	[45] [46]
Venerupis decussate	Adult	F	33		15		0.01	0.105 / 40 / 0.01	2 LC₅₀ No effect / 40–75 / 75 + 35 Clean	Burrowing reduced 90% / Burrowing restored	[47]
								0.01	<20	No effect on burrowing	[46]

(Continued)

TABLE 8.9 (Continued)
Toxicity of Copper to Saltwater Life

Species	Life Stage	Static (S) or Flowing (F)	Salinity (g kg⁻¹)	pH	Temperature (°C)	Chemical Form	Nominal (N) or Observed (O) LC₅₀	Concentration (mg L⁻¹)	Duration (days)	Effect	Reference
Bullia digitalis	Adult		Seawater		15			0.01	20–50	Burrowing eliminated	
						+1 mg L⁻¹ EDTA		0.01	110	Burrowing unaffected	
						+1 mg L⁻¹ EDTA		0.01	110	Burrowing unaffected	
								0.1	4	No effect	
								0.2	4	Burrowing ceases	
								0.3	4	Oxygen uptake reduced 70%	[47]
								0.5	4	LC₅₀	
Haliotis cracherodii	Adult	S	33		14	CuSo₄	O	0.05	4		
Halliotis frufescens	Adult	S	33		14	CuSo₄	O	0.065	4	LC₅₀	
	Larva							0.114	2		

(Continued)

TABLE 8.9 (Continued)
Toxicity of Copper to Saltwater Life

Species	Life Stage	Static (S) or Flowing (F)	Salinity (g kg⁻¹)	pH	Temperature (°C)	Chemical Form	Nominal (N) or Observed (O) LC$_{50}$	Concentration (mg L⁻¹)	Duration (days)	Effect	Reference
Crustacea											
Acartia	Adult	S	Seawater			Cu^{2+}		0.017–0.055	4	LC$_{50}$	[48]
Artemia salina	Larva and adult	S			21	CuCl$_2$	N	⎰ 0.25	14	High mortality	[40]
								⎱ 0.1	14	No mortality	[40]
Carcinus meaenas	Adult				15			109	2		[49]
	Larva				15			0.6	2	LC$_{50}$	
Crangon crangon	Adult				15			30	2		
	Larva				15			0.33	2		
Euphausia pacifica	Adult	S	Seawater			Cu^{2+}	O	0.014–0.03		LC$_{50}$	[34]
Hamarus americanus	Adult	S	30		13	CuSO$_4$		0.056	11	Median period of survival	[50]
	Adult		Seawater			Cu^{2+}		0.100	4	LC$_{50}$	[50]
	Larva		Seawater			Cu^{2+}		0.048	4	LC$_{50}$	[51]
Echinodermata											
Echinometra mathaei	Larva ⎰ Adult ⎱		32–35		28	CuCl$_2$	N	0.050	3	LC$_{50}$	[52]
								0.30	4	LC$_{50}$	[52]
Paracentrotus lividus	Pluteal larva		Seawater			Cu^{2+}		0.010–0.020	4	Growth retarded	[53]

(Continued)

TABLE 8.9 (Continued)
Toxicity of Copper to Saltwater Life

Species	Life Stage	Static (S) or Flowing (F)	Salinity (g kg⁻¹)	pH	Temperature (°C)	Chemical Form	Nominal (N) or Observed (O) LC₅₀	Concentration (mg L⁻¹)	Duration (days)	Effect	Reference
Fish											
Archosargus probatocephalus	Adult	S	30		21–24	CuCl₂	N	4.17	1	LC₅₀	[54]
								3.02	2	LC₅₀	
								2.9	3	LC₅₀	
								1.14	4	LC₅₀	
								0.05	4	No mortality	
Arius felis	27 cm	S	30		21–23	CuCl₂	N	5.43	1	LC₅₀	[55]
								4.17	2	LC₅₀	
								3.57	3	LC₅₀	
								2.4	4	LC₅₀	
								0.1	3	Hyperactivity	
								0.01	3	Turning behaviour reduced	
Fundulus heteroclitus	Adult	S	23.6		20		O	0.61	14	LC₅₀	[56]
			5.5		20		N	3.1	4	LC₅₀	[57]
			20	8.0	20		N	2.0	4	LC₅₀	[57]
							N	78.0	4	LC₅₀	[58]
							N	1.0	4	LC₅₀	[59]
							N	0.16	9	LC₅₀	[60]

(Continued)

TABLE 8.9 (Continued)
Toxicity of Copper to Saltwater Life

Species	Life Stage	Static (S) or Flowing (F)	Salinity (g kg⁻¹)	pH	Temperature (°C)	Chemical Form	Nominal (N) or Observed (O) LC₅₀	Concentration (mg L⁻¹)	Duration (days)	Effect	Reference
Lagodon rhomboids	Larva	{ S { S	20	8.0	20		O	0.15	14	LC_{50}	[56]
Lagodon rhomboids	Larva	F									
Lagodon rhomboids	Adult	S	30		21–24	$CuCl_2$	N	7.34	4	LC_{50}	[54]
								4.43	4	LC_{50}	
								4.14	4	LC_{50}	
								2.75	4	LC_{50}	
								0.20	4	No mortality	
Leiostomus		F					O	0.16	14	LC_{50}	[56]
Micropogon undulates	Larva	F					O	0.21	14	LC_{50}	[56]
Micropogon undulates	Adult	S	30		21–24	$CuCl_2$	N	9.15	1	LC_{50}	[54]
								7.55	2	LC_{50}	
								7.20	3	LC_{50}	
								5.66	4	LC_{50}	
								0.40	4	No mortality	
Nectes americanus	Adult	F		8.0	9.4	Cu^{2+}		2.0	4	LC_{50}	[61]

EDTA, ethylenediaminetetraacetic acid.

capacity in San Diego controls copper toxicity by keeping the concentration of $Cu(II)_{aq}$ at nontoxic levels. Comparison of the 4-day LC_{50} values between phyla suggests that all are similar, apart from the annelids, which are more sensitive to copper. In the crustacea, the young-life-stage larvae are considerably more sensitive than the adults, with 4-days LC_{50} values of $6.33\,mg\,L^{-1}$ for adult *Crangon crangon* larvae and $30\,mg\,L^{-1}$ for adult *Crangon crangon*. This alone applied to bivalve molluscs but not the annelid. For *Neanthes arenaceodentata*, both the 4-day and 28-day LC_{50} values are similar.

In acute studies of the fish *Trachinotus*, low salinity increased the toxicity of copper of $250\,mg\,L^{-1}$ collected from a clean environment (1-day LC_{50} of $250\,mg\,L^{-1}$), suggesting acclimation of the *Nereis* from the contaminated sediments.

8.3.4 CADMIUM

Reduced salinity and higher water temperature both increase the toxicity of cadmium to invertebrates [15,16]. Planktonic crustaceans have a 4-day LC_{50} of $60–380\,\mu g\,L^{-1}$ [17]. Young life stages of invertebrates are sensitive to cadmium. Adult crustaceans are also susceptible to $60\,\mu g\,L^{-1}$ of cadmium causing 30% mortality in a 60-day exposure [18]. Fish are relatively resistant to cadmium with a 4-day LC_{50} of $6400–16,400\,\mu g\,L^{-1}$. Fiddler crabs *Uca annulipes* and *Uca triangularis*, when exposed to cadmium, gave 4-day LC_{50} values of 43,230 and $48,210\,\mu g\,L^{-1}$.

Planktonic crustaceans are very sensitive to cadmium with 4-day LC_{50} values ranging from 0.06 to $0.38\,mg\,L^{-1}$ for adult *Penaeus indicus* and 3.5 to $13.3\,mg\,L^{-1}$ for adult *Marinogammarus obtusatus*. Young life stages are also sensitive with a comparable response for both zoeae of *Cancer magister* and *Pagurus longicarpus* after a 60-day exposure [18]. The same concentration reduced the growth of *Argopecten irradiens* in a 42-day exposure, while $0.25\,mg\,L^{-1}$ caused a 10% mortality. The 11-day LC_{50} was considerably higher at $1.0\,mg\,L^{-1}$ [18].

Short-term acute toxicity tests with fish indicate that they are relatively resistant to cadmium with 4-day LC_{50} values of 6.4 and $60.0\,mg\,L^{-1}$, respectively, for *Menidia menidia* and *Cyprinodon variegates* [58]. *Pseudopleuronectes americanus* suffered 12% mortality after an 8-day exposure to $0.56\,mg\,L^{-1}$, rising to 90% at $1.09\,mg\,L^{-1}$. In contrast, a 63-day exposure of *Platichthys flesus* at $0.5\,mg\,L^{-1}$ caused occasional body spasms but no mortality [57].

Comparison of the 4-day LC_{50} between phyla suggests that all are similar, apart from the annelids, which are more sensitive to copper. On the crustacea, the young-life-stage larvae are considerably more sensitive than the adults with 4-day LC_{50} values of $6.33\,mg\,L^{-1}$ for adult *Crangon crangon* larvae and $30\,mg\,L^{-1}$ for adult *Crangon crangon*. This alone applies to bivalve molluscs, but not the annelid.

For *Neanthes arenaceodentata*, both the 4-day and 28-day LC_{50} values are similar.

In acute studies of the fish *Trachinotus*, low salinity increased the toxicity of copper.

For fish, the response of young life stages is such that 4-day LC_{50} values for embryos ($0.028\,mg\,L^{-1}$) and larvae ($0.136\,mg\,L^{-1}$) are less than the majority of results for adult fish ($0.129–0.510\,mg\,L^{-1}$). The only adverse response at a lower concentration was a disturbance of the behaviour of *Arius felis* after 3 days at $0.01\,mg\,L^{-1}$, with hyperactivity at $0.\,mg\,L^{-1}$.

In contrast, the annelids, which are consistently sensitive in 4-day LC_{50} tests, evidenced little increase in sensitivity when exposure increased to 28 days.

The response of crustaceans is confusing, with one study of *Artemia* reporting no mortality arising from 11 days at $0.1\,mg\,L^{-1}$. In contrast, a median period of survival of 11 days at $0.056\,mg\,L^{-1}$ has been reported for *Homarus*.

The bivalve molluscs are among the most sensitive species reported, with adverse effects occurring between 0.003 and $0.01\,mg\,L^{-1}$ after long exposure. The longest exposure spanned 630 days, during which the growth of *Mytilus edulis* was reduced, but no mortalities until 540 days had passed. However, in another study, $0.01\,mg\,L^{-1}$ caused 20% mortality after only a 14-day exposure of *Crassostrea gigas*. Abnormal development of the larvae of two species was observed after only 2 days at $0.005\,mg\,L^{-1}$ (EC_{50}).

The most sensitive species tested is the hydrozoan *Companularia flexuosa*, for which colony growth was stimulated by copper concentrations of 0.001 and $0.01\,mg\,L^{-1}$ but reduced by concentrations of 0.015 and $0.025\,mg\,L^{-1}$ after a 16-day exposure. An exposure of 8 days at $0.05\,mg\,L^{-1}$ halted growth of the colonies completely.

In one study, the annelid *Nereis diversicolor* was collected from two locations, one of which had sediments contaminated with copper. These animals were more resistant (12.5-day LC_{50} of $250\,mg\,L^{-1}$) collected from a clean environment (1-day LC_{50} of $250\,mg\,L^{-1}$), suggesting acclimation of the *Nereis* from the contaminated sediments.

8.3.5 CHROMIUM

The relative LC_{50} values of annelids, molluscs, crustacea and fish when exposed to trivalent chromium (exposure period not stated) are, respectively, 2200–8000, 14,000–105,000, 2000–98,000 and $12,400–91,000\,\mu g\,L^{-1}$. Reduction in salinity from 35 to $15\,g\,kg^{-1}$ reduced the 4-day LC_{50} from 640,000 to $190,000\,\mu g\,L^{-1}$.

The toxicity data are almost exclusively confined to acute exposure of 16 days or less (Table 8.10). The results for each phylum show similar ranges of 4-day LC_{50} values and, with the exception of the annelids, span more than an order of magnitude:

Phylum	Range of 4-day LC_{50} Values ($mg\,L^{-1}$)
Annelids	2.2–8.0
Molluscs	14–105
Crustaceans	2.0–98.0
Fish	12.4–91.0

These intraphyletic variations appear to be independent of test conditions. However, one study of the three species *Nereis diversicolor*, *Corophium volutator* and *Macoma balthica* demonstrated that salinity affected toxicity. For *Macoma*

TABLE 8.10

Toxicity of Chromium to Saltwater Life

Species	Life Stage	Static (S) or Flowing (F)	Salinity (g kg⁻¹)	pH	Temperature (°C)	Chemical Form	Nominal (N) or Observed (O) LC₅₀	Concentration (mg L⁻¹)	Duration (days)	Effect	Reference
Annelida											
Capitella capitata	Larva / Adult	S	Seawater			Chromium trioxide	O	N	8.0	4	[65]
	Adult							5.0	4	LC₅₀	
								0.55	8		
Ctenodrilus serratus		S	Seawater			Chromium trioxide	N	4.3	4	LC₅₀	[66]
Neanthes arenaceodentata	Adult	S	Seawater			Chromium trioxide	N	0.5	28	LC₅₀	[65]
							N	3.1	4	LC₅₀	[67]
						Potassium dichromate	O	2.2–4.3	4	LC₅₀	[68]
	Life cycle						O	0.030		Reduced reproductive success	[69]
							N	10.0	8	LC₅₀	[70]
							O (T)	80.0	4		
								22	4		
								9.5	4		
								34	4		[71]
								7.5	4		
								27	8		
								8	8		
								3.3	8		

(Continued)

TABLE 8.10 (Continued)
Toxicity of Chromium to Saltwater Life

Species	Life Stage	Static (S) or Flowing (F)	Salinity (g kg⁻¹)	pH	Temperature (°C)	Chemical Form	Nominal (N) or Observed (O) LC₅₀	Concentration (mg L⁻¹)	Duration (days)	Effect	Reference
Nereis virens	Adult	S	Seawater			Potassium dichromate	N	2.0	4	LC₅₀	[72]
Ophryotrocha diadoma		S	Seawater			Chromium trioxide	N	7.5	4	LC₅₀	[66]
Mollusca											
Crassostrea gigas	Larva	S	33.8	8.1	20	K₂CrO₇	O	4.54 ± 0.72	2	EC₅₀ developed abnormal	[77]
Macoma balthica	Adult	S	15		5	Cr(VI)	O (T)	190	4	LC₅₀	[71]
			35		5			36	8		
					15			640	4		
								180	8		
								15	16		
								110	4		
								34	8		
Mya arenaria	Adult	S	Seawater			Potassium chromate	N	57	4	LC₅₀	[72]
Mytilus edulis	Larva	S	33.8	8.1	17	Potassium chromate	O	44.47 ± 0.74	2	EC₅₀ developed	[77]
Perna perna	7 cm	S				Chromium trichloride	N	2.0 (1.8–2.3)	1 h	Filter rate reduced 50%	[78]
						Sodium dichromate	N	1.6 (1.0–2.0)	1 h	Filter rate reduced 50%	[78]

(Continued)

TABLE 8.10 (*Continued*)
Toxicity of Chromium to Saltwater Life

Species	Life Stage	Static (S) or Flowing (F)	Salinity (g kg^{-1})	pH	Temperature (°C)	Chemical Form	Nominal (N) or Observed (O) LC$_{50}$	Concentration (mg L^{-1})	Duration (days)	Effect	Reference
Rangia cuneate	Adult	S	Seawater			Potassium dichromate	N	14.0–35.0	4	LC$_{50}$	[73]
Nassarius obsoletus	Adult	S	Seawater			Potassium chromate	N	105.0	4	LC$_{50}$	[72]
Echinordermata											
Asterias forbesi	Adult	S	Seawater			Potassium chromate	N	32.0	4	LC$_{50}$	[72]
Chelon labrosus	Adult	F	34	8.3	12.5	Potassium dichromate	O (D)	140.2	1	LC$_{50}$	[79]
								90.0	2		
								67.5	3		
								47.2	4		
Menidia menidia	Larva Juvenile	S	Seawater			Potassium dichromate	N	12.4–14.3	4	LC$_{50}$	[79]
								20.1	4	LC$_{50}$	[81]
Nitrocra spinipes	Adult	S	7	7.8	21	Potassium dichromate	N	16 (11–23)	4	LC$_{50}$	[76]

(*Continued*)

TABLE 8.10 (Continued)
Toxicity of Chromium to Saltwater Life

Species	Life Stage	Static (S) or Flowing (F)	Salinity (g kg^{-1})	pH	Temperature (°C)	Chemical Form	Nominal (N) or Observed (O) LC$_{50}$	Concentration (mg L^{-1})	Duration (days)	Effect	Reference
						Crustacea					
Acartia clause	Adult autumn generation	S	Seawater	14		Na$_2$CrO$_4$ 18	N	16.99 ± 2.38	2	LC$_{50}$	[74]
								11.47 ± 3.87	2	LC$_{50}$	
				22				8.83 ± 2.1	2	LC$_{50}$	
				14		CrO$_3$	N	16.37 ± 0.18	2	LC$_{50}$	
				14		Na$_2$CrO$_4$	N	19.27 ± 2.8	2	LC$_{50}$	
	Winter generation			14			N	12.26 ± 2.62	2	LC$_{50}$	
				22				1.0	1	Feeding reduced	
	Summer generation			14		Cr(NO$_3$)$_3$	N	1.0–17.0	2	No mortality, all precipitated	
	Adult	S	Seawater			Potassium dichromate	N	6.60	4	LC$_{50}$	[79]
Allorchetes compressa	Adult	F	33.0	8.0	20	Potassium	O (T)	5.56 (4.75–6.50)	4	LC$_{50}$	[80]
						Potassium		6.34 (5.43–7.39)	4	LC$_{50}$	[80]

(*Continued*)

TABLE 8.10 (Continued)
Toxicity of Chromium to Saltwater Life

Species	Life Stage	Static (S) or Flowing (F)	Salinity (g kg⁻¹)	pH	Temperature (°C)	Chemical Form	Nominal (N) or Observed (O) LC₅₀	Concentration (mg L⁻¹)	Duration (days)	Effect	Reference
Callinectes sapidus	Adult	S	Seawater	8.1	15	Potassium dichromate	N	89.0–98.0	4	LC₅₀	[75]
Cancer magister	Zocae	S	33.8		5	K₂CrO₇	O	3.44	4	LC₅₀	[77]
Corophium volutator	Adult	S	20		5	Cr(VI)	O (T)	90	1		
								50	2		
			35		5	Cr(IV)	O (T)	17	4	LC₅₀	[71]
					10	Cr(IV)	O (T)	7.5	8		
					15	Cr(IV)	O (T)	3.2	16		
			5		5	Cr(IV)	O (T)	40	4		
								36	4		
								5.8	4		
								5.0	4		
Mysidopsis bahia	Adult	S	Seawater			Potassium dichromate	N	2.0	4	LC₅₀	
Mysidopsis bigelowi	Adult	S	Seawater			Potassium dichromate	N	4.40	4		
Pagurus longicarpus	Adult	S	Seawater			Potassium dichromate	N	10.0	4	LC₅₀	[80,81]
Pseudololeptus coranatus	Adult	S	Seawater			Potassium dichromate	N	3.65	4		
Tigriopus japonicas	Adult	S	Seawater			Potassium dichromate	N	17.2	4		
Fish											
Alburnus	Adult	S	7	7.8	10	Potassium dichromate	N	240.0 (194–297)	4	LC₅₀	[76]

balthica, a reduction of salinity from 35 to 15 g kg^{-1} at 5°C caused a change in 4-day LC$_{50}$ from 540 to 1,900 mg L^{-1}, a substantial increase in toxicity.

An increase in temperature had the opposite effect, increasing the toxicity of chromium; for instance, the 4-day LC$_{50}$ for *Corophium volutator* was reduced from 40 mg L^{-1} at 5°C to 5.8 mg L^{-1}, both at a salinity of 35 g kg^{-1}.

The longer exposure data indicate that the adverse effect concentrations become progressively lower with a 28-day LC$_{50}$ of 0.5 mg L^{-1} for *Capitella capitata* compared with a 4-day LC$_{50}$ of 1 mg L^{-1}.

A similar 28-day LC$_{50}$ was recorded for *Neanthes arenaceodentata*; however, prolonged exposure to concentrations as low as 0.03 mg L^{-1} disrupted the reproductive success of this species, emphasising the importance of long-term exposures in assessing potential adverse effects.

The available information for trivalent chromium, as opposed to hexavalent chromium, is too limited to allow an objective comparison of the relative toxicities, nor do the data permit comments on the relative toxicity of chromium to different life stages.

8.3.6 LEAD

In one study of the toxicity of lead to *Chelon labrosus*, it was observed that the lead, added as a nitrate, was insoluble in seawater at concentrations in excess of 4.5 mg L^{-1}, raising doubts about the validity of other studies reporting effects at higher concentrations. Perhaps because of this chemical complication, the adverse effects reported are varied and disparate (Table 8.11). It is clear, however, that the larvae of molluscs and the zoeae of *Cancer magister* are particularly sensitive to lead, with abnormal development occurring after a 2-day exposure to concentrations in excess of 0.45 mg L^{-1}, these being the lowest reported adverse effect concentrations.

The available data are too limited to allow comment on the selective sensitivity of toxicity or on the possible effect of salinity or temperature.

However, studies of organolead compounds clearly demonstrate the greater toxicity of tetramethyl and tetraethyl lead compounds with inorganic forms of lead or, indeed, with other organolead compounds, such as trimethyl lead. Mollusc larvae are particularly sensitive to lead, with abnormal development occurring upon a 2-day exposure to 400 µg L^{-1} lead.

Couture et al. [76] studied the seasonal variations in lead sources in eastern north Atlantic mussels.

The concentration of lead and its stable isotope composition were measured in 216 composite samples of 50 blue mussels (*Mytilus edulis*) collected quarterly between 1985 and 2005 at three sites along the French Atlantic coast, one site in the La Fresnaye Bay and another in the Loire and Seine River Estuaries. Depending on the sites and the time periods, lead concentrations were 5–66 times higher than the natural background value for the North Atlantic. Even for the samples with the lowest lead concentrations, the isotopic signature of lead is very different to that of the regional natural lead, suggesting that most of the bioaccumulated lead is anthropogenic in origin. Stable lead isotope ratios measured in the mussels differ markedly

TABLE 8.11
Toxicity of Lead to Saltwater Life

Species	Life Stage	Static (S) or Flowing (F)	Salinity (g kg⁻¹)	pH	Temperature (°C)	Chemical Form	Nominal (N) or Observed (O) LC$_{50}$	Concentration (mg L⁻¹)	Duration (days)	Effect	Reference
						Annelida					
Capitella capitella	Larva		Seawater					1.2	4		
	Adult							1.0	28		
Ctemodrilus serratus								>20.0	4		
Neanthes arenaceodentata	Juvenile							2.5	28	LC$_{50}$	[28]
	Adult							2.2			
Phyryotrocha diadema								11.0	4		
Ophryotrocha labronica								>1.0	25		
						Mollusca					
Crassostrea gigas	Larva	S	33.8	8.1	20	Pb(NO$_3$)$_2$	O	0.758 ± 0.02	2	EC$_{50}$ abnormal development	[82]
Mercenaria mercenaria	Embryo							0.78	2	EC$_{50}$	[85]
Mya arenaria								8.8	7	LC$_{50}$	
Mytilus	Adult							>5.0	39	LC$_{50}$	

(Continued)

TABLE 8.11 (Continued)
Toxicity of Lead to Saltwater Life

Species	Life Stage	Static (S) or Flowing (F)	Salinity (g kg⁻¹)	pH	Temperature (°C)	Chemical Form	Nominal (N) or Observed (O) LC₅₀	Concentration (mg L⁻¹)	Duration (days)	Effect	Reference
Mytilus edulis	Adult	F	32.8		9.9	Pb(C$_2$H$_3$O$_2$)	N	0.2	8	Growth unaffected	[84]
	Adult	F	Seawater			Tetramethyl Pb		0.27	4	LC$_{50}$	[86]
						Tetraethyl Pb		0.10	4		
						Trimethyl PbCl		0.5	4		
						Trimethyl PBCl		1.1	4		
	Larva	S	33.8	8.1	17	Pb(NO$_3$)$_2$	O	0.476 ± 0.001	2	EC$_{50}$ abnormal development	[82]
Bullia digitalis	Adult		Seawater		15			0.5	4	No effect	
								1.0	4	Burrowing halted	
								5.0	4	Decreased oxygen uptake	
Crustacea											
Artenia salina	Zoeae	S	33.8	8.1	15	lead nitrate	O	1.0	24	LC$_{50}$	[85]
Cancer magister								0.575 ± 0.192	4	LC$_{50}$	[82]
Crangon crangon		F	Seawater			Tetramethyl Pb		0.11	4	LC$_{50}$	[86]
						Tetramethyl Pb		0.02	4		
						Trimethyl PbCl		8.8	4		
						Trimethyl PbCl		5.8	4		

(Continued)

TABLE 8.11 (Continued)
Toxicity of Lead to Saltwater Life

Species	Life Stage	Static (S) or Flowing (F)	Salinity (g kg⁻¹)	pH	Temperature (°C)	Chemical Form	Nominal (N) or Observed (O) LC$_{50}$	Concentration (mg L⁻¹)	Duration (days)	Effect	Reference
						Fish					
Chelon labrosus	Adult	F	34.7	7.5	12.2	Lead nitrate	O (D) (limited by solubility)	>4.5	1–4	LC$_{50}$	[85]
Limanda limanda	Adult	F	Seawater			Tetramethyl Pb		0.05	4		
						Tetraethyl Pb		0.23	4		
						Trimethyl PbCl		24.6	4		
						Trimethyl PbCl	1.7	4	4	LC$_{50}$	[86]
						Dimethyl PbCl$_2$		300	4		
						Dimethyl PbCl$_2$	75	4			

from the lead emitted in western Europe as a result of leaded gasoline combustion, which was still a dominant source of contaminant lead to the atmosphere during most of the study period. The isotope composition of lead in the mussels was instead more typical of that of the lead released to the environment by water treatment plants, municipal waste incinerators and industries such as metal refineries and smelters.

Continental runoff, rather than atmospheric deposition, is therefore identified as the lead along the French Atlantic coast. From the strong seasonal variations in $^{206}Pb:^{208}Pb$ ratios in the mussels from the Seine Estuary site, it was concluded that the resuspension of contaminated sediments, triggered by high river runoff events, is a chief factor affecting the bioaccumulation of lead in *Mytilus edulis*. The value of this organism as a biomonitor of coastal contamination is thus further demonstrated.

Further information on the toxicity of lead to invertebrates is given in Table 8.12.

8.3.7 Mercury

Crustaceans appear to be as sensitive to mercury as fish (see Table 8.12), while the zoeae of the crab, *Cancer magister*, have an LC_{50} of $0.008\,mg\,L^{-1}$. In contrast, $0.006\,mg\,L^{-1}$ had no effect after a 28-day exposure of *Penaeus indicus*, although the 4-day LC_{50} was $0.015\,mg\,L^{-1}$ [86].

Molluscan larvae are as sensitive as the crab zoeae with EC_{50} values for abnormal development of 0.0067 and $0.0058\,mg\,L^{-1}$, respectively, for *Crassostrea gigas* and *Mytilus edulis* [79].

Exposure tests employing inorganic and organic mercury do not suggest any obvious differences in their toxicity to marine fauna.

The 4-day LC_{50} for the embryos of a fish was $0.07\,mg\,L^{-1}$. However, exposure for 1 day followed by 3 days in clear water produced an LC_{50} value of $0.09\,mg\,L^{-1}$, indicating a rapid and irreversible impact on the organisms [19]. A similar response was evident in longer exposure to mercury in which the hatching and posthatching vitality of offspring were examined.

In the latter case, a 5-day exposure produced an EC_{50} value of $6.032\,mg\,L^{-1}$ compared with $0.061\,mg\,L^{-1}$ from a 1-day exposure and 31-day exposure in clean water [19].

Minganti et al. [89] studied the variations of mercury and selenium in *Adamussium* cod taken from Terra Nova Bay, Antarctica, during a 5-year period. Gaden et al. [103] determined mercury trends in ringed seals. This is associated with the length of the ice-free season.

8.3.8 Nickel

Nickel is relatively nontoxic to marine organisms with 4-day LC_{50} values in excess of $10\,mg\,L^{-1}$ (Table 8.13). However, marine planktonic crustaceans, bivalve larvae and embryonic echinoderms are more sensitive. In the latter organisms, abnormal development results from exposures of 20 h to concentrations between 0.05 and $0.58\,mg\,L^{-1}$ [97], in contrast to the 4-day LC_{50} for adult *Asterias forbesi*, which is $150\,mg\,L^{-1}$ [67].

TABLE 8.12
Toxicity of Mercury to Saltwater Life

Species	Life Stage	Static (S) or Flowing (F)	Salinity (g kg⁻¹)	pH	Temperature (°C)	Chemical Form	Nominal (N) or Observed (O) LC₅₀	Concentration (mg L⁻¹)	Duration (days)	Effect	Reference
Annelida											
Limnodriloides lerrucsus	Adult	S	20	7.0 / 7.0 / 6.0 / 8.0	10 / 1 / 10 / 10	HgCl₂	N	0.12 / 0.15 / 0.14 / 0.13	4 / 4 / 4 / 4	LC₅₀	[90]
Monopylophorus cuticulatus	Adult	S	20 ... 10	7.0 / 7.0 / 7.0 / 6.0 / 8.0 / 7.0	10 / 1 / 20 / 10 / 10 / 10	HgCl₂	N	0.23 / 0.25 / 0.32 / 0.16 / 0.14 / 0.415	4 / 4 / 4 / 4 / 4	LC₅₀	[90]
Coelenterata											
Bunodsoma cavernata	Adult	S	26		Room	HgCl₂	N	1.2	7	No effect on amino acid balance	[92]
Mollusca											
Bullia digitalis	Adult	Seawater			15			0.5 / 2.0 / 5.0	4 / 4 / 4	No effect / Burrowing ceases / 20% decrease in oxygen uptake	
Crassostrea gigas	Larva	S	33	8.1	20	HgCl₂	O	0.0067 ± 0.0016	2	EC₅₀ developed abnormal	[77]
Mytilus edulis	Larva	S	33.8	8.1	17	HgCl₂	O	0.0058 ± 0.001	2	EC₅₀ developed abnormal	[77]

(Continued)

TABLE 8.12 (Continued)
Toxicity of Mercury to Saltwater Life

Species	Life Stage	Static (S) or Flowing (F)	Salinity (g kg⁻¹)	pH	Temperature (°C)	Chemical Form	Nominal (N) or Observed (O) LC$_{50}$	Concentration (mg L⁻¹)	Duration (days)	Effect	Reference
	Adult	F	33.8		8.6	HgCl$_2$	N	00.003 / 0.0016 / 0.025	5 / 4 / 1	Growth reduced / Growth halted / Some mortalities	[84]
Perna perna	7 cm	S				Mercuric chloride	N	0.025	1 h	50% reduction in filter rate	
						Methyl mercuric chloride	N	0.050	1 h	50% reduction in filter rate	
						Ethyl mercuric chloride	N	0.030	1 h	Filter rate reduced 50%	[83]
						Phenyl mercuric acetate	N	0.020	1 h	Filter rate reduced 50%	
						Sodium acetate	N	1.0	1 h	(No effect)	
Pernaperna						Mercuric chloride + selenium chloride	N	0.040	1 h	Filter rate reduced 50%	[78]
Crustacea											
Allorchestes compressa	Adult	F	36	8.0	20	Mercuric chloride	O (T)	0.08 (0.06–0.098)	4	LC$_{50}$	
Cancer magister	Zoeae	S	33.8	8.1	15	Mercuric chloride	O	0.0082 ± 0.0037	4	LC$_{50}$	

(Continued)

TABLE 8.12 (Continued)
Toxicity of Mercury to Saltwater Life

Species	Life Stage	Static (S) or Flowing (F)	Salinity (g kg⁻¹)	pH	Temperature (°C)	Chemical Form	Nominal (N) or Observed (O) LC₅₀	Concentration (mg L⁻¹)	Duration (days)	Effect	Reference
Palaeomonetes pugio	Adult	S	25	8.3–8.7	22	Mercuric acetate	N	0.06 (0.03–0.09)	4	LC₅₀	[89]
						Mercuric thiocyanate	N	0.09 (0.01–015)	4		
						Mercuric acetate	O	0.075	4		
						Mercuric thiocyanate	O	0.120	4		
Penaeus indicus	Adult	F	35		19–24	Mercuric chloride	O	0.0153	4	LC₅₀	[91]
								0.0161	2		
	Postlarval	F	35		19–24	Mercuric chloride	O	0.006	28	No effect	
Fish											
Aphanius	Fingerlings					HgCl₂		1.0	4	Plasma Na and K increased, serum Ca and Mg increased, Cl decreased	[93]
								1.0	30	Plasma Na and K above normal, serum Ca and Mg above normal, Cl below normal	

(Continued)

TABLE 8.12 (*Continued*)
Toxicity of Mercury to Saltwater Life

Species	Life Stage	Static (S) or Flowing (F)	Salinity (g kg⁻¹)	pH	Temperature (°C)	Chemical Form	Nominal (N) or Observed (O) LC₅₀	Concentration (mg L⁻¹)	Duration (days)	Effect	Reference
Fundulus heteroclitus	Embryo	S	20		25	Mercuric chloride	N	0.0674	4	LC₅₀	[88]
								0.069 (+2 days in clean water)	2		
								0.0806 (+3 days in clean water)	1		
						Methyl	N	0.0511 (continuous)	4		
								0.050 (+2 days in clean water)	2		
								0.0583 (+3 days in clean water)	1		
Fundulus heteroclitus	1-day embryo	S	20		25	Mercuric chloride					

TABLE 8.13
Toxicity of Nickel to Saltwater Life

Species	Life Stage	Static (S) or Flowing (F)	Salinity (g kg⁻¹)	pH	Temperature (°C)	Chemical Form	Nominal (N) or Observed (O) LC₅₀	Concentration (mg L⁻¹)	Duration (days)	Effect	Reference
Annelidai											
Capitella capitate	Adult		Seawater			Nickel chloride		>50.0	4	LC_{50}	[95]
Cienodrilus serratus			Seawater			Nickel chloride		17.0	4		
Neanthes arenaceodentata			Seawater			Nickel chloride		49.0	4		
Nereis virens			Seawater			Nickel chloride		25.0	4	LC_{50}	[82]
Mollusca											
Crassostrea gigas	Larva	S	33.8	8.1	20	Nickel sulphate	O	0.349 ± 0.05	2	EC_{50} abnormal development	[82]
Crassostrea virginica	Larva / Larva		Seawater			Nickel chloride		1.18	4	LC_{50} EC_{50} larval growth	[43]
Macoma balthica	Adult	S	25		5	Nickel chloride	O (T)	1.21	12		
					10			180.0	8		
			15		15			150.0	8	LC_{50}	
			25		10			80.0	8		
			25		10			95.0	4		
					10			360.0	4		
								750.0	4		

(Continued)

TABLE 8.13 (Continued)
Toxicity of Nickel to Saltwater Life

Species	Life Stage	Static (S) or Flowing (F)	Salinity (g kg⁻¹)	pH	Temperature (°C)	Chemical Form	Nominal (N) or Observed (O) LC$_{50}$	Concentration (mg L^{-1})	Duration (days)	Effect	Reference
Mercenaria mercenaria	Larva		Seawater			Nickel chloride		0.310 / 5.71	4 / 8–10	LC$_{50}$ / EC$_{50}$ larval growth	[96]
Mya arenaria	Adult		Seawater			Nickel chloride		320.0	4	LC$_{50}$	[72]
Mytilus edulis	Larva	S	33.8	8.1	17	Nickel sulphate	O (T)	0.891 ± 0.21	2	EC$_{50}$ abnormal development	[82]
	Adult	F	32.2		6.8	Nickel chloride	N	0.2	8	Growth unaffected	
Perna perna	7 cm	S					N	0.7 (0.6–1.0)	1 h	Filter rate reduced 50%	[78]
Crustacea											
Acartia clause	Adult		Seawater			Nickel chloride		2.08	4	LC$_{50}$	[97]
Allorchestes compressa	Adult	F	35	8.1	21	Nickel chloride	O (T)	34.68 (32.02–37.53) / 12.28 (10.86–13.88)	4 / 7	LC$_{50}$	[98]
Cancer magister	Zoeae	S	33.8	8.1	15	Nickel sulphide		4.36 ± 2.21	4	LC$_{50}$	[82]
Carcinus	Adult		Seawater			Nickel					[98]

(*Continued*)

TABLE 8.13 (Continued)
Toxicity of Nickel to Saltwater Life

Species	Life Stage	Static (S) or Flowing (F)	Salinity (g kg⁻¹)	pH	Temperature (°C)	Chemical Form	Nominal (N) or Observed (O) LC$_{50}$	Concentration (mg L⁻¹)	Duration (days)	Effect	Reference
Corophium volutator	Adult	S	25		5	Nickel chloride	O (T)	15.0	8	LC$_{50}$	
			25		10			8.5	8		
			25		15			5.2	8		
			10		5			21.0	4		
			35		5			54.0	4		
					10			52.0	4		
					15			34.0	4		
			10		15			16.0	4		
Crangon crangon	Adult		Seawater			Nickel chloride		150.0	2	LC$_{50}$	[98]
Eurytemora affinis			Seawater			Nickel chloride		9.67	4	LC$_{50}$	[97]
Heteromysis formosa	Adult		Seawater			Nickel chloride		0.152	4	LC$_{50}$	[97]
Mysidopsis bahia			Seawater			Nickel chloride		0.508	4	LC$_{50}$	[97]
Mysidopsis bigelowi			Seawater			Nickel chloride		0.634	4	LC$_{50}$	[97]
Nitrocra spinipes			Seawater			Nickel chloride		0.60	4	LC$_{50}$	[99]
Pagurus longicarpus			Seawater			Nickel chloride		47.0	4	LC$_{50}$	[72]
Tigriopus japonicus			Seawater			Nickel chloride		6.36	4	LC$_{50}$	[97]

(Continued)

TABLE 8.13 (Continued)
Toxicity of Nickel to Saltwater Life

Species	Life Stage	Static (S) or Flowing (F)	Salinity (g kg⁻¹)	pH	Temperature (°C)	Chemical Form	Nominal (N) or Observed (O) LC$_{50}$	Concentration (mg L⁻¹)	Duration (days)	Effect	Reference
Echinodermata											
Asteris forbesi	Adult		Seawater			Nickel chloride		150.0	4	LC$_{50}$	[72]
Arbacia punctalata	Embryo	S	Seawater			Nickel chloride		17.0	2	>50% mortality	[100]
Lytechinus pictus	Embryo	S	Seawater			Nickel chloride		0.580	1	Abnormal development	[101]
Fish											
Chelon labrosus	Adult	F	34.5	8.2	12.2	Nickel nitrate	O (D)	271.0 149.0 118.3	2 3 4	LC$_{50}$	
Fundulus heteroclitus			Seawater			Nickel chromium		350.0	4	LC$_{50}$	[72]
Menidia menidia			Seawater			Nickel chromium		7.96	4	LC$_{50}$	[97]

Nickel is relatively nontoxic to fish. It is less toxic in saline water (4-day $LC_{50} = 35,000\,\mu g\,L^{-1}$) than in nonsaline water (4-day $LC_{50} = 10,000\,\mu g\,L^{-1}$). Toxicity is greater at higher water temperatures.

A similar difference exists between larval and adult bivalve molluscs, but the data are less consistent. The reported 4-day LC_{50} for larvae of *Mercenaria mercenaria* is $0.31\,mg\,L^{-1}$, but in a separate study the EC_{50} for growth of the same species over 8–10 days was much higher at $5.71\,mg\,L^{-1}$ with no mortality.

One study has examined the effects of salinity and temperature on the toxicity of nickel to two species, *Corophium volutator* and *Macoma balthica* [24]. For both species, reduction in salinity increased the toxicity, as reflected by the LC_{50} values, and increases in temperature increased toxicity. In Corophium, a salinity reduction from 35 to $10\,g\,kg^{-1}$ altered the 4-day LC_{50} from 54 to $12\,mg\,L^{-1}$, while a rise in temperature from 5°C to 10°C altered the 8-day LC_{50} from 15.0 to $5.2\,mg\,L^{-1}$.

There is a complete absence of long-term studies of the toxicity of nickel to either fish or invertebrates, but in short-term tests, adult fish are less sensitive than the invertebrates tested.

Nickel is relatively non-toxic to marine organisms (4-day LC_{50}: $10,000\,\mu g\,L^{-1}$), and planktonic crustacea and bivalve molluscs larvae are more sensitive (4-day LC_{50}: $50–60\,\mu g\,L^{-1}$).

8.3.9 SELENIUM

Information on the toxic effects of selenium to marine fauna is reviewed in Table 8.14.

Selenium is not particularly toxic, with the 4-day LC_{50} ranging from 2.88 to greater than $10\,mg\,L^{-1}$. The lowest adverse effect concentration was $0.2\,mg\,L^{-1}$, which reduced the filtering rate of the bivalve *Perna perna* in a 1 h exposure [75].

The lowest adverse effect concentration was $200\,\mu g\,L^{-1}$.

In recent years, the physiological role of selenium as a trace element has created considerable speculation and some controversy. Selenium has been reported as having carcinogenic as well as toxic properties; other authorities have presented evidence that selenium is highly beneficial as an essential nutrient. Its significance and involvement in the marine biosphere is not known. A review of the marine literature indicates that selenium occurs in seawater as selenite ions (SeO_3^{2-}) with a reported average of $0.2\,\mu g\,L^{-1}$.

8.3.10 SILVER

All the reported studies relate to tests employing silver nitrate (Table 8.15) so that the significance of the chemical form of the silver cannot be assessed, although this was of major importance in determining the toxicity of silver to freshwater fish.

Eggs and embryos of the fish *Pseudopleuronectes americanus* are relatively insensitive to silver with concentrations of $0.092\,mg\,L^{-1}$, having no effect after 9 days, while $8.18\,mg\,L^{-1}$ caused reduced growth, deformities and 30% mortality. In contrast, $0.010\,mg\,L^{-1}$ caused a 90% reduction in the successful development of *Crassostrea gigas* embryos after only a 2-day exposure. This was in low saline water ($16.5\,mg\,kg^{-1}$), and an increase in salinity reduced the salinity toxicity of silver.

TABLE 8.14
Toxicity of Selenium to Saltwater Life

Species	Life Stage	Static (S) or Flowing (F)	Salinity (g kg⁻¹)	pH	Temperature (°C)	Chemical Form	Nominal (N) or Observed (O) LC$_{50}$	Concentration (mg L⁻¹)	Duration (days)	Effect	Reference
Mollusca											
Bullia digitalis	Adult		Saltwater		15		N	1.0	4	No effect	[82]
								4.0	4	Burrowing ceases	
								7.0	4	Irreversible stress	
								5.0	4	15% decrease in oxygen uptake	
Crassostrea gigas	Larva	S	33.8	8.1	20	SeO$_2$	O	10.0	2	EC$_{50}$ not reached	[82]
Mytilus endulis	Larva	S	33.8	8.1	17	SeO$_2$	O	10.0	2	EC$_{50}$ not reached	[82]
Notocallista sp.	Juvenile	F	36	8.0	15.5	Sodium selenite	O (T)	2.88 (2.75–3.01)	4	LC$_{50}$	[102]
Perna perna	7 cm	S				Selenium	N	0.2 (0.1–0.3)	1 h	Filter rate reduced 50%	[78]
Crustacea											
Allorchestes compressa	Adult	F	35	7.7	20	Sodium selenite	O (T)	6.17 (5.38–7.07)	4	LC$_{50}$	[102]
				7.5	20	Sodium selenite	O (T)	4.77 (4.24–5.35)	4	LC$_{50}$	[102]
Cancer magister	Zoeae	S	33.8	8.1	15	SeO$_2$	O	10.0	4	LC$_{50}$ not reached	[77]
Cyclaspis usitatum	Adult	F	35	8.1	20	Sodium	O	6.12 (5.03–7.45)	4	LC$_{50}$	[102]

TABLE 8.15
Toxicity of Silver Nitrate to Saltwater Life

Species	Life Stage	Static (S) or Flowing (F)	Salinity (g kg⁻¹)	pH	Temperature (°C)	Chemical Form	Nominal (N) or Observed (O) LC50	Concentration (mg L⁻¹)	Duration (days)	Effect	Reference
Mollusca											
Crassostrea gigas	Larva	S	33.8	8.1	20		O	0.022 ± 0.013	2	EC$_{50}$ abnormal development	[82]
	Embryo	S	33	8.2	20		O (initial)	0.010	2	Development unaffected	[103]
								0.0135	2	Development success	[103]
			16.5	8.2	20		N	0.002	2	Development normal	
								0.0056	2	Development normal	
								0.01	2	7% normal development	[104]
								0.0135	2	2% normal development	
								0.018	2	No normal development	
			22.7	8.2	20		N	0.010	2	Development normal	
								0.0135	2	20% normal development	
								0.0155	2	5% normal development	
								0.018	2	1% normal development	
			33.0	8.2	20		N	0.002–0.018	2	Development unaffected	

(Continued)

TABLE 8.15 (Continued)
Toxicity of Silver Nitrate to Saltwater Life

Species	Life Stage	Static (S) or Flowing (F)	Salinity (g kg^{-1})	pH	Temperature (°C)	Chemical Form	Nominal (N) or Observed (O) LC$_{50}$	Concentration (mg L^{-1})	Duration (days)	Effect	Reference
Mytilus edulis	Larva	S	33.8	8.1	17		O	0.014 ± 0.002	2	EC$_{50}$ abnormal development	[82]
	Juvenile adult	F	25		2.8–24		N	0.010	630	Growth unaffected	
								0.050	182	No growth	
								0.050	365	Smaller size, no mortality	
								0.025	365	Growth unaffected	
Crustacea											
Cancer magister	Zoeae	S	33.8	8.1	15		O	0.055 ± 0.021	4	LC$_{50}$	[84]
							O	0.174	8	No effect	
								0.167	8		
								0.166	8		

Fish embryos and eggs are relatively sensitive to silver, and a 9-day exposure to $90 \mu g L^{-1}$ silver had no adverse effect [20]. At $180 \mu g L^{-1}$, silver growth deformities and 30% mortality were observed [21]. An increase in salinity reduced the toxicity of silver.

8.3.11 VANADIUM

The information on the effects of vanadium on marine organisms is scant and is limited to acute toxicity studies of 9 days or less (Table 8.16). The results suggest little difference for the four species tested, although each is from a different phylum and all the LC_{50} values exceed $10 mg L^{-1}$. In addition, it should be noted that tunicates may accumulate extremely high concentrations of vanadium in their tissues, even in uncontaminated environments. The reasons for this are not known [100,101], but the concentrations in the tissues have not reported adverse effects.

Vanadium has a tendency towards concentrating in the environment for reasons not yet understood. Environment mobilisation of vanadium and its compounds occurs by a number of means in the net transport of vanadium into the oceans. Some of these transport processes include terrestrial runoff, industrial emissions, atmospheric washout (vanadium in the air comes only from industry, as there are no significant natural sources), river transport and oil spills, resulting in a complex ecological cycle. The possibility of vanadium deposition due to oil spillage has been discussed, but no evidence is available to confirm the release of vanadium from oil. Since crude oils are rather rich in vanadium (50–200 ppm), it is not inconceivable that some vanadium should be released upon the contact of oil with seawater. The LC_{50} of vanadium is greater than $10,000 \mu g L^{-1}$.

8.3.12 ZINC

Marine life is relatively resistant to zinc at all life stages. Crustacea, bivalve molluscs and worms undergo damage or fatalities upon a 1- to 2-week exposure to $340 \mu g L^{-1}$ zinc [22,23]. A decrease in salinity increased the toxicity of zinc to invertebrates and fish sevenfold [42,80–83]. Fiddler crabs *Uca annulipes* and *Uca triangularis*, when exposed to zinc, gave a 4-day LC_{50} value of $66,420–76,950 \mu g L^{-1}$.

The toxicity information for zinc indicates that marine fish are relatively resistant to zinc at all life stages, compared with the crustaceans, bivalves, molluscs and larval polychaete worms. The responses of adult polychaetes and gastropods are comparable with those of fish. An increasing duration of exposure from 4 to 12 days has little impact on the observed toxicity, and in a number of species LC_{50} values below $1 mg L^{-1}$ have been recorded, while in one study of *Mytilus edulis* lasting 22 days a nominal concentration of $0.01 mg L^{-1}$ reduced the growth rate. In another study, a 12-day LC_{50} of $0.34 mg L^{-1}$ was reported for the bivalve *Mercenaria mercenaria*.

The toxicity of zinc, chromium, copper and cadmium to both *Macoma balthica* and *Corophium volutator* species increased from $1.1 mg L^{-1}$ at 15°C to $3.0 mg L^{-1}$ at 5°C [29]. This same study also demonstrated that decreasing salinity had the effect of increasing the toxicity of zinc to these estuarine species. In the case of *Macoma*, the 4-day LC_{50} of $35 g kg^{-1}$ at $380 mg L^{-1}$ decreased to $60 mg L^{-1}$ at $15 g kg^{-1}$, both at

TABLE 8.16
Toxicity of Vanadium to Saltwater Life

Species	Life Stage	Static (S) or Flowing (F)	Salinity (g kg⁻¹)	pH	Temperature (°C)	Chemical Form	Nominal (N) or Observed (O) LC$_{50}$	Concentration (mg L⁻¹)	Duration (days)	Effect	Reference
Annelida											
Nereis diversicolor	Adult		Seawater			NaVO$_3$	N	10.0	9		
Mollusca											
Mytilus edulis	Adult		Seawater			NaVO$_3$	N	65.0	9	LC$_{50}$	[107]
Crustacea											
Carcinus maenas	Adult		Seawater			NaVO$_3$	N	35.0	9		
Fish											
Limanda limanda	Adult	F	34.7	8.1	12.5	NH$_4$VO$_3$	O (D)	44.0	1	LC$_{50}$	[83]
								30.7	2		
								27.8	3		
								27.8	4		

TABLE 8.17
Toxicity of Zinc to Saltwater Life

Species	Life Stage	Static (S) or Flowing (F)	Salinity (g kg⁻¹)	pH	Temperature (°C)	Chemical Form	Nominal (N) or Observed (O) LC_{50}	Concentration (mg L⁻¹)	Duration (days)	Effect	Reference
Annelida											
Capiella capiata	Larva		Seawater			Sulphate		1.7	4	LC_{50}	[65]
	Adult		Seawater			Sulphate		3.5	4	LC_{50}	[65]
Ctendrilus serratus	Adult		Seawater			Sulphate		7.1	4	LC_{50}	[35]
Neanthes arenaceodentata	Juvenile		Seawater			Sulphate		0.90	4	LC_{50}	[65]
	Adult							1.80	4	LC_{50}	[65]
Neris virens	Adult		Seawater			Chloride		8.1	4	LC_{50}	[72]
Mollusca											
Crassostrea gigas	Larva	S	33.8	8.1	20	Sulphate	O	0.119 ± 0.012	2	EC_{50} abnormal development	[43]
Crassostrea virginica	Embryo	S	Seawater			Sulphate	O (T)	0.310	2	LC_{50}	[90]
Macoma balthica	Adult	S	35		5			750.0	4		
					10			950.0	4		
			25		15			380.0	4		
			15		15			180.0	4		
			25		15			60.0	4		
					5			65.0	8		
Mytilus edulis	Larva	S	33.8	8.1	17	Sulphate	O	0.175	2	EC_{50} abnormal development	
	Adult	F	33		17	Chloride	N	0.05	1	Growth reduced	[84]
								0.025	2		
								0.010	22		

(Continued)

TABLE 8.17 (Continued)
Toxicity of Zinc to Saltwater Life

Species	Life Stage	Static (S) or Flowing (F)	Salinity (g kg⁻¹)	pH	Temperature (°C)	Chemical Form	Nominal (N) or Observed (O) LC$_{50}$	Concentration (mg L⁻¹)	Duration (days)	Effect	Reference
Nassarius obsoletus	Adult		Seawater			Chloride		50.0	4	LC$_{50}$	[72]
								7.4	7	LC$_{50}$	[72]
Artemia salina	Larva		Seawater			Chloride		0.100	6	LC$_{50}$	[39]
	Adult							1.00	14	LC$_{50}$	[39]
Cancer magister	Zoeae	S	33.8	8.1	15	Sulphate	O	0.456 ± 0.174	4	LC$_{50}$	[77]
Carcinus maenas	Larva		Seawater			Chloride		1.0	4	LC$_{50}$	[50]
Corophium volutator	Adult	S	25		5	Sulphate	O (T)	3.0	8		
					10			2.7	8		
					15			1.1	8		
			25		5			16.0	4		
					10			15.0	4	LC$_{50}$	[90]
					15			3.6	4		
			10		10			1.6	4		
			15		10			8.5	4		
			25		10			11.0	4		
Pagurus longicarpus	Adult		Seawater			Chloride		0.400	4	LC$_{50}$	[32]
								0.200	7	LC$_{50}$	[32]
Tigriopus japonicus	Adult		Seawater			Chloride		2.160	4	LC$_{50}$	[42]
Fish											
Asterias forbesi	Adult		Seawater			Chloride		3.90	4	LC$_{50}$	[72]
								2.30	7	LC$_{50}$	[72]
Fundulus heteroclitus	Adult		Seawater			Chloride		60.0	4	LC$_{50}$	[72]
								52.0	7	LC$_{50}$	[72]
								120.0	7	100% mortality	

TABLE 8.18
Effects of Chromium, Zinc and Cadmium Contents of Fish Organs on Mortality

	Laboratory Tests on Rudd Fish				Maximum Concentrations Found in Organs of a Wide Variety of Fish		Minimum Concentrations Found in Organs of a Wide Variety of Fish	
Exposure Time	10 Weeks	3 Weeks	12 h	<12 h	mg kg⁻¹ Dry Weight	Comments	mg kg⁻¹ Dry Weight	Comments
					Element: Chromium			
Concentration (µg L⁻¹) of metal in water	3	16	20	80–145				
Condition of animal (rudd fish)	Good	Good	Good	100% mortality				
Concentration of Metal in Organ (mg kg⁻¹ Dry Weight)								
Opercle	<0.2	<2	8.3	20–26	26	Mortalities during 12 h exposure	8.3	No mortalities in 12 h exposure
Liver	<0.2	<2	5.6	15–18	18.4	Mortalities during 12 h exposure	5.6	No mortalities in 12 h exposure
Kidney	<0.2	<2	10.3	24–27	23.7	Mortalities during 12 h exposure	10.3	No mortalities in 12 h exposure
					Element: Zinc			
Concentration (µg L⁻¹) of metal in water	180	800	1600	7500–18,000				
Condition of animal	Good	Good	Good	100% mortality				

(Continued)

TABLE 8.18 (Continued)
Effects of Chromium, Zinc and Cadmium Contents of Fish Organs on Mortality

Exposure Time	Laboratory Tests on Rudd Fish				Maximum Concentrations Found in Organs of a Wide Variety of Fish		Minimum Concentrations Found in Organs of a Wide Variety of Fish	
	10 Weeks	3 Weeks	12 h	<12 h	mg kg⁻¹ Dry Weight	Comments	mg kg⁻¹ Dry Weight	Comments
Concentration of Metal in Organ (mg kg⁻¹ Dry Weight)								
Opercle	48	20	115.3	91–174	120	Mortalities during 12 h exposure	No data	
Liver	120	15	42.5	34–63	150	Mortalities during 12 h exposure	13	No mortalities
Kidney	29	28	154.6	92–216	57		No data	
Element: Copper								
Concentration (µg L⁻¹) of metal in water	11	50	—	250–1600				
Condition of animal	Good	Good	—	100% mortality				
Concentration of Metal in Organ (mg kg⁻¹ Dry Weight)								
Opercle	12	31	—	52–104	12.4	Mortalities during 12 h exposure	No data	
Liver	7	20	—	22–40	62	Mortalities during 12 h exposure	1.7	No mortalities
Kidney	6	28	—	30–100	6	Mortalities during 12 h exposure	0.7	No mortalities

(Continued)

TABLE 8.18 (Continued)
Effects of Chromium, Zinc and Cadmium Contents of Fish Organs on Mortality

	Laboratory Tests on Rudd Fish				Maximum Concentrations Found in Organs of a Wide Variety of Fish		Minimum Concentrations Found in Organs of a Wide Variety of Fish	
Exposure Time	10 Weeks	3 Weeks	12 h	<12 h	mg kg⁻¹ Dry Weight	Comments	mg kg⁻¹ Dry Weight	Comments
				Element: Cadmium				
Concentration ($\mu g\,L^{-1}$) of metal in water	3	250	—	1100–11,000				
Condition of animal	Good	Good	—	100% mortality				
Concentration of Metal in Organ (mg kg⁻¹ Dry Weight)								
Opercle	9.5	9		6–29	9.5	Mortalities during 12 h exposure	No data	
Liver	5	10		4–12	9	Mortalities during 12 h exposure	0.2	No mortalities
Kidney	4	14		14–28	7.1	Mortalities during 12 h exposure	3	No mortalities

15°C. This effect of salinity has also been reported for two species of isopod and for two species of salmonid fish.

The toxicity of zinc to fish and various invertebrates is reviewed in Table 8.17, as well as the toxicity of fish organs.

The toxic effect of four metals, namely, chromium, zinc, copper and cadmium, on the organs of fish and their mortality is reviewed in Table 8.18.

REFERENCES

1. F. Van Hoof and H. Van Son, *Chemosphere*, 1981, 10, 1127.
2. D.I. Mount and C.E. Stephen, *Journal of Wildlife Management*, 1967, 31, 168.
3. M. Martin, M.D. Stephenson and J.H. Martin, *Californian Fisheries and Game*, 1977, 63, 95.
4. T. Kariya, Y. Haga, T. Hoga and T. Tsuda, *Bulletin of Japanese Society Scientific Fisheries*, 1967, 33, 818.
5. W.A. Brings, E.N. Leonard and J.A. McKim, *Journal of the Fisheries Research Board of Canada*, 1973, 30, 583.
6. S.E. Papoutsoglou and D. Abel, *Bulletin of Environmental Contamination and Toxicology*, 1988, 41, 404.
7. H. Segner, *Journal of Fish Biology*, 1987, 30, 423.
8. D.J. Lauren and D.G. McDonald, *Canadian Journal of Fisheries and Aquatic Science*, 1987, 44, 105.
9. T.S. Presser and H.M. Ohlendorf, *Environmental Management*, 1987, 11, 805.
10. A.L. Hilmy, N.A. El-Domiatry, A.Y. Daabees and H.A.A. Latik, *Biochemistry and Physiology*, 1987, 86C, 263.
11. D.J. Madigan, Z. Baumann and E. Olwyn, *Environmental Science and Technology*, 2013, 47, 2287.
12. A. Zirino and S. Yammamoto, *Limnology and Oceanography*, 1972, 17, 661.
13. D.R. Turner, M. Whitfield and A.G. Dickson, *Geochemica and Cosmochemica Acta*, 1981, 45, 855.
14. A. Calabrese, J.R. MacInnes, D.A. Nelson, R.A. Grieg and P.P. Yevich, *Marine Environmental Research*, 1984, 11, 253.
15. D.J.H. Phillips, Toxicity and accumulation of cadmium in marine and estuarine biota, in J.O. Nriagu (ed.), *Cadmium in the Environment Part 1: Ecological Cycling*, Wiley, New York, 1980, p. 426.
16. D. Taylor, A review of the lethal and sublethal effects of cadmium on aquatic life, in *Proceedings of the Third International Cadmium Conference*, Miami, FL, 1981, p. 75.
17. H. Thede, N. Schlotz and H. Fascher, *Marine Ecology Programme Series*, 1979, 1, 13.
18. G. Pesch and G. Stewart, *Marine Environmental Research*, 1980, 33, 145.
19. J.R. Sharp and J.M. Neff, *Marine Environmental Research*, 1980, 3, 195.
20. R.A. Voyer, J.A. Cardin, J.F. Heltsche and G.L. Hofman, *Aquatic Toxicology*, 1982, 2, 223.
21. G. Klein-MacPhee, J.A. Cardin and W.J. Berry, *Transactions of the American Fisheries Society*, 1984, 113, 247.
22. T. Stromgren, *Marine Biology*, 1982, 72, 69.
23. A. Calabrese, J.R. MacInnes, D.A. Nelson and J.E. Miller, *Marine Biology*, 1977, 41, 179.
24. V. Bryant, D.M. Newberry, D.S. McLusky and R. Campbell, *Marine Ecology Progress Series*, 1985, 24, 139.
25. M.B. Jones, *Marine Biology*, 1975, 30, 13.
26. D.M.E. Herbert and A.C. Wakeford, *International Journal of Air and Water Pollution*, 1964, 8, 251.

27. A.R.D. Stebbing, *Journal of the Marine Biological Association of the United Kingdom*, 1976, 56, 977.
28. L. Karbe, *Marine Biology*, 1976, 12, 316.
29. W.R. Reeve, A controlled environmental bulletin experiment (CEPEX) and its usefulness in the study of larger marine zooplankton under toxic stress, in P.M. Lockwood (ed.), *Effects of Pollutants on Aquatic Organisms*, Cambridge University Press, New York, 1976, p. 145.
30. D.J. Reish, *Revue Internationale d'Oceanographic Medicale*, 1978, 49, 99.
31. C.E. Pesch and D. Moreau, *Water Research*, 1978, 12, 747.
32. G.W. Bryan and L.G. Hummerstone, *Journal of the Marine Biological Association of the United Kingdom*, 1977, 51, 845.
33. L.H. Jones, N.V. Jones and A.J. Radlet, *Estuarine and Coastal Marine Science*, 1976, 4, 107.
34. B. Brown and M. Ahsanullah, *Marine Pollution Bulletin*, 1971, 2, 182.
35. L.J. Saliba and M. Ahsanullah, *Marine Biology*, 1973, 23, 297.
36. D.S. McLusky and C.N.K. Phillips, *Estuarine and Coastal Marine Science*, 1975, 3, 103.
37. J.E. Raymond and J. Shields, *Water Pollution Research*, 1964, 3, 275.
38. A. Calabrese, R.S. Collier, D.A. Nelson and J.R. MacInnes, *Marine Biology*, 1973, 18, 162.
39. E.F. Mandelli, *Water Research*, 1975, 9, 287.
40. R.K. Ozaki, *Bulletin of Environmental Contamination and Toxicology*, 1974, 16, 658.
41. R.R. Stephenson and D. Taylor, *Bulletin of Environmental Contamination and Toxicology*, 1975, 144, 304.
42. A.C. Brown, *Bulletin of Transactions of Royal Society*, 1982, 44, 555.
43. S. Sosnowski and J.H. Gentile, *Journal of the Fisheries Research Board of Canada*, 1978, 35, 1366.
44. P.M. Connor, *Marine Pollution Bulletin*, 1972, 3, 190.
45. D.W. McLeese, *Journal of the Fisheries Research Board of Canada*, 1974, 31, 1949.
46. M.W. Johnson and J.H. Gentile, *Bulletin of Environmental Contamination and Toxicology*, 1979, 22, 258.
47. G.A. Heslinga, *Marine Biology*, 1976, 35, 135.
48. P. Bongis, *Comptes Rendus des Séances de L'académie des Sciences*, 1965, 260, 2929.
49. C.W. Steel, *Marine Pollution Bulletin*, 1983, 14, 425.
50. C.W. Steel, *Marine Pollution Bulletin*, 1983, 14, 168.
51. D.W. Engel, W.G. Sunda and R.M. Thoutee, *Environmental Health Perspectives*, 1976, 17, 288.
52. D. Dorfmann, *Bulletin of the New Jersey Academy of Science*, 1977, 22, 21.
53. R. Eisler and G.R. Gardner, *Journal of Fish Biology*, 1973, 5, 131.
54. G.R. Gardner and G. La Roche, *Journal of the Fisheries Research Board of Canada*, 1973, 30, 363.
55. J. Gentile, Semi-annual report, Environmental Research Laboratory, Narrangansett, RI, 1975.
56. J.T.P. Baker, *Journal of the Fisheries Research Board of Canada*, 1969, 26, 2785.
57. A. Larsson, B. Bengtsson and C. Haux, *Aquatic Toxicology*, 1981, 1, 19.
58. M.H. Roberts, H. Warinner, C.F. Tsai, D. Wright and L.E. Cronin, *Archives of Environmental Science and Toxicology*, 1982, 11, 681.
59. I. Rivera-Duarte, G. Rosen, D. Lapota, D.P. Chadwick, L. Kea-Padilla and A. Zirino, *Environmental Science and Technology*, 2005, 39, 1542.
60. D.R. Reish, J.M. Martin, F.M. Piltz and J.O, Word, *Water Research*, 1976, 10, 299.
61. D.J. Reish and R.S. Carr, *Marine Pollution Bulletin*, 1978, 9, 24.
62. A.J. Mearns, P.S. Oshida, M.J. Sherwood, D.R. Young and D.J. Reish, *Water Pollution Control Federation*, 1976, 48, 1929.

63. P.S. Oshida, A.J. Mearns, D.J. Reish and C.W. Word, *Neanthes arenaceodentata* (Polychaete: Annelida), Southern California Coastal Water Research Project, T.M. 225, El Segundo, CA, 1976.

64. P.S. Oshida, A safe level of hexavalent chromium for a marine polychaete, Southern California Coastal Water Research Project, Annual report, El Segundo, CA, 1977.

65. G.W. Bryan, Recent trends in research on heavy metal contamination of the sea, *Hegloander Meersuntersuchungen*, 1980, 33, 6.

66. V. Bryant, D.S. McLusky and K. Roddie, *Toxicity of Chromium to Three Species of Estuarine Invertebrates*, Report 697-M, Water Research Centre, Marlow, Bucks, UK, 1984.

67. R. Eilser and R. Henneky, *Archives of Environmental Contamination and Toxicology*, 1977, 6, 315.

68. K.R. Olson and R.C. Harrel, *Rangia Cueata: Contributions to Marine Science*, 1973, 17, 9.

69. M. Moraitou-Apostolopoulou and G. Verriopoulos, *Hydrobiologia*, 1982, 19, 121.

70. P.M. Frank and P.B. Robertson, *Bulletin of Environmental Contamination and Toxicology*, 1979, 21, 74.

71. E. Lindem, B.-E Bengtsson, O. Svanberg and G. Sundstorm, *Chemosphere*, 1979, 8, 843.

72. M. Martin, K.E. Osborn, P. Billig and N. Glickstein, *Marine Pollution Bulletin*, 1981, 12, 305.

73. H.R. Watling and R.J. Watling, *Bulletin of Environmental Contamination and Toxicology*, 1982, 29, 651.

74. USEPA, *Ambient Water Quality Criteria for Chromium*, EPA 440/5–80–305 Washington, DC, 1980.

75. M. Ahsanullah, *Australian Journal of Marine and Freshwater Research*, 1982, 33, 465.

76. R.M. Couture, J.F. Chiffolean, D. Anger, D. Claisse, G. Gobeit and D. Lorsa, *Environmental Science and Toxicology*, 2010, 44, 1211.

77. M. Martin, K.E. Osborn, P. Billig and N. Gliickstein, *Marine Pollution Bulletin*, 1981, 121, 305.

78. S.J. Hugman and G. Mance, *The Acute Toxicity of Priority List H Substances for Marine Fish Report*, Report 616-M, Water Research Centre, Marlow, Bucks, UK, 1983.

79. T. Stromgren, *Marine Biology*, 1982, 72, 69.

80. R. Lloyd, Effects of lead on marine organisms, Annex B of Technical Report TR 208, Water Research Centre, Marlow, Bucks, UK, 1984, p. 41.

81. B.G. Maddock and D. Taylor, The acute toxicity and bioaccumulation of some lead alkyl compounds in marine animals, in *International Experts Discussion Meeting, Lead – Occurrence, Fate and Pollution in the Marine Environment*, Rovinj, Yugoslavia, 1977.

82. M. Ahsanullah, *Australian Journal of Marine Freshwater Research*, 1982, 33, 465.

83. J.R. Sharp and J.M. Neft, *Environmental Biology Fish*, 1982, 7, 277.

84. M.W. Curtis, T.L. Copeland and C.H. Ward, *Water Research*, 1979, 13, 137.

85. P.M. Chapman, M.W. Farrell and R.O. Bringman, *Water Research*, 1982, 16, 1405.

86. T.P. McClurg, *Water, South Africa*, 1984, 10, 40.

87. M.R. Kasschau, M.M. Skaggs and E.C.M. Chen, *Bulletin of Environmental Contamination and Toxicology*, 1980, 25, 873.

88. A.M. Hilmy, M.B. Shabana and M.M. Saied, *Water, Air, and Soil Pollution*, 1982, 18, 467.

89. V. Minganti, R. Capelli, F. Florentino, R. De Pellegrinn and M. Vacchi, *International Journal of Environmental Analytical Chemistry*, 1995, 61, 239.

90. S.M. Petrich and D.J. Reish, *Bulletin of Environmental Contamination and Toxicology*, 1979, 23, 698.

91. A. Calabrese and D.A. Nelson, *Bulletin of Environmental Contamination and Toxicology*, 1974, 2, 92.

92. USEPA, *Ambient Water Quality Criteria for Nickel*, EPA 440/5–80–060, Washington, DC, 1980.
93. J.E. Portman, *Helgolander Meersuntersuchungen*, 1968, 17, 1–4.
94. B.E. Bengtsson, *Marine Pollution Bulletin*, 1978, 9, 238.
95. A.J. Waterman, *Biology Bulletin*, 1937, 73, 401.
96. H. Timourian and G. Watchmaker, *Journal of Experimental Zoology*, 1972, 182, 379.
97. M. Abdullah and D.H. Palmer, *Journal of Freshwater and Marine Research*, 1980, 31, 795.
98. M.P. Coglianese and M. Martin, *Marine Environmental Research*, 1981, 5, 13.
99. M.P. Coglian, *Archives of Environmental Contamination and Toxicology*, 1982, 11, 297.
100. K. Kustin, K.V. Ladd and G.C. McLeod, *Journal of General Physiology*, 1975, 65, 15.
101. J.H. Swinhart, W.R. Biggs, D.J. Halko and N.C. Shcroeder, *Biological Bulletin*, 1974, 146, 302.
102. P. Miraman and M. Unsal, *Chemosphere*, 1975, 10, 827.
103. A. Gaden, S.H. Ferguson, L. Harwood, H. Melling and G.A. Stern, *Environmental Science and Technology*, 2009, 43, 3646.

9 Toxic Effects of Organic Compounds on Fish and Invertebrates

Compared with the metal and organometallic compounds, there are fewer published toxicity data on the effects of organic compounds on freshwater or oceanic sea creatures. It is seen in Table 9.1 that 4-day LC_{50} values obtained with fish and creatures other than fish exposed to various organic compounds range from $10 \, \mu g \, L^{-1}$ (i.e. very toxic) to $106 \, \mu g \, L^{-1}$ (i.e. less toxic). As seen in Table 9.2, of the organic compounds discussed in various creatures found in the seas, the chlorophenols, chloroaromatics and chlorinated insecticides are the most toxic ($LC_{50} < 10 \, \mu g \, L^{-1}$), while compounds such as alcohols and surface-active agents are the least toxic.

Further data for the effect of organic substances in seawater are given in Table 9.3.

The toxicity of various types of organic compounds on the injury or mortality of fish and invertebrates is now discussed in more detail.

9.1 ALIPHATIC HYDROCARBONS

In many areas, oil has become the most frequently encountered water pollutant. This reflects the expanding and widespread consumption of petroleum products, a consumption that will continue to increase in the foreseeable future. Oil pollution has harmful effects on aquatic life and in the oceans and lowers the aesthetic appearance of inland water. Occasionally, it necessitates the closure of waterworks intakes. Although marine oil pollution has received much attention in recent years, this has not been the case with oil pollution of inland waters, and for some time now there has been the need for an assessment of the analytical related problems in this field. These problems are discussed below. Particular attention is given to the identification of the polluting oil.

Wherever oil is produced, stored or transported by vehicle or pipeline, there exists a potential source of oil pollution, either directly by surface drains and surface runoff or indirectly by seepage into the ground. Unlike some pollutants, oil pollution is generally unpredictable as to location and time, and usually exists as a surface phenomenon. Heavy pollution is obviously unwelcome, but even thin ephemeral films representing only small amounts of oil may cause complaints and require investigation.

Practically the total range of petroleum products is encountered as pollutants in water. Crude oils are very seldom found in inland waters, but petrol, paraffin, gas oils, heating fuels, lubricating oils, transformer oils and cutting fluids have caused widespread problems. In most areas, heating fuel, due to its widespread

TABLE 9.1
LC$_{50}$ Values of Organic Compounds in Non-Saline and Saline Waters

Compound	Organism	Water Type	LC$_{50}$ Value	LC$_{50}$ Test Duration (days)	Reference
Diethyl hexyl phosphate 1-octanol	Daphnia and fish	ns	10–1000 mg kg^{-1}	4	[20,25]
	Rainbow trout (*Salmo gairdneri*)	ns	15.84 mg L^{-1}	4	[21]
Sodium dodecyl sulphonate	Lugworm (*Arenicola marina*)	ns	15.2 mg L^{-1}	4	[22]
Triton X-100			15.2 mg L^{-1}	4	[22]
Sodium dodecyl benzene sulphonate			12.5 mg L^{-1}	4	[22]
Ethylene dibromide	*Hydra oligatis*	ns	50 mg L^{-1}	3	[23]
Methylene dichloride	Juvenile fathead minnow (*Pimphales promelas*)	ns	502 mg L^{-1}	2	[24]
1,2,4-Trichlorobenzene	Fathead minnow (*Pimphales promelas*)	ns	(a) 7.8 mg L^{-1} (b) 2.76 mg L^{-1}	4 4	[25]
1,4-Dichlorobezene		ns	1.10 mg L^{-1}	4	[25]
1,2,3,4-Tetrachlorobenzene		ns	4.2 mg L^{-1}	4	[25]
Pentachlorophenol	*Selenastrum capricornatum*		ns(a) 0.11–0.15 mg L^{-1} (soft water) (b) 0.76 mg L^{-1} (hard water)	4	[26]
	Roach (*Rutilus rutilus*)	ns	0.028 mg L^{-1}	4	[27]
	Rainbow trout (*Salmo gairdneri*)	ns	0.09 mg L^{-1}	4	[21]
2,4-Dichlorophenol	Rainbow trout (*Salmo gairdneri*)	ns	4.64 mg L^{-1}	4	[21]
2,4,6-Trichlorophenol	Roach (*Rutilus rutilus*)	ns	0.05 mg L^{-1}	4	[27]
2,3,4,6-Tetrachlorophenol	Roach (*Rutilus rutilus*)	ns	0.071 mg L^{-1}	4	[27]

(*Continued*)

TABLE 9.1 (*Continued*)
LC$_{50}$ Values of Organic Compounds in Non-Saline and Saline Waters

Compound	Organism	Water Type	LC$_{50}$ Value	LC$_{50}$ Test Duration (days)	Reference
	Rainbow trout (*Salmo gairdneri*)	ns	0.09 mg L^{-1}	4	[21]
2,4-Dichlorophenol	Rainbow trout (*Salmo gairdneri*)	ns	4.45 mg L^{-1}	4	[21]
2,4,6-Trichlorophenol	Roach (*Rutilus rutilus*)	ns	0.05 mg L^{-1}	4	[27]
2,3,4,6-Tetrachlorophenol	Roach (*Rutilus rutilus*)	ns	0.071 mg L^{-1}	4	[27]
Polychlorobiphenyl (Archlor 1254)	Rainbow trout (*Salmo gairdneri*)	ns	30 mg L^{-1}	4	[28]
Endosulphan	Crab (*Oziotelphusa senex senex*)	s	6.2 mg L^{-1} (sublethal) 18.62 mg L^{-1}	4	
Kepone	Lamprey (*Petromyzon marinus*)	s	414–444 mg L^{-1}	4	[29]
Carbaryl	Lugworm (*Arenicola marina*)	ns	7.2 mg L^{-1}	3	[25]
	Catfish (*Clarius batrachus*)	ns	46.9–107.7 mg L^{-1}	4	[30]
Parathion ethyl	Lugworm (*Arenicola marina*)	ns	2.7 mg L^{-1}	3	[25]
Mirex	Rainbow trout (*Salmo gairdneri*)	ns	5.0 mg L^{-1}	4	[28]
Malathion	Rainbow trout (*Salmo gairdneri*)	ns	1.73 mg L^{-1}	4	[31]
Roundup herbicide	Rainbow trout (*Salmo gairdneri*), Chinook Coho salmon	ns	7.4–12 mg L^{-1}	4	[32]
Rodeo herbicide	Rainbow trout (*Salmo gairdneri*), Chonook Coho salmon	ns	580 mg L^{-1}	4	[32]

(*Continued*)

TABLE 9.1 (*Continued*)
LC$_{50}$ Values of Organic Compounds in Non-Saline and Saline Waters

Compound	Organism	Water Type	LC$_{50}$ Value	LC$_{50}$ Test Duration (days)	Reference
Bromacil	Fathead minnow (*Pimphales promelas*)	ns	182 mg L^{-1}	4	[33]
Diuron	Fathead minnow (*Pimphales promelas*)	ns	14.2 mg L^{-1}	4	[33]
Lindane	Teleost fish (*Anguilla anguilla*)	ns	0.32–0.68 mg L^{-1}	4	[34]
Hexazinone	Juvenile pacific salmonid	s	276 mg L^{-1}	4	[35]
Methylene bis(thiocynanate)	*Chlorella pyrenozdosa*	ns	0.042 mg L^{-1}	4	[36]
	Poecilia reticulate	ns	0.39 mg L^{-1}	4	
Cyanogen chloride	*Daphnia magna*	ns	0.065 mg L^{-1} (adult) 0.029 mg L^{-1} (juvenile)	2	[37]
3,4-Dichloro-aniline	Fathead minnow (*Pimphales promelas*)	ns	6.99–8.06 mg L^{-1}	4	[38]
Pronone 109	Fish	s	904 mg L^{-1}	4	[35]
Varpar L	Fish	s	1989 mg L^{-1}	4	[35]
3-Fluoromethyl	Walleye (*Stizostedium vitreum*)	s	LC$_{25}$ 4.1 mg L^{-1} (gametes)	0.5	[39]
3-Fluoromethyl	Larval sea lamprey (*Petromyzon marinus*)	s	LC$_{99.9}$ 1 mg L^{-1}	8 h	[39]

Note: s = saline water; ns = non-saline water.

TABLE 9.2

Relative 4-day LC_{50} Values for Organic and Organometallic Compounds in Water Creatures

$0.01–10\,\mu g\,L^{-1}$	Most Toxic			Least Toxic
	$0.01–1\,mg\,L^{-1}$	$1–10\,mg\,L^{-1}$	$10–100\,mg\,L^{-1}$	$100–1000\,mg\,L^{-1}$
Organotin compounds	Pentachlorophenol	1,2,4-Trichlorobenzene	1-Octanol	Methylene dichloride
	2,4,6-Trichlorophenol	1,4-Dichlorobenzene	Sodium dodecyl sulphonate	Kepone
	2,3,4,6-Tetrachlorophenol	1,2,3,4-Tetrachlorobenzene	Trion X-100	Rodeo bromacil hexazinone
	Lindane	2,4,-Dichlorophenol carbaryl	Sodium dodecyl benzene sulphonate	Pronone 109
	Methylene bis(thiocyanate)	Parathion	Ethylene dibromide	Varpar L
	Cyanogen chloride	Mirex, malathion, Roundup, 3,4-dichloroaniline, 3-fluoro-4-methyl nitrophenol	Polychlorobiphenyls, endosulphan, carbaryl, diuron	

TABLE 9.3

Concentration of Organic Substances in Seawater and Effect on Fish

Substance	Fish Type	Water Type	Exposure pH	Toxicity Concentration	Index	Adverse Effects	Reference
Hydrocarbons	*Myagropsis myagroides* Fendsholt	Sea	—	$10–10,000\,mg\,L^{-1}$	—	Allometric growth inhibited by $10,000\,mg\,L^{-1}$ or $10–10,000\,mg\,0.1\,L^{-1}$ dispersant and $10–10,000\,mg\,L^{-1}$ oil plus dispersant	[1]
	Pink salmon (*Onchorhychus gorbuscha*)	Sea	—	$0.7–2.4\,mg\,L^{-1}$	—	Exposure to $0.7–2.4\,mg\,L^{-1}$ oil for 30 days reduced yolk reserves	
	Plaice	Sea in vicinity of *Amoco Cadiz* spill	—	—	—	Polyaromatics in sediments $100\,mg\,kg^{-1}$ adverse effect on reproduction	[2]
	Baltic herring (*Clupea horengus*)	Sea	—	$3–200\,\mu g\,L^{-1}$	—	$180–200\,\mu g\,L^{-1}$ hydrocarbon toxic to developing fish	[41]
Phthalate esters	Atlantic cod (*Gadus morhua*)	Sea	—	$7–25\,\mu g\,L^{-1}$	—	$7–25\,\mu g\,L^{-1}$ phthalate esters toxic to developing fish	[41]
Hexazinone	Juvenile Pacific salmonida	Sea	—	96h LC_{50} Hexazinone = $276\,mg\,L^{-1}$ Pronone 109 = $904\,mg\,L^{-1}$	—	—	[35]

(Continued)

TABLE 9.3 (*Continued*)
Concentration of Organic Substances in Seawater and Effect on Fish

Substance	Fish Type	Water Type	Exposure pH	Toxicity Concentration	Index	Adverse Effects	Reference
3-Trifluoromethyl-nitrophenol lampricide	Walleye (*Stizostedium vitreum*)	Sea	—	Velpar L = 1686 mg L^{-1}; Pronone carrier = 4330 mg L^{-1}; Velpar carrier = 20,000 mg L^{-1}	12 h LC$_{25}$	Adverse effects on eggs and fry	[39]
					4.1 mg L^{-1} (gametes)		
					2.6 mg L^{-1} (eggs)		
	Larval sea lampreys (*Petromyzon marinus*)	Sea	—	—	8 h LC$_{99.9}$		
					1 mg L^{-1}		
Roundup herbicide	Coho salmon smolts (*OncorhynchusKisutch*)	Sea	—	0.03–2.78 mg L^{-1}	—	10-day exposure to 0.03–2.78 mg L^{-1}; Roundup had no effect on survival or growth	[53]
Garlon 3A, Garlon 4A, trichloropyrester,3,5,6-trichloro-2-pyridinol, 2-methoxy-3,5,6-trichlorophyridine	Juvenile Pacific salmonids	Sea	—	—	24–96 h LC$_{50}$ reported	Garlon 4 and 3,5,6-trichloro-2-pyridinol highly toxic	[54]

industrial and domestic utilisation, and diesel fuel are the most commonly occurring oil pollutants. Petrol, although used in greater quantity than most other petroleum products, does not often pollute inland waters. Probably this is due to its high volatility on water surfaces, the strict regulations concerning its storage and the general public awareness of its dangers. Its relatively high water solubility may be a lesser factor. Lubricating oils cause pollution problems especially in highly industrialised areas. Since lubricating oils are seldom stored or used in large quantities, pollutions tend to be of a smaller nature, but are often responsible for intermittent surface films on inland waters and the oceans, which are regarded as insignificant. However, the increase in concern over the aesthetic appearance of our inland waters may soon render this significant. Hydrocarbons can occur in seawater as a result of oil spills at sea. These can have adverse effects on fish and invertebrates, which include allometric growth inhibition [1], mortality and reduced yolk reserves in fish [2] and reduced pumping activity and increased mucus production in marine bivalves [3]. Oil dispersants have been used to attempt to disperse oil spillages. These dispersants themselves can have adverse effects on fish and invertebrates such as allometric growth inhibition and reduced growth rate [4,7].

During and after the *Deepwater Horizon* oil spill, the oyster *Crassostrea virginica* absorbed oil [141].

Porto et al. [161] carried out a temporal study of the hydroaccumulation in mussels, dam cockles and oysters collected between 6 and 34 months after an oil spill.

9.2   POLYAROMATIC HYDROCARBONS

Many polyaromatic hydrocarbons in trace quantities have been shown to be directly carcinogenic to mammals (Table 9.4). These are attributed to particular materials which may be present in water samples and are also water soluble to some extent, so that their occurrence in the environment has caused widespread concern. At least a hundred compounds of this type have been detected and characterised in environmental samples. The basic molecular structure consists of benzene rings either fused together or bridged by methylene side chains. Alkyl substituents also occur.

The compounds can be produced by biochemical degradation of other organic compounds under suitable conditions. They may occur in the environment from the combustion of materials such as wood or leaves. Other sources of aromatic materials from which polyaromatic hydrocarbons may be derived include crude oil, which can contain 20% by weight of dicyclic and higher polyaromatic hydrocarbons and high-grade petrol, the aromatic content of which is more than 50%. Unsaturated fatty acids, terpenoids and steroids may also be potential polyaromatic hydrocarbon precursors.

The behaviour and effects of anthropogenic polycyclic aromatic hydrocarbons in aquatic biota in chronically and acutely polluted waterways have been intensely studied for many years [5–10]. Although molluscs have been shown to accumulate polyaromatic hydrocarbons, the question of whether the concentrations of potentially

TABLE 9.4
PAHs Commonly Found in Water

Structure	IUPAC Name	Molecular Weight	Relative Carcinogenicity	Abbreviation
	Benzo[*ghi*]perylene	276	—	B[*ghi*]P
	Chrysene	228	—	Ch
	Fluoranthene	202	—	Fl
	Indeno(1,2,3-*cd*)pyrene	276	+	IP
	Phenanthrene	178	?	Ph
	Perylene	252	—	Per
	Pyrene	202	—	Pyr
	Anthracene	178	?	An
	Benzo[*a*]anthracene	228	+	B[*a*]A
	Benzo[*b*]fluoranthene	252	++	B[*b*]F
	Benzo[*j*]fluoranthene	252	++	B[*j*]F

(*Continued*)

TABLE 9.4 (*Continued*)
PAHs Commonly Found in Water

Structure	IUPAC Name	Molecular Weight	Relative Carcinogenicity	Abbreviation
	Benzo[*k*]fluoranthene	252	—	B[*k*]F
	Benzo[*a*]pyrene	252	+++	B[*a*]P
	Benzo[*e*]pyrene	252	+	B[*e*]P

Note: +++ = active; ++ = moderate; + = weak; ? = unknown; — = inactive.

toxic and carcinogenic polyaromatic hydrocarbons are magnified through the food chain is not yet resolved.

The analytical chemistry of polyaromatic hydrocarbons in tissues can provide an important part of the answer to the biomagnification question, but it must be improved by new technology and the modification of existing analytical procedures to the point where unambiguous, detailed and reproducible data can be obtained on a routine basis. Furthermore, few papers dealing with the analytical methodology for determining polycyclic aromatic sulphur heterocycles and polycyclic aromatic nitrogen heterocycles in fish tissue are to be found in the literature. Considering that the heterocyclic fractions are at least as biologically active as the polyaromatic hydrocarbons [11–17], it is clearly desirable that techniques be developed which will provide accurate and quantitative and qualitative data on the sulphur and nitrogen heterocycles in aquatic biota. Polyaromatic hydrocarbons have been shown to reduce the reproduction rate of plaice in seawater [18].

It has been postulated that polyaromatic hydrocarbons cause liver, lip and skin tumours in brown bullhead trout (*Ictalurus nebulosus*) [19]. The concentrations of polyaromatic hydrocarbons, including benzo[*a*]anthracene and benzo[*a*]pyrene found in organisms so affected, were high (up to $16 \mu g \, kg^{-1}$ wet weight of benzo[*a*] anthracene and up to $6.4 \mu g \, kg^{-1}$ wet weight of benzo[*a*]pyrene).

Shahabi et al. [133] studied the distribution of polycyclic aromatic hydrocarbons in green mussels (*Perna viridis*) from various sites in coastal waters of Peninsular Malaysia between August 2004 and January 2007, in order to assess contamination by petroleum hydrocarbons. The range of total polycyclic aromatic hydrocarbons detected in mussels was from 766 to 110,500 ($ng \, g^{-1}$ lipid weight). High concentrations of polycyclic aromatic hydrocarbons were found in mussel tissues collected near Penang Bridge. The ratios of methyl phenanthrenes to phenanthrene (ΣMP/P) for Penang, Kampung Pasir Puteh and Tebing Runtuth (Johore Straights) were greater than 2, indicating extensive input of petrogenic polycyclic aromatic hydrocarbons. The results indicated that male individuals had more considerable concentrations of polycyclic aromatic hydrocarbons in their soft tissues than female individuals. The results of an independent sample t-test showed that there were no significant differences ($p > 0.05$) between male and female mussels analysed in the Pasir Panjang station. Negative significant correlations ($r = 0.89$, $p < 0.01$; $r = 0.0655$, $p < 0.05$) were found between the weight and total polycyclic aromatic hydrocarbons in female and male species, respectively. This indicated that the body weight of each individual was not affected by the polycyclic aromatic hydrocarbon concentrations. This present study proposed the use of soft tissue of *Perna viridis* as a biomonitor of perylene bioavailability and contamination in coastal waters of Peninsular Malaysia.

Garrigues et al. [134] used biochemical indices based on enzymatic activities in a study of fish and mussels taken at various locations in the Mediterranean Sea.

Preliminary results show a good agreement between biochemical measurements in marine organisms and chemical analyses of polycyclic aromatic hydrocarbons present in sediments. The results obtained suggest the use of biochemical indices for application in chemical contaminant biomonitoring.

9.3   DI-2-ETHYL HEXYL PHTHALATE

This ester causes increased surfacing behaviour in *Daphnia magna* [40] as well as deterioration of reproductive capacity and the immune system and carcinogenic activity in *Daphnia* and fish. Diethyl hexyl phthalate causes mortalities of young Baltic herring (*Clupea lapengus*) and Atlantic cod (*Gadus marina*) [40].

Between 0.002 and 0.02 mg kg^{-1} diethyl hexyl phthalate, PVC and plasticiser have been found in fish.

9.4   SURFACE-ACTIVE AGENTS

Sodium dodecyl sulphate, Triton X-100 nonionic detergents and sodium dodecyl benzene sulphonate surface-active agents have adverse effects on gills and epidermic receptors in the non-saline water lugworm (*Arenicola marina*) [41].

9.5   ACRYLATES AND METHACRYLATES

These esters cause respiratory and metabolic inhibition and neurotoxicity in juvenile fathead minnow (*Pimphales promelas*) [22].

9.6   PHENOL

Phenol above certain concentrations in non-saline waters causes immobilisation, paralysis and mortality in *Ascellus aquatics* [42].

9.7   POLYBROMINATED DIPHENYLETHERS

Oros et al. [132] determined the levels and distribution of polybrominated diphenyle-thers in samples of water, sediments and bivalves taken from San Francisco Estuary.

9.8   PERFLUOROOCTANE SULPHATES

High accumulation of this compound has been found in marine tucuxi dolphins (*Solatia guianensis*) along the Brazilian coast [44].

9.9   DOMOIC ACID

A recent outbreak of poisoning resulting from the consumption of cultured blue mussels (*Mytilus edulis L.*) from a localised area in eastern Canada has been attributed to the presence of domoic acid, a relatively rare neurotoxic amino acid, previously found only in some algae of the family Rhodomelaceae [135]. Studies on aqueous extracts of shellfish tissue indicated that the toxin and several of its isomers could be separated and isolated in sufficient amounts of subsequent structural identification by reversed-phase high-performance liquid chromatography with ultraviolet diode array detection. Aqueous acetonitrile containing 0.1% v/v trifluoroacetic acid was used as mobile phase. As the retention time and characteristic UV absorption spectrum of demoic acid 1 (λ_{max} = 242 nm) permit unequivocal identification, the high-performance liquid chromatography diode array detection procedure was refined with a microbore column to provide a rapid (5 min), sensitive (0.3 ng detection limit) and reproducible assay method for the determination of domoic in shellfish tissue. Extraction was accomplished by boiling homogenised shellfish tissue for 5 min with distilled water. Extracts were taken through an octadecyl silica solid-phase extraction clean-up prior to high-performance liquid chromatography. This method has been applied to a variety of shellfish and phytoplankton.

9.10   DITHIOPHENE

Dithiophene and related methylated derivatives are known to be among the most persistent and probably the most toxic dithiophenes in the marine environment. Their analysis and their fate by photo-oxidation and biodegradation were studied by Berthou and Vignier [136]. The methylated dibenzothiophene isomers, provided that they are resolved by high-resolution gas chromatography, were used as organic markers of oil pollution in oysters. The determination of the relative distribution of the four monomethyldithiophenes made it possible to characterise the source of pollution in an oyster area in North Brittany, France. The fate of methylated dithiophene compounds was studied in controlled seawater enclosures where Arabian light oil had been spilled. Analysis of the weathered oil showed that (1) oil was degraded by photo-oxidation at a rate of 0.004% day and (2) the half-life of photolysis of methylated dibenzothiophene was dependent on the number of methyl groups on the aromatic nucleus: 8 days for dibenzothiophene and

20 days for trimethylated dibenzothiophene. Compounds solubilised in the water column include methyl dibenzothiophene as methyl substituted dibenzothiophene sulphoxides and sulphones which can be determined by high-performance liquid chromatography with synchrofluorescence and gas chromatography–flame photometric detection.

The metabolic pathway of dibenzothiophene was established *in vitro*. Rat microsomes transformed this substrate to dibenzothiophene-5-oxide and subsequently to dibenzothiophene-5-dioxide. Such an enzymatic S-oxidation was shown to be principally cytochrome P-450 dependent. It is suggested that the mixed-function oxidase dibenzothiophene oxidase activity of marine species could be evaluated by this S-oxidation.

9.11 DICHLORO-6-NITROPHENOL

It is shown that this compound is formed upon nitration of 2,4-dichlorophenol, which in turn is a transformation intermediate of the herbicide dichloroprop. However, the chemical and spectroscopic characteristics of 2,4-dichloro-6-nitrophenol, as well as its toxicity, are poorly known.

Tognazzi et al. [137] showed that 2,4-dichloro-6-nitrophenol behaves as a diprotic acid in 3.0 ± 0.9 in aqueous solutions. At pH values of less than 3.0, 2,4-dichloro-6-nitrophenol would undergo protonation. The adsorption spectra suggests that anionic 2,4-dichloro-6-nitrophenol, which prevails at pH values of less than 5, would have an orthoquinoid structure that is responsible for the absorption peak centred at 428 nm. Considering that 2,4-dichloro-6-nitrophenol has been detected in the brackish lagoons of the Rhône Delta (southern France), where its levels are comparable to those of the parent herbicide, it is necessary to examine the possible effect of 2,4-dichloro-6-nitrophenol on the species living in that environment. For this reason, the acute toxicity of the anionic form of 2,4-dichloro-6-nitrophenol was assessed for the brine shrimp (*Artemia salina*), a zooplankton species that lives in both brackish and saline aquatic environments. The toxicity test yielded a LC_{20} value of $8 \pm 2\,mg\,L^{-1}$ and a LC_{50} value of $18.7 \pm 0.8\,mg\,L^{-1}$. Such values are safely higher than the maximum detected concentrations of 2,4-dichloro-6-nitrophenol in the Rhône Delta lagoons. Further studies need to be concentrated on the long-term effects of 2,4-dichloro-6-nitrophenol and, in particular, its potential genotoxicity.

9.12 ALLYL FORMATE

Rainbow trout (*Salmo gairdneri*), which picked up a body burden of $100\,\mu g\,kg^{-1}$ of allyl formate, developed severe liver damage [43].

9.13 PENTACHLOROPHENOLS

Pentachlorophenols above certain concentrations in non-saline waters cause low survival rates (e.g. 31.6 h at $90\,\mu g\,L^{-1}$) in rainbow trout (*Salmo gairdneri*) [45].

9.14   CHLOROBENZENES

Guppies (*Poecilia reticulate*) have been exposed to 1,2,3-trichlorobenzenes (1.92, 3.78 and 55.9 μmol L^{-1}), 1,2,3,4-tetrachlorobenzene (1.13 and 1.69 μmol L^{-1}) or pentachlorobenzene (0.40 and 0.54 μmol L^{-1}) in acute flow-through tests. In each experiment, time of death was inversely related to toxicant concentration. Irrespective of test compound or exposure concentration, death occurred when the internal toxicant concentration reached 2.0–26 μmol g^{-1} fish.

Carlson and Kosian [46] studied the toxicity of chlorinated benzenes to fathead minnows (*Pimephales promelas*). Compounds studied were 1,3-dichlorobenzene, 1,4-dichlorobenzene, 1,2,3,4-tetrachlorobenzene, pentachlorobenzene and hexachlorobenzene. The mean tissue residue concentrations were as follows:

Compound	No Effect Concentration (NOEC) (mg kg^{-1})	Lowest Effect Concentration (LOEC) (mg kg^{-1})
1,3-Dichlorobenzene	120	160
1,4-Dichlorobenzene	70	103
1,2,3,4-Tetrachlorobenzene	640	1100

Tissue residue concentrations in fish chronically exposed to maximal test concentrations of pentachlorobenzene and hexachlorobenzene were 380 and 97 mg kg^{-1} respectively.

9.15   CHLOROPHENOLS

An assessment of the sublethal effects on rainbow trout (*Salmo gairdneri*) of 2,4-dichlorophenol and 2,4,6-trichlorophenol has been carried out [47]. Both compounds were accumulated in fish even at the lowest concentration in water tested (5 μg L^{-1}), the greatest amount of chlorophenol being accumulated in liver, adversely affecting liver enzyme activity. Rogers and Hall [48] determined the levels of polychlorodiphenol in herring muscle in the range 0.01–2.2 mg kg^{-1}.

9.16   *p*-CHLOROANILINE

p-Chloroaniline above certain concentrations (100,000 μg L^{-1}) kills embryos in the South African clawed toad (*Xenobus laevis*) [49].

9.17   METHYL BROMIDE

Methyl bromide at concentrations above 100,000 μg L^{-1} for a period of 1–3 months causes paralysis in the guppy *Poecilia reticulate*, and at concentrations above 1800 μg L^{-1} for 4 days causes degenerative changes in the gills as well as oral mucosa [50].

9.18   ORGANOBROMINE COMPOUNDS

Covaci et al. [91] have discussed the occurrence of anthropogenic and naturally occurring organobrominated compounds in deep-sea species from the Mediterranean sea.

9.19 TETRACHLORO-1,2-BENZOQUINONE

This compound, when present in seawater, causes skeletal abnormalities in the fourhorn skulphin (*Hyoxecephalus quadricornis*) [51].

9.20 POLYCHLOROBIPHENYLS

Since their introduction, polychlorinated biphenyls (PCBs) have caused much eco-logical damage and are harmful to humans. One aspect of these toxicants is that they have been shown to have a very severe adverse effect on wildlife by causing thin eggshells and, consequently, a poor reproductive rate in the laying season.

Reijnders [52] has reported that seals that fed on polychlorobiphenyl-contami-nated fish undergo reproductive failure.

Arochlor causes severe weight reduction and liver degeneration in rainbow trout (*Salmo gairdneri*) and inhibition of reproduction at the $50–100\,\mu g\,L^{-1}$ level in non-saline waters in *Daphnia puliccaria*. Levels in herring muscle between 1979–1986 were $0.3–2.2\,\mu kg^{-1}$ reducing to $0.01–0.017\,\mu kg^{-1}$ in 1988. Polychlorobiphenyl levels in cod liver between 1979 and 1986 were 0.3 to $3.7\,mg\,kg^{-1}$, reducing to 0.013 to $0.19\,mg\,kg^{-1}$ in 1988.

Nicholson and Moore [64] compared the distribution of polychlorobiphenyl con-geners and other chlorinated compounds in fish in coastal areas.

These reduced to $0.01–0.017\,\mu g\,kg^{-1}$ in 1988. Polychlorobiphenyl levels in cod liver between 1979 and 1986 were $03.7–0.37\,\mu g\,kg^{-1}$ reducing to 0.013 to $0.19\,\mu g\,kg^{-1}$ in 1988, compared with the distribution of polychlorobiphenyl congeners and other chlorinated compounds in fish in coastal areas.

A total of 28 polychlorobiphenyl congeners, *p,p'*-DDD, *p,p'*-dichlorodiphenyl-trichloroethane (*p,p'*-DDT), HCB and gamma-hexachlorocyclohexane (γ-HCH) were determined in late trouts (*Salmo trutta*) from two isolated lakes in the Pyrenees and in red mullets (*Mullus barbatus*) from the Mediterranean coast. The lower concentrations, in the range of $1.9–6.0\,ng\,g^{-1}$ (wet weight) for *p,p'*-1,1-dichloro-2,2-bis-(*p*-chlorobiphenyl)-ethylene (*p,p'*-DDE) and $0.1–0.8\,ng\,g^{-1}$ (wet weight) for the other individual compounds ($2.6–7.2\,ng\,g^{-1}$ for the sum of polychlorobiphenyl congeners), were found in the remote lake samples, where the atmosphere was the only source for these anthropogenic com-pounds. These samples were also relatively enriched in γ-HCH, consistent with the predominant atmospheric transport pathway of this compound.

The polychlorobiphenyl congener patterns exhibited substantial differences between both sets of samples. The trouts were enriched in the tri- to pentachloro-biphenyl congeners, whereas the mullets were dominated by the hexachlorobiphe-nyl isomers, reflecting the predominance of atmospheric and aquatic point-source inputs, respectively. On the other hand, the congeners with adjacent unsubstituted positions were clearly depleted in the marine fish, indicating a relatively enhanced metabolic activity of these chronically exposed organisms. All these features illus-trate the advantage of the analysis of individual polychlorobiphenyls for a better understanding of their environmental transport and fate.

Hela and Papadopoulos [56] carried out an estimation of uncertainty using matrix solid-phase extraction methodology for the determination of chlorinated compounds in fish. The methodology was applied for the determination of selected organochlorine pesticides and PCBs in real fish samples, from selected study locations in western Greece (Mesologgi lagoon and Trichonida Lake) based on chromatographic techniques. The sources of uncertainty are presented along with the calculated combined uncertainty terms derived from the analytical procedure and the chromatographic process.

Roscales et al. [57] determined the geographical patterns of the distribution of polychlorobiphenyls and DDT in *Colonectris* sp. across the northeast Atlantic and the Mediterranean archipelago. It was demonstrated that lower chlorinated polychlorobiphenyls are more volatile and easily transported to remote locations, whereas highly chlorinated congeners are more persistent and not easily metabolised.

Morgan and Lohmann [58] studied the dietary uptake from contaminated sediments as a source of polychlorobiphenyls in migratory fish and invertebrates in an urban estuary.

Wells and Echarri [141] described a method for the determination of 20 undivided chlorobiphenyls in marine mammals, including rake, dolphins, porpoise and seal tissue. An estimate was made of the toxic equivalence concentration for the non-*ortho* and monochlorobiphenyls.

9.21 *p*-DIOXINS AND DIBENZOFURANS

Polychlorinated dibenzo-*p*-dioxins, polychlorinated dibenzofurans and *ortho*-unsubstituted PCBs (non-*ortho* PCBs) are three structurally and toxicologically related families of anthropogenic chemical compounds that have in recent years been shown to have the potential to cause serious environmental contamination [28,60,62–66,96,103].

The substances are trace-level components or by-products of several large-volume and widely used synthetic chemicals, principally polychlorobiphenyl and chlorinated phenols [67–70]. They can also be produced during combustion processes and by photolysis [71,72]. In general, polychlorinated dibenzo-*p*-dioxins, polychlorinated dibenzofurans and non-*ortho* PCBs are classified as highly toxic [73], although the toxicities are very dependent on the number and positions of the chlorine substituents [74]. About 10 individual members of a total of 216 polychlorinated dibenzo-*p*-dioxins, polychlorinated dibenzofurans and non-*ortho* PCBs are among the most toxic man-made or natural substances to a variety of animal species. The toxic hazards posed by those chemicals are exacerbated by their propensity to persist in the environment [64] and to readily bioaccumulate [75,87,88], and although the rate of metabolism and elimination is strongly species dependent [76–78], certain high-toxicity isomers have been observed to persist in the human body for more than 10 years.

The majority of scientific concern for the hazards of these compounds has been directed towards the disposition in the environment of the single most toxic isomer, 2,3,7,8-tetrachlorodibenzo-*p*-dioxin (2,3,7,8-TCDD) [28,59–67]. More recently, however, investigations into the formation and occurrence of polychlorinated dibenzofurans suggest that this family of toxic compounds may also commonly occur at comparable levels and could possibly pose a greater hazard than polychlorinated dibenzo-*p*-dioxins. Polychlorinated dibenzofurans are often found as co-contaminants in and are more readily produced from pyrolysis of PCBs [70,79,80]. Most important, polychlorinated dibenzofurans produced from pyrolysis of PCBs are predominantly the most toxic isomers, those having a 2,3,7,8-chlorine substitution pattern [65]. A number of fires involving electrical transformers and capacitators have demonstrated the potential for formation of hazardous levels of polychlorinated dibenzofurans from the pyrolysis of PCBs [70,80–83].

In the light of these findings and because of a dearth of data pertaining to the occurrence of these compounds in the environment, polychlorinated dibenzofurans and non-*ortho* polychlorinated dibenzo-*p*-dioxins were included as target compounds in a survey of important US rivers and lakes for polychlorinated dibenzo-*p*-dioxins. The decision to include as many chlorinated dibenzo-*p*-dioxin isomers as possible was based on several facts: (1) several other polychlorinated dibenzo-*p*-dioxins isomers are also extremely toxic [74]; (2) pentachlorophenol, a large-volume fungicide and wood preservative, contains relatively high levels of hexa-, hepta- and octachlorodibenzodioxins and essentially no tetrachlorodibenzo-*p*-dioxins [28,89,90]; and (3) incineration of materials containing chlorophenols readily produces mixtures of polychlorinated dibenzo-*p*-dioxins, but 2,3,7,8-TCDD is a minor component. On the other hand, the highly toxic 1,2,3,7,8-penachloroisomer is a major component of polychlorinated dibenzo-*p*-dioxin incineration products of pentachlorophenol [69,84,85,149].

Component-specific analyses can be a crucial link to the sources of contamination because different sources of polychlorinated dibenzo-*p*-dioxins and polychlorinated dibenzofurans usually produce mixtures of distinctly different relative component abundances [86]. On the other hand, the preferential accumulation of certain isomers in animals may prevent source identification from analyses of biological samples.

Haglund et al. [92] studied the temporal variation of polybrominated dibenzo-*p*-dioxin and methoxylated diphenylether concentrations in fish. Large differences in exposure and metabolic stability were observed.

9.22   MALATHION

Above certain concentration levels, malathion reduces the survival time of freshwater teleosts (*Channa punctatis*). It also produces mortalities upon a 5-day exposure at 44,000 µg L^{-1} in toad embryos (*Bufo arenarum*) [116].

9.23   BROMACIL

Bromacil reduces the growth and survival time and deforms fry in fathead minnow (*Pimephales promelas*) [117].

9.24   PHOSPHAMIDON

Phosphamidon causes glycogen depletion in muscles, that is, reduced mobility, in freshwater prawn (*Macrobrachum lamarrei*) [118].

9.25 FENITROTHION

Fenitrothion causes growth abnormalities of follicle and epithelium in freshwater murrel (*Channa punctatis*) [119].

9.26 CARBARYL

Carbaryl above certain concentration levels reduces the survival time of catfish (*Clarias leatrachus*) [30].

9.27 AMINOCARB

Fingerling rainbow trout (*Salmo gairdneri*) exposed for up to 4 days to $10 \mu g L^{-1}$ of aminocarb in water at pH 4.6–8.2 picked up $9.1 \, mg \, kg^{-1}$ aminocarb in fish tissue in the first 6 h of exposure [120]. At pH 8.2, whole body aminocarb increased to $12 \, mg \, kg^{-1}$ in 1 h and remained elevated until the fish died in 72 h.

9.28 FENTHION

Ram and Sathyaneson [128] determined LC_{50} values obtained when mysid (*Mysidopsis bahia*), grass shrimps (*Palaemonetes pugio*), pink shrimps (*Penaeus duorarum*) and sheepshead minnow (*Cyprinodon variegates*) were exposed to spray applications (336 g/ha) of fenthion to water on an estuarine shoreline. Mortalities and non-lethal effects occurred in these species.

9.29 CHLORINATED INSECTICIDES

Persistent chlorinated hydrocarbons of agricultural and non-agricultural interest, such as 1,1,1-trichloro-2,2-bis-(*p*-chlorophenyl)ethane (DDT), PCBs and hexachlorobenzene, now have a global distribution and can be detected in wildlife samples in variable amounts. Polychlorobiphenyls together with DDE are the main types of chlorinated hydrocarbons found in Norwegian avian fauna and in fish along the Norwegian coast [93–95].

In Friefjorden, a fjord in southeast Norway, heavy local contamination with chlorinated hydrocarbons of industrial origin has been detected. The contaminants most often found in fish in this area are hexachlorobenzene, octachlorostyrene and decachlorobiphenyl. In addition, complex mixtures of PCBs and chlorinated naphthalenes have been detected [71,97]. Decachlorobiphenyl has previously been found in arctic fox (*Alopex lagopus*) from Norheim [98], and octachlorostyrene was first detected in birds in the Netherlands [99–101].

Various other workers have reported on concentrations of chlorinated insecticides found in fish and fish tissues. These include benzene hexachloride (BHC), HCH, heptachlor, aldrin, DDT and polychlorobiphenyl, toxaphene, polychlorobiphenyls, DDT and hydrocarbons [104,111], polychlorobiphenyls and *p,p'*-DDE [102] and lindane [105].

In a monitoring programme over 6 years, the above chlorinated lindane, endo-sulphan and trichlorophon insecticides have been shown to cause erratic swimming behaviour, hyperventilation and mortalities in invertebrates.

Endosulphan in seawater increases the body weight, haemolymph volume and hydration at sublethal concentrations ($6200\,\mu g\,L^{-1}$) when the crab *Oziotelphusa senex* is exposed to it [106,107].

Quintel et al. [108] have studied the worldwide occurrence and distribution of endosulphan in water, sediment and fish tissue.

Kawano et al. [109] reported on the concentrations of chlordane compounds present in fish, seabirds, invertebrates and mammals. The metabolite oxychordane, which is much more toxic than the parent compounds and very persistent, was found in higher concentrations in seabirds than in marine mammals.

Karlsson et al. [110] have reported that the enantiomer ratios of chlordane conge-ners at the pictograms per liters range are gender specific in cod, gudus, morhuan, herring and salmon, from the Barents Sea.

Trim [130] has discussed the results obtained in static 96 h toxicity tests with malathion, endosulphan and fenvalerate on estuarine waters on the mummichog (*Fundulas heteroclitus*). All three insecticides were highly toxic to estuarine and coastal water fish.

Kolankaya [140] determined organochlorine pesticide residues and ascertained their toxic effect in organisms in Turkey.

Persistent toxic substances appear in most urbanised coastal areas of the world, accumulating in water, sediments and biota, where they cannot be eliminated effi-ciently, and their derivatives can be highly toxic. Like many other countries, Turkey is also facing problems concerning pesticides and other persistent toxic substance residues. BHC and alpha, beta, gamma and delta isomers of HCH were determined in 16 samples of surface and ground waters and mussels in the Middle Black Sea Region, and the concentrations of polychlorobiphenyls and organochlorine pes-ticide residues were analysed in the eastern Aegean Sea water and fish samples. Polychlorobiphenyls were not detected in *Mullus barbatus* living in the Aegean Sea, and its derivatives were detected in low levels. Thirteen organochlorine pesticides were determined in water, sediment, fish and water birds in the Göksu Delta and analysed in the Sariyar Dam Lake, the Sakarya Basin, five lakes in Central Anatolia, and the Meriç Delta. These analyses were conducted on water sediment and fish samples. Some of these pesticides were found in more or less all of the samples. Microorganisms in soil and some saprophyte fungi detoxify chlorinated hydrocar-bons by de-chlorinating them. The toxicity of degraded metabolites of these insec-ticides, for which the chlorine amounts have been reduced, decreased consequently, and toxic effects also reduced in mice when chlorine was removed. Mixed cultures of microorganisms isolated from agricultural soils were used to degrade endosulphan, carbaryl and malathion. This result indicates that these chemicals break down in soil, resulting in a significant decrease in their toxic effects. Toxic substances cause acute, subchronic and chronic effects in the environment on biota. The hermit ibises and endemic migratory bird species no longer visit the Birecik district in Turkey. The reason for this is thought to be the usage of organochlorine pesticide sprays in agricultural areas of the district. In recent years, studies have been conducted on the activation of persistent organic substances and cytochrome P-450 monooxygenase enzymes in indicator fish species. One of these studies has been conducted on the bottom-feeding fish that live in Izmir Bay. The outcomes showed a significant decrease in the ethoxyresorufin-*O*-deethylase (EROD) activity in fish (mullet) and benthic fish (common sole), when these are used as biomonitors for polycyclic aromatic hydrocarbons and polychlorobiphenyl-type pollutants, that have been harvested from six different stations along the bay from the city borders towards the inner parts of the Aegean Sea. The effects of organochlorine pesticides have been determined by other studies in Sariyar Dam Lake and Meriç Delta. An increase in the EROD activity has been determined in fish living in these environments.

Dela Torre et al. [130] found concentrations of dechlorane and related compounds in the Franciscana dolphin (*Pontoporia blainvillei*).

Yamashita et al. [112] found organochlorine pesticides in waters off the southeast coast of Brazil.

Samples of river lake waters and suspended matter and fish samples were collected during 1993 and from the Nile River and Manzala Lake in Egypt. Of the different organochlorine pesticides analysed, *p,p'*-DDE was the most predominant in fish (7.6–67 ng/g wet weight), sediments (3.2–432 ng/g dry weight) and suspended solids (5.3–138 pg/L). However, in the dissolved phase of water samples, HCH compounds predominated (α-HCH, 71–2815 pg/L). Concentrations or organochlorine pesticides, except chlordane, were higher in Manzala Lake than in the Nile River. Concentrations of organochlorine pesticides in fish corresponded with those in sediments from each location. Comparison of organochlorine concentrations in the Nile River water with those reported in earlier studies suggested a decrease in the concentrations during the last decade. However, concentrations of *p,p'*-DDE have increased in fish. It appears that the release of this metabolite from contaminated sediment is the major source of *p,p'*-DDE in fish in recent years.

Hallou et al. [113] determined levels of a series of organochlorine pesticides, including PCBs in muscle, liver and gonad of yellowtail flounder (*Pleuronectes ferruginea*), collected in waters of the northwest Atlantic, offshore of the island of Newfoundland. Comparisons were made to concentrations in tissues of fish maintained in tanks of the Northwest Atlantic Fishers Center, located in the city of St John's. Polychlorobiphenyl and DDE concentrations in the liver of inshore females were more than 10 times more elevated than in offshore females. Concentrations of several organochlorine insecticides were also significantly more elevated in the liver, muscle and gonads of inshore than offshore fish. The polychlorobiphenyl congener fingerprint displaced the predominance of PCB in all samples: 153 > 138/163/164 > 118. Twenty-one polychlorobiphenyl congeners were detectable in more than 80% of liver samples, eight were detectable in muscle and four in gonads, while no fingerprint differences were apparent between sexes or inshore and offshore samples.

Erkman and Kolankaya [114] determined organochlorine pesticides in water, sediment and fish samples from Meriç Delta, Turkey. This study was carried out between May 2002 and August 2003 in Meriç Delta, which is located at a site where the Meriç River falls to the Aegean Sea. Residues of organochlorine pesticides in surface water, sediment and fish (*Cyprinus carpio*) samples from Meriç Delta were analysed by gas chromatography. The results of the study showed that all analysed 20 organochlorine pesticides and their residues are widespread throughout the study area. According to the results, it was found that the concentrations of these selected organochlorine pesticides in fish (*Cyprinus carpio*) samples were higher than in water and sediment samples. Because of the low water solubility of the organochlorine compounds, it is expected that any organochlorine pesticides present in the study area will be preferably adsorbed to sediment or bioaccumulated in fish. The α- and β-HCH were the predominant HCH isomers in all analysed fish samples and ranged between 319.5 and 968.15 ng g^{-1} and between 397.5 in BHC and 876.4 ng g^{-1} in β-BHC, respectively. The concentration levels of *p,p′*-DDT (ranging from 2.68 to 52.45 ng g^{-1}) in all analysed fish samples were consistently higher than its metabolite *p,p′*-DDE, indicating a recent use of this organochlorine pesticide in the area. Erkman and Kolankaya [114] analysed the distribution characteristics of individual organochlorine pesticides and found that α-HCH, β-HCH, *p,p′*-DDT, *p,p′*-DDE, β-endosulphan, heptachlore epoxide and endrin ketone were the most common organochlorine pesticide contaminants in the study.

9.30 PERMETHRIN

This insecticide causes nerve poisoning and blocking of anaerobic and aerobic metabolism in the snail *Hymnaea acuminata* [115].

9.31 VINEYARD PESTICIDES

Sondykl [121] carried out a genotoxicity assessment of two vineyard pesticides used as vineyard chemicals, diuron and azoxystrobin.

This study deals with the use of a chronic exposure scenario of zebrafish (*Danio rerio*) in laboratory conditions to evaluate the genotoxic potential of these two. Adult male zebrafish were exposed for 3 weeks in the semistatic mode in four 20 L aquaria. Treatments allowed to each aquarium were negative control (untreated), positive control (methyl methane sulphonate 1 µM), diuron 4.3 nM and azoxystrobin 1.2 nM. Once per week, genotoxicity was assessed (six fish/treatment) by the use of two complementary biomarkers; the primary DNA damage was evaluated in somatic (liver) and germ (spermatozoa) cells by the alkaline version of the comet assay and the micronucleus formation assessed in erythrocytes. Very low basal DNA damage was obtained with both biomarkers in the negative control during the three consecutive weeks, and a significant genotoxic response was obtained in 1 µM methyl methyl sulphonate exposed fish, in both liver and germ cells with the comet assay and in erythrocytes with the micronucleus test, starting after 1 and 3 weeks. With this chronic exposure scenario, both diuron and azoxystrobine revealed a genotoxic potential at realistic environmental concentrations and a significant response was obtained in all cell types investigated and with both biomarkers used, mainly after 7 or 14 days, this stressing the interest of a long-term exposure scenario. Further studies need to be undertaken in order to evaluate whether DNA damage observed in spermatozoa of fish exposed to an environmental concentration of pesticides could lead to subsequent reproductive disorders.

9.32   TOXAPHENE

Musial and Uthe [122] have presented chromatographic and chemical confirmatory evidence to confirm the presence of residues of toxaphene, a polychlorinated camphene pesticide, in herring (*Clupea harengus harengus*) and cod (*Gadus morhua*) from widely separated areas of the Canadian east coast. Toxaphene residues were not detected in a sample of deep-sea scallops (*Placopecten magellanicus*). Toxaphene was determined by capillary gas chromatography following a combination of chromatography and fuming nitric-concentrated sulphuric acid clean-up, a procedure which greatly simplified the capillary gas chromatograms and eliminated many co-extractives. Concentrations in the fish tissues ranged from 0.4 to 1.1 µg/g on a wet weight basis and from 2.4 to 12 µg/g on a fat weight basis. These data indicate widespread contamination of the marine environment by chlorinated camphenes.

9.33   NITROGEN-CONTAINING COMPOUNDS

In seawater, the compound fluoro-methyl-4-nitrophenol lampricide damages the eggs and fry of walleye (*Stizostedium vitreum*) [123,124].

Many *N*-nitrosamines are toxic and carcinogenic, and furthermore, the carcinogenic action exhibits a high degree of organ specificity. Nitrosamines are formed by interaction between a nitrite and an amine with varying ease, depending on the nature of the amine and the prevailing conditions. The reaction is not restricted to secondary amines but also occurs with primary and tertiary amines and even quaternary ammonium salts. Thus, the precursors are widespread as naturally occurring compounds and nitrosamines are generated in many commercial and industrial processes. It is therefore conceivable that trace amounts may be present in air and water in the vicinity of industrial sites [82].

Eggs (late blastula) or 406-day larvae of striped bass (*Morone saxatilis*) or sheepshead minnow (*Cyprinodon variegates*), when exposed to cyclophosphamide or *N*-methyl-*N*-nitro-*N*-nitroguanidine at concentrations of 1–1000 µM for 1–4 days, show a close dependent relationship between aberration frequency of chromosomes and the concentration of toxicant in eggs and larvae of both species [131].

9.34   BIOGENIC AMINEODORANTS

Biogenic amineodorants such as monotoxin microcystin are known to accumulate in the tissues of diverse aquatic biota. Wood et al. [126] studied several variations in microcystin production of fish and shellfish in a river estuary.

Gaseous biogenic amines such as putrescine, spermidine, aniline and trimethylamine are important biomolecules that play many crucial roles in metabolism and medical diagnostics. A chemodosimetric detection assay has been developed by Chow et al. [124] for these gaseous amines by Ru^{II}–Eu^{III} heterobimetallic complexes, $K\{[Ru^{II}(^tBubpy)(CN)_4]_2Eu^{III}(H_2O)_4\}$ (where $^tBubpy = 4,4'$-di-tert-butyl-2,2'-bipridine). Synthesis, X-ray crystal characterisation and spectroscopic properties of this Ru^{II}–Eu^{III} heterobimetallic complex were reported. Binding properties of the Ru^{II}–Eu^{III} complex with common gases revealed that this complex is very selective to gaseous amine molecules. Sensitivity of this complex towards the amines was found as $\sim\log k = 4.5$–4.8. Real-time monitoring of gaseous biogenic amines was applied to Atlantic mackerel by studying spectrofluorimetric responses of the Ru^{II}–Eu^{III} complex towards different biogenic amine concentrations. Gas chromatography–mass spectrometry was used as a reference for the studies. A linear spectrofluorimetric response was found towards biogenic amine concentrations in real fish samples. This complex was found to respond specifically to these biogenic amines down to 10 ppb.

9.35   2-METHYL-3-METHOXY-4-PHENYLBUTANOIC ACID

Microcystins are widespread cyanobacterial toxins in freshwater systems, and have been linked to both acute and chronic health effects. A growing number of studies show that they can bioaccumulate in food webs. Although several methods (e.g. enzyme-linked immunosorbent assay [ELISA] and liquid chromatography–mass spectrometry [LC-MS]) have been developed for analysis of microcystins in water, extraction (for subsequent analysis) of the toxin biological matrices (i.e. animal tissues) is impeded owing to covalent binding of toxins and active sites of their cellular targets, that is, protein phosphatises. As an alternative approach, chromatographic methods for analysis of a unique marker, 2-methyl-3-methoxy-4-phenylbutanoic acid, the product of the Lemieux oxidation of microcystins, have been previously developed and shown to measure total bound and unbound microcystins. Application of this method, however, has been limited by poor recovery of the analyte. Suchy and Berry [125] have described an improved recovery method, namely the use of solid-phase microextraction. The 2-methyl-4-methoxy-4-phenylbutanoic acid, analogue 4-phenylbutanoic acid and oxidised microcystin were used to develop methods. Specifically, Suchy and Berry [125] investigated several solid-phase microextraction fibres and postoxidation steps.

Specifically, a method employing postoxidation methyl esterification, followed by headspace solid-phase microextraction of 2-methyl-4-methoxy-4-phenylbutanoic acid, was developed and subsequently applied to analysis of environmental samples, namely fish tissue that contain microcystins. This method shows high linearity for both water and tissues spiked with microcystins. This method has a limit of detection of 9 ng g^{-1}.

Evaluation of field samples by solid-phase microextraction with gas chromatography–mass spectrometry detection has detected considerably higher levels of microcystin than was detected by conventional methods such as ELISA, and it is proposed that the technique reveals microcystin, particularly in the bound form, that is not detected by these methods. These results indicate that this method provides improved detector capability for microcystins in biological metrices and will enhance our ability to understand bioaccumulation in food webs and monitor exposure.

9.36   CIGUATERA TOXINS

Ciguatera fish poisoning is a human foodborne intoxication caused by ingestion of tropical fish contaminated with the potent polyether toxins known as ciguatoxins. These toxins are issued from *Gambierdiscus* species of dinoflagellates. Herbivorous fish accumulate these toxins in their musculature and viscera after ingesting dino-flagellates. Epidemiological studies carried out by Otero et al. [127] showed that ciguatoxins has been present in areas between 35° north and 35° south latitude, mainly, Indo-Pacific and Caribbean areas, but not in waters closer to the European and African continents. A specimen of *Seriola dumerili* weighing 70 kg and a smaller *Seriola fasciata* specimen, captured in waters belonging to the Selvagens islands (Madeira Arquipelago), were analysed. Fish from this genus were impli-cated in previous suspended ciguatera poisoning outbreaks in the Portuguese Madeira Arquipelago in the North Atlantic Ocean. Analysis was performed by two approaches, a functional method using cerebellar granule cells and the ultraperfor-mance LC-MS method. The study was carried out in one portion of the tail muscle of *Seriola fasciata* and five parts of the body of *Seriola dumerili* (tail muscle, head, ventral muscle, midmuscle and liver). The functional method consisted of the modi-fication of the inward sodium current in cerebellar granule cells, and the chemical method was a high-resolution chromatography procedure which allowed elucidat-ing the toxin profile in the sample. In addition, ultraperformance LC-MS was opti-mised and used for detecting and quantifying ciguatera for the first time. After fish extraction and clean-up, the chromatograms revealed the presence of ciguatera-1B at 1111.6 *m/z*, ciguatera-3C at 1023 *m/z*, a ciguatera analogue at 1040.6 *m/z* and a cigua-tera from the Caribbean or Indic waters at 1141.6 *m/z*. Therefore, the results obtained in this chapter for both methods confirm for the first time the presence of ciguatera in fish from Madeira Arquipelago.

9.37   TETRODOTOXIN AND 5,6,11-TRIDEOXY TETRODOTOXIN

Rodriguez et al. [138] have reported on the toxicity of these compounds in the trum-pet shell (*Charonia lampas lampas*).

Tetrodotoxin is one of the most potent toxins already isolated, and it occurs in a wide variety of animals. In this work, the occurrence of tetrodotoxin and analogues was examined using mass spectrometry, confocal microscopy, LC-MS and mouse bioassay in a trumpet shell (*Charonia lampas lampas*) and in the fluids of a patient poisoned by consuming this shell. Retention time data in the LC-MS system within the enhanced mass spectrum mode indicated the presence of tetrodotoxin and the analogue 5,6,11-trideoxy tetrodotoxin; the enhanced product ion mode confirmed the existence of both toxins with the formation of characteristic daughter ions from the fragment patterns of each molecule. Tetrodotoxin and 5,6,11-trideoxy tetrodo-toxin were only detected in the digestive gland of the trumpet shell and the urine and serum of the patient. The concentration of 5,6,11-trideoxy tetrodotoxin checked in the samples by the LC-MS system was three times higher than that for tetrodotoxin. However, the results obtained by mouse bioassay showed that the analogue is much

TABLE 9.5
Toxicity of Various Organic Compounds on Fish, Crustacea and Mammals

Organic Compound	Creature	Reference
Halogenated 1-methyl-1,2′-biphenyl bipyroles	Marine biota	[140,143]
Perfluoroalkyl acids	Others	[144]
Polychlorinated dibenzo-p-dioxins	Bivalves	[145]
Phthalate esters	Marine biota	[146]
Perfluorinated compounds	Aquatic biota	[147]
Fluorinated organic compounds	Marine biota	[148]
Imidacloprid	Burrowing shrimp	[149]
Perfluorooctane sulphonate	Marine biota	[150]
4-Nitrophenol, nonylphenol	Marine biota	[151]
Polychlorobiphenyl	Killer whales	[152]
Polychlorobiphenyl	Marine biota	[153]
Polychlorobiphenyl, polybromobiphenyl	Killer whale	[154]
Hexabromocyclododecane polybrominated diphenylethers and polychlorobiphenyl	Ringed seal	[155]
2-Ethyl hexyl 2,3,4,5-tetrabromobenzoates, 2-ethyl hexyl tetrabromophthalate, 1,2-bis-2,4,6-ethane decabromodiphenyl ethane	Bivalves, gastropod, elimieproxine	[156]
Hexabromocyclodecane	Biota	[157,160]
Polybrominated hexahydroxanthene	Sponges	[158]
Methoxylated diphenylethers	Marine biota	[159]
Polybrominated diphenylethers	Bivalves	[46,163,164]

less toxic than tetrodotoxin. *In vitro* toxicity was checked using cerebellar cells; in these experiments, the trumpet shell sample showed high toxicity, but the level was lower than *in vivo* results, probably due to some competition between analogues. Rodriguez et al. [138] showed for the first time the presence and toxicity of tetrodotoxin and 5,6,11-trideoxy tetrodotoxin in a trumpet shell collected on the European coasts. The LC-MS method is a useful tool to confirm the presence of tetrodotoxin and further identify tetrodotoxin analogues.

Further information, reviewed in Table 9.5, is available on the toxic effects of fish, crustacea and mammals of a wide variety of other organic compounds.

REFERENCES

1. L. Jong-Hwa, *Bulletin of Fisheries Science Institute*, 1987, 3, 11.
2. A. Moles, M.M. Babcock and S.D. Rice, *Marine Environmental Research*, 1987, 21, 49.
3. V. Axiate and J. George, *Marine Biology*, 1987, 94, 241.
4. T. Stromgren, *Marine Environmental Pollution*, 1987, 21, 239.
5. *Petroleum in the Marine Environment*, National Academy of Sciences, Washington, DC, 1975.

6. GESAMP, *Impact of Oil on the Marine Environment*, Rep. Stud., GESAMP, No. 6, Group of Experts on the Scientific Aspects of Marine Environmental Protection, London, 1977.

7. J.M. Neff, in *Polycyclic Aromatic Hydrocarbons in the Aquatic Environment, Sources, Fates and Biological Effects*, Applied Science Publishers, London, 1979.

8. J.W. Farrington, J. Albaiges, K.A. Burns, B.P. Dunn, P. Eaton, J.L. Laseter, P.L. Parker and S. Wise, in *The International Mussel Watch*, National Academy of Sciences, Washington, DC, 1980, chap. 2.

9. J.W. Howard and T. Fazio, *Journal of the Association of Official Analytical Chemists*, 1980, 63, 1077.

10. D.W. Connel and G.J. Miller, *CRC Critical Reviews in Environmental Control*, 1980, 11, 37, 105.

11. K.R. Davies, T.W. Shultz and J.N. Dumont, *Archives of Environmental Contamination and Toxicology*, 1981, 10, 371.

12. T.M. Dillon, J.M. Neff and J.S. Warner, *Bulletin of Environmental Contamination and Toxicology*, 1978, 22, 320.

13. J.N. Dumont, T.W. Schultz and R.D. Jones, *Bulletin of Environmental Contamination and Toxicology*, 1979, 22, 159.

14. J.M. Giddings, *Bulletin of Environmental Contamination and Toxicology*, 1979, 23, 360.

15. B.R. Parkhurst, A.S. Bradshaw, J.L. Forte and G.P. Wright, *Bulletin of Environmental Contamination and Toxicology*, 1979, 23, 349.

16. G.R. Southworth, C.C. Keffer and J.J. Beauchamp, *Environmental Science and Technology*, 1980, 14, 1529.

17. B.W. Wilson, R.A. Pelrov and J.T. Cresto, *Mutation Research*, 1980, 79, 193.

18. T. Brule, *Journal of Marine Biological Association*, 1987, 67, 227.

19. P.C. Baumann, W.D. Smith and W.K. Parkland, *Transactions of the American Fisheries Society*, 1987, 116, 79.

20. C.D. Knowles, H.J. McKee and D.U. Palawski, *Environmental Toxicology and Chemistry*, 1987, 6, 201.

21. J.M. McKim, P.K. Schmeider, R.W. Carlson, E.P. Hunt and G.J. Niemi, *Environmental Toxicology and Chemistry*, 1987, 6, 295.

22. E. Conti, *Aquatic Toxicology*, 1987, 10, 325.

23. C.O. Herring, J.A. Adams, B.A. Wilson and S. Pollard, *Bulletin of Environmental Contamination and Toxicology*, 1988, 17, 731.

24. D.C. Dill, P.G. Murphy and M.A. Mayes, *Bulletin of Environmental Contamination and Toxicology*, 1987, 39, 869.

25. S. Nota, C. Vernassi, A. Compare and N. Sannolo, *Journal of Chromatography*, 1979, 173, 238.

26. L. Maros, M. Kaldy and S. Inaz, *Analytical Chemistry*, 1989, 61, 733.

27. S.C. Foti, Report of the Atomic Energy Commission, USA AD 734, 1971, 383.

28. G.B. Cleland, P.N. McElroy and R.A. Sonstegard, *Aquatic Toxicology*, 1988, 12, 141.

29. J. Mallatt and H.C. Barron, *Archives of Environmental Contamination and Toxicology*, 1988, 17, 73.

30. G. Tripathi and G.P. Shulka, *Ecotoxicology and Environmental Safety*, 1988, 15, 277.

31. J.M. McKim, P.K. Schmeider, J. Niemi, R.W. Carlson and T.R. Henry, *Environmental Toxicology and Chemistry*, 1987, 6, 313.

32. D.G. Mitchell, P.M. Chapman and T.J. Lang, *Bulletin of Environmental Toxicology*, 1987, 39, 1028.

33. D.J. Call, L.T. Brooke, R.J. Kent, M.L. Knuth, S.H. Poirier, J.M. Huot and A.R. Lima, *Archives of Environmental Contamination and Toxicology*, 1987, 16, 607.

34. M.D. Ferrando, M.M. Almar and E. Andreu, *Journal of Environmental Science and Health*, 1988, B23, 45.
35. M.T. Wan, R.G. Watts and D.J. Moul, *Bulletin of Environmental Contamination and Toxicology*, 1988, 41, 609.
36. J.L. Maas Diepeveen and C.J. Van Leeuwen, *Bulletin of Environmental Contamination and Toxicology*, 1988, 40, 517.
37. D.W. Konouen, *Bulletin of Environmental Contamination and Toxicology*, 1988, 41, 371.
38. D.J. Call, S.H. Poirier, M.L. Knuth, S.L. Harting and C.A. Londberg, *Bulletin of Environmental Contamination and Toxicology*, 1987, 38, 352.
39. J.G. Seelye, L.L. Marking, E.L. King, L.H. Hanson and T.D. Bills, *North American Journal of Fisheries Management*, 1987, 7, 598.
40. C.L. Russon, R.A. Drummond and A.O. Hoffman, *Bulletin of Environmental Contamination and Toxicology*, 1988, 41, 589.
41. R.M. Kocan, H. Van Westernhagen, M.L. Landolt and G. Furstenberg, *Marine Environmental Research*, 1987, 23, 291.
42. D.W. Green, K.A. Williams, D.R.L. Hughes, E.A.R. Shaik and D. Pascoe, *Water Research*, 1988, 22, 225.
43. B.F. Troy and D.E. Hinton, *Marine Environmental Research*, 1988, 24, 259,
44. C.S. Giam, H.S. Chau and T.S. Nelt, *Analytical Chemistry*, 1975, 47, 2225.
45. G.B. Cleland, P.N. McElroy and R.A. Sonstegard, *Aquatic Toxicology*, 1988, 12, 141.
46. A.R. Carlson and P.A. Kosian, *Archives of Environmental Contamination and Toxicology*, 1987, 16, 129.
47. A. Tang, *Water Science and Technology*, 1988, 20, 77.
48. I.H. Rogers and K.J. Hall, *Water Pollution Research Journal of Canada*, 1987, 22, 197.
49. K. Dampert, *Ecotoxicology and Environmental Safety*, 1987, 13, 324.
50. P.W. Webster and J.H. Canton, *Journal of Aquatic Toxicology*, 1988, 12, 323.
51. B.E. Bengisson, *Water Science and Technology*, 1988, 20, 87.
52. P. Reijnders, in *Fundamental Protection of the North Sea*, Research Institute for Nature Management, London, March 24–27, 1987, p. 5.
53. D.G. Mitchell, P.M. Chapman and T.J. Lang, *Environmental Toxicology and Chemistry*, 1987, 6, 875.
54. M.T. Wan, D.J. Moul and R.C. Watts, *Bulletin of Environmental Contamination and Toxicology*, 1987, 39, 721.
55. J. Sanchez, M. Sole and J. Albaiges, *International Journal of Environmental Analytical Chemistry*, 1993, 50, 269.
56. D.G. Hela and V.D. Papadopoulos, *International Journal of Environmental Analytical Chemistry*, 2012, 93, 1676.
    Proceedings of the 13th Symposium on the Chemistry and Fate of Modern Pesticides and 7th European Conference on Pesticides and Related Organic Micropollutants in the Environment.
57. J.L. Roscales, J. Munoz Arnaaz, J. Gonzales-Solis and B, Jiminez, *Environmental Science and Technology*, 2010, 44, 2328,
58. E.J. Morgan and R. Lohmann, *Environmental Science and Technology*, 2010, 44, 5444,
59. S.D. Bridgham, *Archives of Environmental Contamination and Toxicology*, 1988, 17, 731.
60. M. Stephenson and G.L. MacKie, *Aquatic Toxicology*, 1986, 9, 243.
61. J.S. Weis and A. Ma, *Bulletin of Environmental Contamination and Toxicology*, 1987, 39, 224.
62. R.D. Kimborough (ed.), *Hologenated Biphenyls, Terphenyls, Naphthalenes, Dibenzodioxins and Related Products*, Elsevier/North-Holland, New York, 1980.
63. O. Hutzinger, R.W. Frei, E. Merian and F. Pocchiari (eds.), *Chlorinated Dioxins and Related Compounds: Impact on the Environment*, Pergamon Press, New York, 1982.

64. W.J. Nicholson and J.A. Moore (eds.), *Health Effects of Halogenated Aromatic Hydrocarbons*, New York Academy of Sciences, New York, 1979.
65. D.H.K. Lee and H.L. Falk (eds.), *Environmental Health Perspectives*, Experimental Issue No. 5, US Department of Health, Education and Welfare Publication No. (NTH) 74-218, Washington, DC, September 1973.
66. J.R. Huff, J.A. Moore, D.R. Saracci and L. Tomatis, in D.P. Rail (ed.), *Environmental Health Perspectives*, US Department of Health and Human Services Publication No. 80-218, No. 36, Washington, DC, 1980, p. 221.
67. M.P. Espositio, T.O. Tiernan and F.E. Dryden, *Dioxins*, EPA-600/2-80-197, Environmental Protection Agency, Washington, DC, November 1980.
68. K. Olie, P.L. Vermeullen and O. Hutzinger, *Chemosphere*, 1977, 6, 455.
69. B. Ahling, A. Lindskog, B. Jannson and G. Sundstrom, *Chemosphere*, 1977, 6, 455.
70. H.R. Buser, H.P. Bosshardt, C. Rappe and R. Lindahl, *Chemosphere*, 1978, 7, 419.
71. D.G. Crosby and A.S. Wong, *Chemosphere*, 1976, 5, 327.
72. L.L. Lamparski, R.H. Stehl and R.L. Johnson, *Environmental Science and Technology*, 1980, 14, 196.
73. E.E. McConnell, in R.D. Kimbrough (ed.), *Halogenated Biphenyls, Terphenyls, Naphthalenes, Dibenzodioxins and Related Products*, Elsevier/North-Holland, New York, 1980, p. 109.
74. J.A. Golwstein, in R.D. Kimburgh (ed.), *Halogenated Biphenyls, Terphenyls, Naphthalenes, Dibenzodioxins and Related Products*, Elsevier/North-Holland, New York, 1980, p. 151.
75. M.H. Bickel and S. Muhelback, in O. Hutzinger et al. (eds.), *Chlorinated Dioxins and Related Compounds: Impact on the Environment*, Pergamon Press, New York, 1982, p. 3036.
76. A. Di Domenico, G. Viviona and G. Zapponi, in R.D. Kimbrough (ed.), *Halogenated Biphenyls, Terphenyls, Naphthalenes, Dibenzodioxins and Related Products*, Elsevier/North-Holland, New York, 1980, p. 105.
77. C.T. Ward and F. Matsumura, *Archives of Environmental Contamination and Toxicology*, 1978, 7, 349.
78. A.L. Young, in W.J. Nicholson and J.A. Moore (eds.), *Health Effects of Halogenated Aromatic Hydrocarbons*, Academy of Sciences, New York, 1979, p. 173.
79. J.G. Vos, J.H. Keoman, H.L. Van der Maas, M.C. Ten Noever de Brauw and R.H. de Vos, *Food and Cosmetics Toxicology*, 1970, 8, 625.
80. R.A. Smith, P.W. O'Keefe, D.R. Hiker, B.L. Jelus-Tyror and K.M. Aldous, *Chemosphere*, 1982, 11, 715.
81. B. Jansson and G. Sundstrom, in O. Hutzinger et al. (eds.), *Chlorinated Toxins and Related Compounds: Impact on the Environment*, Pergamon Press, New York, 1982, p. 201.
82. C. Rappe, S. Markland, P.A. Bergqvist and M. Hansson, *Chemica Scripta*, 1982, 20, 56.
83. A.I. Mills and M. Alexander, *Journal of Environmental Quality*, 1976, 5, 437.
84. D. Firestone, J. Rees, N.L. Brown, R.P. Barron and J.N.J. Damico, *Journal of the Association of Analytical Chemists*, 1972, 55, 85.
85. C. Rappe, S. Markland, H.R. Buser and H.P. Bosshardt, *Chemosphere*, 1978, 3, 269.
86. C. Rappe, H.R. Buser, H.P. Bosshardt, W.J. Nicholson and J.A. Moore (eds.), *Health Effects of Halogenated Aromatic Hydrocarbons*, Academy of Sciences, New York, 1979, p. 1.
87. A.R. Isensee, *Biological Bulletin*, 1978, 27, 255.
88. D.M. Decad, L.S. Biraum, H.,B. Matthews and G. Hatzinger (eds.), *Chlorinated Dioxins and Related Compounds: Impact on the Environment*, Pergamon Press, New York, 1982, p. 307.
89. J.W. Farrington, J. Albaiges, K.A. Burns, B.P. Dunn, P. Eaton, J.L. Laseter, P.L. Parker and S. Wise, *The International Mussel Watch*, National Academy of Science, Washington, DC, 1980, chap. 2.

90. J.W. Howard and T. Fazio, *Journal of the Association of Official Analytical Chemists*, 1980, 63, 1077.
91. A. Covaci, S. Hosada, L. Ravens, W. Vetter, F. Santos, H. Neels, A. Storelli and M.M. Stonelli, *Environmental Science and Technology*, 2008, 42, 8654.
92. P. Haglund, K. Loftsrand, A. Malmvarm, A. Bignert and L. Asplund, *Environmental Science and Technology*, 2010, 44, 2466.
93. G. Holt, A. Froslie and G. Northeim, *Acta Veterinaria Scandinavica Supplementum*, 1979, 70, 28.
94. E.M. Brevik, J.E. Bjerk and N.J. Kveseth, *Bulletin of Environmental Contamination and Toxicology*, 1978, 20, 715.
95. N.J. Kveseth, J.E. Bjerk, N. Fimreite and J. Stenerson, *Archives of Environmental Contamination and Toxicology*, 1979, 8, 201.
96. G. Lunde and E. Bauman Ofstad, *Fresenius Zeitschrift für Analytische Chemie*, 1976, 282, 395.
97. E. Bauman Ofstad, G. Lunde, K. Marinsen and B. Rygg, *Science of the Total Environment*, 1978, 10, 219.
98. G. Norheim, *Acta Pharmacol*, 1978, 42, 7.
99. J.H. Koeman, M.C. Noever de Brauw and R.H. de Vos, *Nature (London)*, 1969, 221, 1126.
100. M.C. Ten Noever de Brauw and J.H. Koeman, *Science of the Total Environment*, 1973, 14, 27.
101. R.E. Tucker, A.L. Young and A.P. Gray (eds.), *Human and Environmental Risks of Chlorinated Dioxins and Related Compounds*, Plenum Press, New York, 1983.
102. J.W. Gooch and F. Matsumara, *Archives of Environmental Contamination and Toxicology*, 1987, 16, 349.
103. J. Albaiges, A. Farran, M. Soler and A. Gallifer, *Marine Environmental Research*, 1987, 22, 1.
104. A. Devaux and G. Monod, *Environmental Monitoring and Assessment*, 1987, 9, 105.
105. M. Cossarini-Dunier, G. Monod, A. Demael and D. Lepot, *Ecotoxicology and Environmental Safety*, 1987, 13, 339.
106. K. Rajeswari, V. Kalarani, D.C. Reddy and R. Ramamyrthi, *Bulletin of Environmental Contamination and Toxicology*, 1988, 40, 212.
107. P. Vijakumari, D.C. Reddy and R. Ramamuratti, *Bulletin of Environmental Contamination and Toxicology*, 1987, 38, 742.
108. N. Quintel, J. Castro, A. Fernandez, I.M. Zamara-Ley, G.M. Rand and G.R. Gardini, *Journal of Agriculture and Food Chemistry*, 2013, 61, 11881.
109. M. Kawano, T. Ingue, T. Wada, H. Hidake and R. Tasinkava, *Environmental Science and Technology*, 1988, 22, 792.
110. H. Karlsson, M. Oehme, S. Skopp and I.E. Burkiew, *Environmental Science and Technology*, 2000, 34, 2126.
111. T.C. Svendsen, K. Vorkamp, M. Frederiksen, B. Rensholdt and J. Frier, *Environmental Science and Technology*, 2007, 41, 5980.
112. N. Yamashita, Y. Urushigawa, S. Masunga, M.I. Walash and A. Miyazaki, *International Journal of Environmental Analytical Chemistry*, 2000, 77, 289.
113. J. Hallou, G. Mercer and I.L. Fancey, *International Journal of Environmental Analytical Chemistry*, 1995, 61, 275.
114. B. Erkman and D. Kolankaya, *International Journal of Environmental Analytical Chemistry*, 2006, 86, 161.
115. D.K. Singh and R.A. Agarwal, *Science of the Total Environment*, 1987, 67, 263.
116. B.S. Khangarot and P.K. Ray, *Archive für Hydrobiologie*, 1988, 113, 465.
117. D.J. Call, L.T. Brooke, R.J. Kent, M.L. Knuth, S.H. Poirier, J.M. Huot and A.R. Lima, *Contamination and Toxicology*, 1987, 16, 607.

118. A. Upadhyaya and C.S. Shuka, *Environmental Research*, 1986, 41, 591.
119. P.K. Saxena and K. Mani, *Environmental Pollution*, 1988, 55, 97.
120. K.G. Doe, W.R. Ernst, W.R. Parken, G.R.J. Julien and P.A. Hennigar, *Canadian Journal of Fisheries and Aquatic Science*, 1988, 45, 287.
121. A.B. Sondykl, *International Journal of Environmental Analytical Chemistry*, 2010, 90, 421.
122. C.J. Musial and J.F. Uthe, *International Journal of Environmental Analytical Chemistry*, 1983, 14, 117.
123. J.G. Seelye, L.L. Marking, E.L. King, L.H. Hanson and T.D. Bills, *North American Journal of Fisheries Management*, 1987, 7, 598.
124. C.F. Chow, H.K. Kiong, S.W. Leung, B.K.W. Chiu, C.K. Koo, E.N.Y. Lei, M.H.W. Lam, W.T. Wong and W.Y. Wang, *Analytical Chemistry*, 2011, 83, 289.
125. P. Suchy and J. Berry, *International Journal of Environmental Analytical Chemistry*, 2012, 92, 1443.
126. J.D. Wood, R.B. Franklin, G. Garman, S. McInisch, A.J. Porter and P.A. Bukaveckas, *Environmental Science and Technology*, 2014, 24, 21.
127. P. Otero, S. Perez, A. Alfonso, C. Vale, N.N. Gonvela, I.N. Gonvela, J. Regado, P. Vale, M. Hirama, Y. Ishihara, J. Molog and L.M. Botana, *Analytical Chemistry*, 2010, 82, 6032.
128. R.N. Ram and A.G. Sathyaneson, *Ecotoxicology and Environmental Safety*, 1987, 13, 185.
129. A.H. Trim, *Bulletin of Environmental Contamination and Toxicology*, 1987, 38, 681.
130. A. Dela Torre, M.B. Alonso, E.J. Reiner, J. Hailson-Brito, J.M.S. Torres, C. Beretozzi, I. Marigo, L. Baribosa, M. Cremer, E. Secchi, O. Maim, E. Eljerrat and D. Barcelo, *Environmental Science and Technology*, 2012, 46, 12364.
131. J.C. Meane, C.B. Daniels and G.M. Baski, *Marine Environmental Research*, 1988, 24, 327.
132. D.R. Oros, D. Hoover, F. Rodigari, D. Crane and J. Sericano, *Environmental Science and Technology*, 2005, 39, 33.
133. A. Shahabi, M.P. Zakaria, C.K. Yap, S. Surif, A.R. Bakhtinuri, K. Chandru, P. Shalpourly, B. Sakairi, A. Mathion, D. Rilura, J.F. Narlonnro and M. Lafurie, *International Journal of Environmental Analytical Chemistry*, 2010, 90, 14.
134. R. Garrigues, C. Raeux, P. Lemaire, A. Mthion, D. Ribera, J.F. Barbonnro and M. Lafaurie, *International Journal of Environmental Analytical Chemistry*, 1990, 38, 379.
135. M.A. Quilliam, P.G. Sim, A.W. McCulloch and A.G. McInnes, *International Journal of Environmental Analytical Chemistry*, 1989, 36, 139.
136. F. Berthou and V. Vignier, *International Journal of Environmental Analytical Chemistry*, 1986, 27, 81.
137. A. Tognazzi, A.M. Dattilo, L. Bracchini, C. Rossi and D. Vione, *International Journal of Environmental Analytical Chemistry*, 2012, 92, 1679.
138. P. Rodriguez, A. Alfonso, C. Vale, C. Alfonso, P. Vale, A. Telles and L.M. Botana, *Analytical Chemistry*, 2008, 80, 5622.
139. R.H. Carmichael, A.L. Jones, A.K. Patterson, W.C. Walton, A. Perez-Hueta, E.B. Overton, M. Dailey and K.L. Willett, *Environmental Science and Technology*, 2012, 46, 12787.
140. D. Kolankaya, *International Journal of Environmental Analytical Chemistry*, 2006, 86, 147.
141. D.E. Wells and I. Echarri, *International Journal of Environmental Analytical Chemistry*, 1992, 47, 75.
142. K.C. Pangalio and C.M. Reddy, *Environmental Science and Technology*, 2010, 44, 5741.
143. A. Rooz, N. Berger, U. Jamberg, J. Van Dijk and A. Bignert, *Environmental Science and Technology*, 2013, 47, 11157.

144. E. Abad, F. Perez, J.J. Lierena, J. Coiyrch and J. Rivera, *Environmental Science and Technology*, 2003, 37, 5090.
145. C.E. Mackintosh, J. Maidanado, J. Hoinywa, N. Hoover, A. Cheng, M.G. Ikonomou and F.A.P.C. Gobas, *Environmental Science and Technology*, 2004, 38, 2011.
146. M. Houde, A.O. De Silva, D.C.G. Muir and R.J. Letcher, *Environmental Science and Technology*, 2011, 45, 7962.
147. G.T. Tomy, W. Budakowski, T. Halidorson, P.H. Heln, G.A. Stem, K. Friesen, K. Pepper, S.H. Titilemier and A.T. Fisk, *Environmental Science and Technology*, 2004, 38, 6475.
148. A.S. Felsot and J.R. Ruppert, *Journal of Agriculture and Food Chemistry*, 2002, 50, 4417.
149. J.A. Armitage, U. Schenker, M. Scheringer, J.W. Martin, M. McLeod and I.T. Cousins, *Environmental Science and Technology*, 2009, 43, 9274.
150. J. Hu, F. Jin, Y. Waa, M. Young, L. An and S. Tao, *Environmental Science and Technology*, 2005, 39, 4801.
151. J.J. Alava, P.S. Ross, C. Lachmuth, J.K.B. Ford, B.E. Hickie and F.A.P.S. Gobas, *Environmental Science and Technology*, 2012, 46, 12655.
152. T.M. Brown, T.A. Sheldon, N.M. Burgess and K.J. Reimer, *Environmental Science and Technology*, 2009, 43, 7635.
153. S. Rayne, M.G. Ikonomau, P.S. Ross, G.M. Ellis and L.G. Barratt-Hennard, *Environmental Science and Technology*, 2004, 38, 4293.
154. K. Vorkamp, F.F. Riget, R. Bossi and R. Dietz, *Environmental Science and Technology*, 2011, 45, 1243.
155. M.J. La Guardia, R.C. Hale, E. Harvess, T.M. Meiner and S. Ciparis, *Environmental Science and Technology*, 2012, 46, 5798.
156. K. Vorkamp, K. Bester and F.F. Riget, *Environmental Science and Technology*, 2012, 46, 10549.
157. Y. Wan, P.D. Jones, S. Wiseman, H. Chang, D. Chamey, K. Kannan, K. Zhany, J.Y. Hu, J.S. Chim, S. Tanebe, M.H. Lam and J.P. Gipsey, *Environmental Science and Technology*, 2010, 44, 6068.
158. B.C. Kelly, M.G. Ikonomon, J.D. Blair and F.A.P.C. Gobas, *Environmental Science and Technology*, 2008, 42, 7009.
159. K. Janati, A. Covaci, S. Voorspoels and G. Becher, *Environmental Science and Technology*, 2005, 39, 1987.
160. D.R. Oros, D. Hoover, T. Rodigari, D. Crane and J. Sericano, *Environmental Science and Technology*, 2005, 39, 33.
161. C. Porto, X. Biobca, D. Pastor, M. Sole and J. Albaiges, *Environmental Science and Technology*, 2000, 34, 5067.
162. P.D. Smith, D.L. Brockway and F.E. Sancil, *Environmental Toxicology and Chemistry*, 1987, 6, 891.
163. A. Dikawa and J. Kukkonen, *Ecotoxical and Environmental Safety*, 1988, 15, 282.

10 Toxic Effects of Organometallic Compounds in Fish and Invertebrates

Many polluting organometallic compounds have been found in fish and crustacea. These include tetraalkyl lead compounds (up to $18.9\,mg\,kg^{-1}$ in carp), methylmercury compounds (up to $11.5\,mg\,kg^{-1}$ in tuna) and alkyl arsenic compounds (up to $23\,mg\,kg^{-1}$ in molluscs). Many of these substances are highly toxic, and their effects on sea creatures and the humans who eat them will be discussed in later chapters.

Details of the occurrence of organolead and organomercury compounds in fish [1] are given in Tables 10.1 and 10.2. These originate predominantly from the use of alkyl lead compounds in petroleum and the methylation of inorganic mercury released into the ecosystem as effluents in the chloroalkali process.

Concentrations of organometallic compounds and the corresponding total ionic metal found in fish are compared in Table 10.3. In the cases of lead and mercury, certainly, the organometallic content of the fish is an appreciable proportion of the total mercury content.

The organs of fish, such as the opercle, kidney and liver, have the property of concentrating metals and organometallic compounds such that the concentration in the organ is appreciably higher than that in the whole fish tissue. This is illustrated in Table 10.4, which shows that the concentrations of metals in the organ can be up to 30 times higher than in the fish tissue.

Richman et al. [2] have discussed the factors that might govern the uptake of mercury by fish in acid-stressed lakes. It was concluded that mercury cycling and uptake in aquatic systems were governed by a variety of interconnecting and sometimes covarying factors, the selective importance of which could differ from lake to lake.

10.1 ORGANOARSENIC COMPOUNDS

Juvenile rainbow trout (*Salmo gairdneiri*) fed for 8 weeks on a diet containing arsenic trioxide ($180-1477\,\mu g$ As g^{-1} diet), disodium arsenate ($137-1054\,\mu g$ As g^{-1} diet), dimethyl arsenic acid ($163-1487\,\mu g$ As g^{-1} diet) or arsanilic acid ($103-1503\,\mu g$ As g^{-1} diet) all experienced adverse effects on growth, food consumption and feeding behaviour when fed with inorganic arsenic compounds, but were unaffected by diets containing the organoarsenic compounds. In all cases, carcass arsenic concentrations of arsenic were related to dietary arsenic concentrations [4].

TABLE 10.1

Concentrations (μg kg^{-1}) of Organolead Compounds Found in Environmental Fish

Compound	Carp	Bass	Small Mouth Bass	Pike Sucker	White	Range, All Types
Me$_4$Pb	140	<1	—	—	—	140
Me$_3$EtPb		—	—	—	—	<1
Me$_2$Et$_2$Pb	1430	—	57	100	—	57–1430
MeEt$_3$Pb	140	—	190–250	150–170	290	140–290
Et$_2$Pb	780–7500	—	1200	1020–1120	2950–4380	780–7500
Me$_3$Pb	0	—	<10	200–210	90–200	0–210
Me$_2$Pb^{2-}	360	—	—	—	—	360
Et$_2$Pb$^+$	90–1210	—	220	53	2170–3430	54–3430
Et$_2$Pb^{2+}	710–1310	—	90–275	—	220–4300	90–4300
Pb^{2+}	1280–4130	—	250	1040–1190	3610–3480	250–4130
Total Pb^{2+}	5090–18940	<1	1760	1520–1650	7700–12600	1630–16650

It has been shown that arsenic is incorporated into both marine and freshwater organisms in the form of both water-soluble and lipid-soluble arsenic compounds [5]. Studies to identify the chemical forms of these arsenic compounds have shown the presence of arsenite (As(III)), arsenate (As(V)), methyl arsenic acid, dimethyl arsenic acid and arsenobetaine [6].

Inorganic arsenic has toxic properties similar to those of lead and mercury as regards its ability to bond to sulphur and inhibit enzyme action such as pyruvate dehydrogenase. It tends to concentrate in the liver, kidneys and lungs of animals exposed to it. Exposure of fish to 4000 μg L^{-1} inorganic arsenic for 39 days causes adverse effects, such as a reduced rate of growth. Marine crustacea are the most sensitive to arsenic and annelids, and insect larvae the least sensitive. Marine organisms contain widely varying concentrations of arsenic in different chemical forms, both organic and inorganic. A knowledge of these chemical forms and their bioavailability is important. Crab is susceptible to arsenic at its larval life stage. In the case of crustacea, toxicity decreases in the order arsenate, organic and arsenite. Fish are less susceptible to arsenic than invertebrates.

For 8 weeks, Cockell and Hilton [4] fed juvenile rainbow trout semi-purified diets containing arsenic trioxide (180–1477 μg g^{-1} arsenic diet), disodium arsenate (137–1053 μg g^{-1} arsenic), dimethyl arsenic acid (193–1503 μg g^{-1} arsenic) or arsanilic acid (193–1503 μg g^{-1} arsenic). Growth, food consumption and feeding behaviour were adversely affected by all dietary concentrations of inorganic arsenicals, but were unaffected by diets containing organic arsenicals. In all treated trout, carcass arsenic concentrations were related to dietary arsenic concentration and dietary arsenic exposure rate (milligrams of arsenic per kilogram of body weight per day). Carcass arsenic concentrations were highest in trout fed inorganic arsenicals, with disodium arsenate and arsenic trioxide yielding the highest carcass arsenic concentrations at low and high exposure levels, respectively.

TABLE 10.2

Concentrations (µg L^{-1}) of Organomercury Compounds Found in Environmental Fish

Compound	Whiting	Sardine	Turbot	Halibut	Coho Salmon	Salmon	Red Tuna	White Tuna	Pike	Trout
MeHg$^+$								930		
EtHg$^+$								<10		
PhHg$^+$								<10		
CH$_3$HgCl	80	30	<10	5650	180–200	60–110	<10–11, 500	10	110–880	60–1460

Compound	Rainbow Trout	Whale	Shark	Swordfish	Octopus	Squid	Rockfish
CH$_3$HgCl	50–1970	560–1090	<10 5410	570–1010 400–3170	<10	<10	100–1000

TABLE 10.3

Concentrations of Organometallic Compounds and Corresponding Inorganic Metal Ions Found in Fish ($\mu g\,k^{-1}$)

Element	Organometal Minimum	Organometal Maximum	Inorganic Metal Minimum	Inorganic Metal Maximum
Lead	1500 (m)	1600 (a)	120 (i)	1360 (i)
	1 (b)	5500 (b)		
	5100 (m)	18,900 (m)		
	7700 (c)	12,600 (c)		
Mercury	100 (d)	900 (d)	90 (j)	2400 (m)
	600 (e)	1500 (e)		
	50 (f)	2000 (f)		
	600 (g)	1100 (g)		
		8400 (h)		
Arsenic	—	—	1100 (k)	2900 (k)
Copper	—	—	390 (l)	3500 (l)
			500 (o)	220 (o)
Nickel	—	—	150 (i)	200 (i)
Selenium	—	—	190 (n)	550 (n)
Manganese	—	—	220 (i)	1630 (i)

Note: (a) Pike, (b) bass, (c) white sucker, (d) whiting, (e) sardine, (f) turbot, (g) halibut, (h) salmon, (i) mullet, (j) chub, (k) herring, (l) flathead, (m) carp, (n) shark, and (o) rainbow trout.

TABLE 10.4

Concentrations of Metallic Ions in Fish Tissue and Organs

Element	Fish Tissue Analysis ($\mu g\,kg^{-1}$) Minimum	Fish Tissue Analysis ($\mu g\,kg^{-1}$) Maximum	Fish Organ Analysis ($\mu g\,kg^{-1}$) Minimum	Fish Organ Analysis ($\mu g\,kg^{-1}$) Maximum	Concentration Factors, Maximum Value in Organ ($\mu g\,kg^{-1}$)
Lead	120 (a)	1360 (b)	120 (A)	3600 (B)	26.4
Mercury	90 (c)	2400 (d)	—	—	—
Arsenic	1100 (c)	2900 (f)	600 (D)	4800 (I, C)	13.7
		530 (i)	2180 (j)	62,000 (m, C)	28.4
Nickel	150 (i)	200 (i)	340 (n, B)	1900 (n, B)	9.4
Selenium	190 (k)	550 (k)	—	—	—
Manganese	220 (a)	1630 (j)	—	—	—

Note: (a) striped mullet, (b) grey mullet, (c) chub, crappie, (d) carp, (e) herring, (f) tuna, (g) flathead, (h) crayfish, (i) rainbow trout, (j) sardine, (k) shark, (l) perch, (m) white fish, (n) trout, (A) muscle, (B) kidney, (C) liver, (D) gill.

TABLE 10.5

Occurrence of Organometallic Compounds and Corresponding Inorganic Metal Ions in Crustacea

Element	Organometallic Compounds ($\mu g\,kg^{-1}$)		Inorganic Compounds ($\mu g\,kg^{-1}$)	
	Minimum	Maximum	Minimum	Maximum
Lead	<10	50	480 (A)	2800 (D)
Mercury	—	—	20 (A)	310 (E)
Arsenic	1400 (G)	23 00 (H)	750 (A)	2650 (A)
Tin	30 (F)	100 (F)	—	—
Selenium	20 (C)	30 (C)	710 (C)	6700 (E)

Note: (A) muscle, (B) king prawn, (C) scallop, (D) crab, (F) whelk, (G) prawn, (H) mollusc.

TABLE 10.6

Comparison of Ranges of Concentrations of Organolead Compounds Found in Crustacea and Fish

Compound	In Crustacea ($\mu g\,kg^{-1}$)	In Fish ($\mu g\,kg^{-1}$)
R_4Pb	<20	1117–9360[a]
R_3Pb^+	<10–50	63–2640[a]
R_2PB^{2+}	<10	450–4660[a]
Pb^{2+}	1100–1800	250–4300

[a] Ethyl plus methyl compounds.

The UK Arsenic in Food Regulations (1959) state that foodstuffs must not contain more than $1000\,\mu g\,kg^{-1}$ of total arsenic. Fish and edible seaweeds are exceptions, as their arsenic contents are normally in excess of $1000\,\mu g\,kg^{-1}$. The UK total diet survey suggests that at least 75% of the total amount of arsenic ingested by humans originates from fish and shellfish in this diet. It is accepted that the arsenic in these foods is mainly organically bound, that is in its least toxic form. If the levels of total inorganic arsenic approaches $1000\,\mu g\,kg^{-1}$, then a knowledge of the proportion of the least toxic arsenite to the most toxic arsenate is important.

Available information on the concentrations of organometallic compounds and the corresponding inorganic metal ions in crustacea is given in Table 10.5. Again, as in the case of fish, determined values vary over a wide range and certainly include in this range concentrations at which adverse effects or mortalities could occur and when the suitability of the creature as an item of human diet would be queried. In particular, the concentrations of organolead compound and inorganic lead found in crustacea are lower than those found in fish (Table 10.6).

Data have been presented on the concentrations of ionic alkyl lead compounds in salt marsh periwinkles (*Littorina irrorata*) collected in Maryland and Virginia [3]. Male periwinkles accumulated higher concentrations of several alkyl lead species than females.

Tributyltin concentrations have been determined in dogwhelk samples taken at various locations in Sullom Voe, Shetland. These ranged from values of $0.1 \, mg \, kg^{-1}$ inside Sullom Voe down to less than $0.03 \, mg \, kg^{-1}$ in Yell Sound. Concentrations of $0.02–0.03 \, mg \, kg^{-1}$ were found in edible tissue of queen scallops inside the Voe, but tin was rarely detected in commercial shellfish outside the Voe. Only very low concentrations of tributyltin ($\sim 2 \, ng \, L^{-1}$) were found in a small proportion of seawater samples taken in the area.

Large amounts of arsenic enter the environment each year due to the use of arsenic compounds in agriculture and industry as pesticides, herbicides and wood and crop preservatives. The main form in which it is introduced is as inorganic arsenic (arsenite and arsenate). Arsenic is easily transformed between its inorganic and organic forms by biological and chemical action. As the toxicity and biological activity of the different species vary considerably, information on the chemical form, that is speciation, is of great importance in environmental studies. Thus, organoarsenic in the trivalent state is very toxic to most organisms, but in the pentavalent state it may be an essential nutrient for crustacea and seems to be innoxious to mammals.

The biological methylation of inorganic arsenic by microorganisms such as moulds and bacteria present in sediments, sludges and muds has been established, although there is no unequivocal evidence of the proposed pathways [5,7–12].

Organoarsenic species are known to vary considerably in their toxicity to humans and animals, as discussed above [13]. Large fluxes of inorganic arsenic into aquatic environment can be traced to geothermal systems [14], base metal smelter emissions and localised arsenite treatments for aquatic weed control. The methylated arsenicals have entered the environment either directly as pesticides or by biological transformation of the inorganic species [6].

Organoarsenical pesticides such as sodium methyl arsenate and dimethyl arsenic acid are used in agriculture as herbicides and fungicides. It is possible that these arsenicals enter soil, plant and consequently humans. On the other hand, arsenic is a ubiquitous element on the earth, and the presence of inorganic arsenic and several methylated forms of arsenic as monomethyl, dimethyl and trimethyl arsenic compounds in the environment has been well documented [15]. The occurrence of biomethylation of arsenic in microorganisms [16] and soil has also been demonstrated. Therefore, further investigation of the fate of arsenicals in the physical environment and living organisms requires analytical methods for the complete speciation of these arsenicals.

Mareda-Piniero et al. [17] applied a matrix solid phase to extract arsenical species (arsenite, As(III); arsenate, As(V); monomethyl arsenic acid; dimethyl arsenic acid [DMA]; arsenobetaine, AsB; and arsenocholine) from seafood. High-performance liquid chromatography coupled to inductively coupled plasma–mass spectrometry was used to separate and detect all arsenic species. Variables affecting matrix solid-phase dispersion, such as the solid support material (dispersing agent), solid support mass/sample mass ratio, elution solvent composition and elution solvent volume, were fully evaluated. Quantitative recoveries for inorganic and organic arsenic species were obtained when using diatomaceous earth or octadecyl-functionalised silica gel (C18) as a solid support material, with a solid support mass/sample mass ratio of 7.0. Elution of arsenical compounds was assessed using 10 ml of 50/50 methanol/ ultrapure water as an elution solvent. The matrix solid-phase dispersion method has been found precise, with relative standard deviations (RSDs) of ~9% for As(III), dimethyl arsenic acid and AS(V), and 3% for arsenobetaine. The developed procedure was tested by analysing different Certified Reference Materials of marine origin, such as DORM-2 and BCR-627, which offer certified contents for some arsenic species. The method has been also applied to assess arsenic speciation in different molluscs, cold water fish and white fish.

10.2   ORGANOLEAD COMPOUNDS

The use of tetraethyl lead as an antiknock additive octane enhancer for automotive gasolines has been reduced in many countries. Another source of organic lead compounds in the environment is the biological methylation of inorganic lead to tetraalkyl lead compounds in lake or oceanic sediments [18–21].

$$R_4 \rightarrow R_3\,Pb \rightarrow R_2\,Pb^2 \rightarrow Pb^2$$

The formation of trialkyl lead salts is probably associated with proteins, and arsenic in tissues from rapid metabolic alkylation of tetraethyl lead is of toxicological importance in evaluations of exposure to tetraalkyl leads. The possibility of biomethylation of inorganic lead or organolead species by microorganisms [19,20], by reversing the decomposition mechanism given above, may add to environmental lead problems, although this area is disputed [22]. The highly polar dialkyl and trialkyl compounds in particular have a high toxicity to mammals and are formed as a result of degradation of tetraalkyl lead in aqueous media [26], as shown above [23]. Tetramethyls and tetraethyl lead compounds are coincidentally more toxic than inorganic lead (1000 times) or di- or trimethyl or di- or trialkyl lead compounds [24]. In general, organolead compounds are more toxic than inorganic lead compounds [25].

The highly polar dialkyl and trimethyl lead compounds in particular have a high toxicity to mammals and are formed as a result of the degradation of tetraalkyl lead in aqueous media.

Wong et al. [21,27] and Reisinger and Stoeppler [22] have demonstrated that microorganisms in lake sediment can transform inorganic and organic lead compounds into volatile tetraalkyl lead.

Organolead compounds have been detected in cod, lobster, mackerel and flounder meal (10%–90% of the total lead burden) and in freshwater fish.

Fairly high concentrations of tetraethyl lead (30 ppm) have been detected in mussels collected at a buoy near the S.S. *Cavtat* incident, where a shipload of tetraethyl lead was sunk in the Adriatic Sea. High organolead concentrations, mainly

TABLE 10.7

Concentrations ($mg\,kg^{-1}$ dry weight) of Organolead Compounds Found in Environmental Fish

Compound	Miscellaneous Fish	Carp	Bass	Small Moult Bass	Pike	White Sucker	Range
Me_4Pb	0.43	0.14					0.14–0.43
Me_3EtPb		<0.001	<0.001				<1.001
Me_2Et_2Pb		1.43		0.057	0.10		0.057–1.43
$MeEt_3Pb$		0.14		0.19–0.25	0.15–0.17	0.29	0.14–0.29
Et_4Pb		0.78–7.5		1.20–1.83	1.02–1.12	2.95–4.38	0.78–7.5
Me_3Pb^+		0.16–2.73		<0.01	0.20–0.21	0.09–0.20	<0.01–2.3
Me_2Pb^{2+}		0.36					0.36
Et_3Pb^+		0.09–1.21		0.22–0.86	0.053	2.17–2.43	0.53–3.43
Et_2Pb^{2+}		0.71–1.31		0.09–2.75		2.2–4.3	0.09–4.3
Pb^{2+}		1.28–4.13		0.25–0.30	1.04–1.19	3.6–3.48	0.25–4.13
Total Pb^{2+}		5.09–18.94	<0.01	1.76–5.55	1.52–1.65	7.70–12.60	

of tetraethyl lead, were also found in mussels in other parts of the Italian seas. The presence of tetraethyl lead in aquatic organisms may indicate that the alkyl lead compounds are not immediately metabolised by living organisms and may remain in the authentic forms in the living tissues for a long time. The occurrence of tetraalkyl lead compounds in aquatic biota is highly significant because of the possibility of their incorporation into the food chain.

The concentrations of organolead compounds found in various fish are listed in Table 10.7.

A renewed interest in the speciation of lead in environmental samples has resulted from several diverse lines of investigation. Organolead compounds have been detected in cod, lobster, mackerel and flounder meal (up to 10%–90% of the total lead burden) [28], and in freshwater fish [29,30].

Between 5 and $18.9\,mg\,L^{-1}$ total mercury was found in carp, 1.7–$5.5\,mg\,kg^{-1}$ in small moult bass, 1.2 to $1.65\,mg\,kg^{-1}$ in white sucker and less than 0.01–$0.05\,mg\,kg^{-1}$ in crustacea.

The concentrations of organolead compounds occurring in coastal and, presumably, seawater are considerably below the 0.1–$1\,\mu g\,L^{-1}$ range, and consequently, no adverse effects due to organolead are to be expected.

10.3   ORGANOMERCURY COMPOUNDS

Marine fish embryos and also marine crustacea and crab undergo adverse effects (damage and poor hatching) when exposed to 30–$70\,\mu g\,L^{-1}$ inorganic mercury in seawater for 4–30 days. Similar effects occur in the case of marine molluscs.

Severe adverse effects (weight reduction and poor spawning) occur when non-marine fish are exposed to as little as $1\,\mu g\,L^{-1}$ methylmercury in water for 30 days; $3\,\mu g\,L^{-1}$ mercury, as methylmercury chloride, in seawater or rivers caused 4%

mortality in fish. The toxic effects of organic and inorganic mercury salts in fish are similar in river water and seawater.

Ram and Sathyanesan [31] have studied the histopathological and biochemical changes in the liver of the teleost fish *Channa punctatus* induced by a mercurial fungicide.

Twenty individuals of *Channa punctatus* from two age groups (adults and young as 3 months old) were divided into two batches of 10 specimens. Batch 1 of both age groups was subjected to a toxicologically safe dose ($0.10 mg^{-1}$ of the commercial formulation of organomercury fungicide, Emisan [methoxyethylmercuric chloride]). Batch 2 of each age group was kept as an untreated control. After a 3-month exposure, the fish were killed and the livers examined for histochemicalaride. In both treated groups, liver histology showed various abnormalities, including hyperplasia, nuclear pycnosis, fatty necrosis and degeneration of hepactyes, leading to a tumour and syncytium formation, which are indicative of carcinogenesis. Young fish also showed blood vessel congestion and oedema. Carcinogenesis biochemical changes (reduction in hepatosomatic index, total protein and lipid, biochemicals in cholesterol and acid and alkaline phosphate) occurred in both these groups, but were more pronounced in young than in adult fish. Emisan, in small concentrations, was capable of inducing severe physiometabolic changes of these, leading to death in *Channa punctatus*.

Kirubagarun and Joy [32] examined the toxic effects of three organomercury compounds on the survival and histology of the kidneys of the catfish *Clarias batrachus*. In acute toxicity tests with *Clarias batrachus* (24, 48, 72 and 96 h LC_{50} values calculated according to Spearman–Karber with trimming at 0% and 10%), methylmercuric chloride, mercuric chloride and Emisan 6 (methoxyethylmercury chloride) had mean 96 h LC_{50} values (0% trimming) of 430, 507 and 432 $\mu g L^{-1}$, respectively. In subsequent experiments with *Clarias batrachus*, concentrations of methylmercuric chloride, mercuric chloride and Emisan 6, respectively, were 200, 125 and 1500 $\mu g L^{-1}$ (14- and 28-day exposures) or 40, 50 and 500 $\mu g L^{-1}$ (90- and 180-day exposures). With the exception of the 180-day treatment groups, kidney histology of mercury-exposed fish revealed increases in the proximal tubule diameter and secretory material.

Czuba et al. [33] exposed cell suspension cultures of *Daucus carota* to methylmercury (0–6 μg mL) for 1, 3 or 24 h. Microtubule arrays were unaffected by 1000 $\mu g L^{-1}$ (all exposures) and severely disrupted by 6000 $\mu g L^{-1}$ (all exposures). The degree of disruption caused by other treatments was dependent on exposure duration and concentration. Experimental evidence indicated that microtubule disruption might be a secondary effect resulting from methylmercury-induced perturbations in protein and carbohydrate metabolism and disruptions in redox energy-related processes.

The concentration of organomercury present in sea and coastal water (up to 0.06 $\mu g L^{-1}$ [34]) probably presents little risk to the health of most types of water creatures, although this cannot be taken as a general rule. Thus, 0.2 $\mu g L^{-1}$ organomercury causes severe mortalities in crayfish during a 3-day exposure. The higher levels in the range 0.006–1.15 $\mu g L^{-1}$ of organomercury in rivers [34–36] will certainly be harmful to fish and many types of water creatures. These higher levels of organomercury, say above 0.05 $\mu g L^{-1}$, are more likely to occur in localised areas, that is, near highly contaminated industrial outfalls to rivers, estuaries and coastal waters, for example, Minimata Bay, Japan.

Regarding concentrations of organomercury in water creature tissues, these have been observed to range from <10 µg kg^{-1} (white tuna, shark, octopus, squid and turbot) to as high as 8300 µg kg^{-1} (red tuna) [37].

The concentration of organomercury that is harmful to water creatures is not necessarily the same as the concentration that is harmful to humans who eat the fish; that is, a fish that might be quite healthy for water creatures, although it has traces of organomercury in its tissues, might be harmful to humans when ingested, or vice versa.

A guide to the effects of organomercury compounds on man is provided by the World Health Organization, the maximum permitted limit of organomercury in fish that may be ingested by humans. This regulation states that the mean daily consumption by humans of organomercury should not exceed 29 µg. Thus, if a man consumes ¼ lb, that is 112 g of fish, daily, then to be within the maximum limit, that is 29 µg organomercury per 112 g fish, the fish should contain no more than 260 µg kg^{-1} organomercury.

Organic compounds of mercury have been found in fish. These originate predominantly from the methylation of inorganic mercury released into the ecosystem as effluents in the chloroalkali process.

Richman et al. [2] have discussed the factors that might govern the uptake of mercury by fish in acid-stressed lakes. It was concluded that mercury cycling and uptake in aquatic systems were governed by a variety of interconnecting and sometimes covarying factors, the relative importance of which could differ from lake to lake.

Concentrations of methylmercury, ethylmercury, phenylmercury and methylmercury chloride found in white tuna were, respectively, 0.93, <0.01, <0.01 and 0.54 mg kg^{-1}. Levels of methylmercury chloride found in pike trout, rainbow trout, walleye, swordfish, octopus, squid, rockfish, whiting, sardine, turbot, halibut, coho salmon and red tuna were usually in the range of <0.01–1.9 mg kg^{-1}. Shark was an exception, containing 8.4 mg kg^{-1} methylmercury chloride.

Certain organomercury compounds, such as HgMe$_2$Cl, are more toxic than elemental mercury [38], and inorganic mercury forms polluted areas [39].

Liang et al. [40] investigated methylmercury and total mercury contamination of molluscs collected at coastal sites in China's Bohai Sea. Cassa et al. [41] studied the influence of bioavailability and the tropic position of growth on the level of methylmercury in hake (*Merluccius merluccius*).

Kirk et al. [42] studied the distribution of total mercury, gaseous elemental Hg(O) (GEM), monomethylmercury and dimethylmercury in marine waters of the Canadian Arctic Archipelago, Hudson Strait and Hudson Bay. Concentrations of the total mercury were low throughout the water column in all regions sampled (mean ± standard deviation, 0.40 ± 0.47 ng L^{-1}). Concentrations of methylmercury were also generally low at the surface (23.8 ± 9.9 pg L^{-1}); however, at mid- and bottom depths, monomethylmercury was present at concentrations sufficient to initiate bioaccumulation of monomethylmercury through Arctic marine foodwebs (maximum 178 pg L^{-1}, 70.3 ± pg L^{-1}). In addition, at mid- and bottom depths, the percentage of total mercury that was monomethylmercury was high (maximum $66\% \pm 16\%$), suggesting that active methylation of inorganic Hg(III) occurs in deep Arctic marine waters.

Interestingly, there was a constant near 1:1 ratio between concentrations of monomethylmercury and dimethylmercury at all sites and depths, suggesting that methylated mercury species are in equilibrium with each other or are produced by similar processes throughout the water column. The results also demonstrate that oceanographic processes, such as water regeneration and vertical mixing, affect mercury distribution in marine waters. Vertical mixing, for example, is likely to transport mono- and dimethylmercury upward from production zones at some sites, resulting in surface waters (up to $68.0 \, pg \, L^{-1}$) where primary production, and thus uptake of monomethylmercury by biota, is potentially highest. Finally, calculated instantaneous ocean atmosphere fluxes of gaseous mercury species demonstrate that Arctic marine waters are a substantial source of dimethylmercury and gaseous elemental mercury to the atmosphere (27.3 ± 47.8 and $130 \pm 138 \, ng \, m^{-2}$, respectively) during the ice-free season.

10.4   ORGANOTIN COMPOUNDS

Minchin et al. [43] studied changes in shellfish populations, including scallops (*Pecten maximus*), flame shells (*Lima hians*), polychaetes, snails, bivalves, mussels (*Mytilus edulis*) and oysters (*Crassostrea virginica*) in Mulroy Bay, Ireland, before, during and after the extensive use of organotin compounds in net dips on salmonid farms in the area. The results were compared with data from other areas on the west and south coasts of Ireland. Tabulated data are included on estimated numbers of scallops in each area, concentrations of tributyltin in molluscan tissues, and shell distortions and tributyltin concentrations in oysters; although use of tributyltin was discontinued in spring 1985, relatively high concentrations were found in adult scallop tissues, but settlements of scallops had reappeared. The concentrations in mussels and flame shells were relatively low, and some mussel settlement had occurred in Mulroy Bay, but shells had failed to settle since 1983, and the remaining population was endangered. Since April 1987, the use of organotin antifouling compounds on boats and other aquatic structures has been illegal in Ireland.

The occurrence and concentrations of organotin compounds in the tissues of scallops (*Pecten maximus*), flame shells (*Lima hians*), bivalves, mussels (*Mytilus edulis*) and oysters (*Crassostrea virginica*) have been studied. Mussel and flame shell populations are adversely affected by organotin compounds [43]. High concentrations of tributyltin have been found in polychaetes, snails and bivalves living in marinas containing 2–$646 \, ng \, L^{-1}$ tributyltin [44], that is, above the environmental quality target for tributyltin of $20 \, ng \, L^{-1}$. San Diego Bay mussels exposed to $0.7 \, \mu g \, L^{-1}$ organotin for 60 days sustained a 50% mortality in the case of mussels and a decline in condition in the case of oysters [45]. Various tissues in the organisms showed tin uptake within 0–30 days.

Roberts [46] demonstrated that acute toxicity with tributyltin chloride (using glacial acetic acid as a carrier) yielded 48 h LC_{50} values of 1.30 and $3.96 \, \mu g \, L^{-1}$ in *Crassostrea virginica* embryos and straight-hinge-stage larvae, respectively, and

1.13 and 1.65 µg L^{-1} in *Mercenaria mercenaria* embryos and larvae, respectively. The 24 LC$_{50}$ (both species) were greater than 1.3 µg L^{-1} in embryos and greater than 4.2 µg L^{-1} in larvae. In the one experiment, which used acetone as a carrier for tributyltin chloride, the 38 h LC$_{50}$ for *Crassostrea virginica* embryos was 0.71 µg L^{-1}. Evidence suggested that tributyltin doses below 48 h LC$_{50}$ delayed clam embryo development, although resultant larvae were not abnormal. In addition to slightly delaying oyster embryo development, tributyltin (0.77 µg L^{-1} and above) caused abnormal shell development (flattened rather than convex shells) in resultant larvae. Comparing these results with those obtained in tests with *Crassostrea gigas*, *Mytilus edulis* and *Mytilus galloprovincialis* indicated that embryos and larvae of the five species of bivalve mollusc had sensitivities similar to those tributyltin and that tolerance increased slightly with increasing larval age.

Weis et al. [47] showed that exposure of adult fiddler crabs to tributyltin concentrations as low as 0.5 µg L^{-1} retarded limb regeneration and ecdysis, and produced morphological abnormalities in regenerated limbs. Deformities included backward curling of the dactyl of the claw, curling and stunting of walking legs and a reduction in the number of setae. In two of the trials, males appeared more sensitive to tributyltin than females, whereas the reverse was true in the third trial. These differences might have been related to seasonal reproductive activity.

It was also observed that the presence of sediment greatly reduced the effects of tributyltin.

In a 13-month experiment conducted by His and Robert [48] in Arachon Bay, oysters were cultivated in trays with wooden sides which had been painted with organotin antifouling paint (tributyltin) or copper oxide antifouling paint (Renaudin La Precieuse). The organotin paint adversely affected oyster weight, length and width compared with controls (in unpainted trays), but did not affect shell height. Shell density (reflective of shell calcification) and dry condition paint factors were markedly decreased. Embryonic and larval viability was unaffected, but larval growth was slightly impaired. Copper oxide paint has no effect on oyster growth or shell calcification. Dry condition factors were reduced, but to a lesser extent than was observed with organotin. Embryonic and larval development were unaffected by the copper oxide paint.

Zischka and Arthur [49] have determined 96 h LC$_{50}$ values of tributyltin compounds to mysids (*Mysidopes bahia*). The age of the fish was an important factor in determining the sensitivity of juveniles to tributyltin compounds.

In chronic toxicity tests [50] carried out in the Chesapeake Bay area in which bioca were exposed to tributyltin, the survival of *Gammarus* sp. was unaffected by 24 h exposure to concentrations up to 0.58 µg L^{-1}, although body weight was reduced by 64% relative to controls. Survival of *Brevootra tyrannus* and larval *Menidia beryllina* was unaffected by a 28-day exposure to concentrations of tributyltin up to 0.49 µg L^{-1}. Extensive variation between individuals made it difficult to discern treatment histological changes in *Brevootra tyrannus*.

In *Mendia beryllina*, growth was reduced by 20%–22% following exposure to 0.09 or 0.49 µg L^{-1} tributyltin, but various morphometric measurements revealed no treatment-related effects. It was considered unlikely that mean environmental concentrations of tributyltin in Chesapeake Bay marines would cause direct mortality to the three test species after 4 weeks of exposure, although sublethal effects could occur.

A 24 h LC_{50} value of 1–3 µg L^{-1} has been reported for adult rainbow trout. The concentrations of tributyltin found in surface microlayers of natural waters were in the range 1.9–473 µg L^{-1}. Consequently, surface swimming rainbow trout could be at risk.

A tributyltin concentration of 0.5 µg L^{-1} caused reduced body weight in samaraus GP, reduced growth rate in *Brevootra tyrannus* and larval *Menidia beryllina* and an adverse effect on weight and length in the oyster *Crassostrea gigas*.

San Diego Bay mussels exposed to 0.7 µg L^{-1} organotin for 60 days sustained a 50% mortality in the case of mussels and a decline in condition in the case of oysters. Various tissues in these organisms showed an uptake within 0–30 days.

Alkyltin compounds have been encountered in invertebrates (oysters and mussels) in the range 0.02–0.1 mg kg^{-1}.

Kannan et al. [53] detected concentrations of mono- and tributyltin in the muscle and liver of fish collected from Australia, Papua New Guinea and the Solomon Islands. Butyltin concentrations ranged from below the limit of detection to 47 ng g^{-1} in muscle and 6.5–570 ng g^{-1} wet weight in liver. Liver was found to accumulate higher concentrations of butyltins than muscle. Butyltin residues in tissues were not positively correlated with lipid content. Monobutyltin was the predominant species in all samples. The daily dietary intake of butyltins by Australians via fish was estimated to be 377–416 ng per person per day, lower than is believed to cause health problems.

Tremolada et al. [54] used a simple model to predict compound loss processes in aquatic ecotoxicological tests for measuring triphenyltin levels in biota and water.

Gomez-Aris et al. [55] used transplanted *Venerupis decussata* to evaluate a pollution of tributyltin in marine clams. Live species of the clam *Venerupis decussata* were suspended in seawater of the Mazagon Marina, located in a heavy metal polluted area at the mouth of Huelva Estuary (southwest Spain). Clams were preserved in plastic cages and subsamples were recovered every 5 days over a period of 40 days. Water from the marina was sampled every 2 days during the time course of the experiment. Clams and water were analysed for metals in organotins. Results showed the accumulation of tributyltin in the bivalves reaching an equilibrium with the surrounding water. Bioconcentration factors ranged from 10^2 (for V) to 4×10^3 (for Cu). Clams also accumulated Al and Pb, but a steady state was not reached. A first-order kinetic model was applied to the data, and results indicated that rates of accumulation differed in relation to clam size class. Clam mortality increased during the experiment and was totalled after 42 days, which was attributed to the high concentration of Cu in seawater.

Navarro et al. [56] carried out a simultaneous determination of mercury and species in the European *Anguilla anguilla* glass eel and yellow eel.

A methodology to simultaneously determine total and methylmercury and butyltin, tributyltin, dibutyltin and monobutyltin compounds in eel samples was assessed and validated using multiple isotopically enriched species. The developed methodology was able to analyse simultaneously the organometal species accurately and precisely and to correct for the potential transformations and degradations of the different species during the various steps of the analytical procedure. Low detection limits were achieved (0.007–0.17 µg kg^{-1} for mercury species and 0.4–0.71 µg kg^{-1} for tin species), allowing analysis of low-mass samples and thus the analysis at the individual organism scale, including glass eels for which the sample dry weight ranged from 60 to 100 mg. The methodology was validated with Certified Reference Materials (BCR-464, BCR-477, BCR-710, DOLT-4 and NIST-2977) and applied to the analysis of these pollutants in two developmental stages of the European eel, *Anguilla anguilla*: individual whole glass eels and muscle tissue from yellow eels. The Adour Estuary (southwest France) was selected to monitor the bioaccumulation of organometal species in these organisms, according to their developmental stage, their morphological parameters and the sampling site. The results suggest that the accumulation of the methylmercury in glass eel tissue is related to weight, with higher concentrations in smaller individuals. Butyltin concentrations were very close to the limit of detection, and no significant differences were detected between glass and yellow eels.

Han and Weber [57] describe speciation of inorganotin and methyltin compounds and butyltin compounds in oyster samples.

The effect has been studied of organotin compounds on marine mussels [58], pinniped cetaceous and cetacean [59] and blue nose dolphins [60,61]. Hu et al. [51] studied the isotopic biomagnification of triphenyltin in a marine food web in the Bohai Sea, northern China.

Crab, burrowing shrimp, short-necked clam, vuned rapa whelk, bay scallop and the fish species weaver, catfish, bartal flathead, white flower croaker, walleye fish and mullet were included in this study.

Rudel et al. [52] carried out retrospective monitoring of organotin compounds in marine biota from 1985 to 1999.

REFERENCES

1. J.K. Chan, P.T.S. Wang, G.A. Bengert and D.L. Dunn, *Analytical Chemistry*, 1984, 86, 271.
2. L.A. Richman, C.D. Wren and P.M. Stokes, *Water, Air and Soil Pollution*, 1988, 37, 465.
3. K. Krishnan, W.D. Marshall and W.I. Hatch, *Environmental Science and Technology*, 1988, 22, 806.
4. K.A. Cockell and J.W. Hilton, *Aquatic Toxicology*, 1988, 12, 73.
5. A.C. Chapman, *Analyst*, 1926, 51, 548.
6. M.O. Andreae, *Analytical Chemistry*, 1977, 48, 820.
7. T.B. Bennett, W.H. McDaniel and R.N. Hemphill, *Advances in Automated Analysis Technical International Conference*, Vol. 8, Media, Inc., Tarrytown, NY, 1972.
8. A.A. Elawady, R.B. Miller and W.J. Carter, *Analytical Chemistry*, 1976, 48, 110.
9. T. Bevan, M.M. Alibers and L. Bokos, *Analytica Chimica Acta*, 1981, 131, 311.
10. J. Story, B. Havlik, J. Preasilova, K. Tretzer and J. Hanausova, *International Journal of Environmental Chemistry*, 1978, 5, 89.

11. T.G. Rains and O. Menis, *Journal of the Association of Official Analytical Chemists*, 1972, 55, 1339.
12. G. Compeau and R. Bartha, *Bulletin of Environmental Contamination and Toxicology*, 1983, 31, 486.
13. J.L. Webb, in *Enzyme and Metabolic Inhibitors*, Vol. 3, Academic Press, New York, 1966, chap. 6.
14. R.E. Stauffer, J.W. Bull and E.A. Jenne, Chemical studies of selected trace elements in hot spring drainage of Yellowstone National Park, Geological Survey Professional Paper 1044F, US Government Printing Office, Washington, DC, 1980.
15. R.S. Braman, in E.A. Woolson (ed.), *Arsenical Pesticides*, Series 7, American Chemical Society, Washington, DC, 1970, p. 108.
16. D.P. Cox, in E.A. Woolson (ed.), *Arsenical Pesticides*, Series 7, American Chemical Society, Washington, DC, 1975, p. 81.
17. A. Mareda-Pineiro, E. Pena-Vazquez, P. Hermedo-Herbello, P. Bermejo-Barrara, J. Mareda-Piniero, E. Alonso-Rodriguez, S. Muniategui-Lorrenzo, P. Lopez-Mahia and D. Prada-Rodriguez, *Analytical Chemistry*, 1008, 80, 9272.
18. A.W.P. Jarvis, R.N. Markell and H.R. Potter, *Nature (London)*, 1975, 255, 217.
19. U. Schmidt and F. Huber, *Nature (London)*, 1976, 259, 159.
20. J.P. Dumas, S. LeRoy Paxdernik, D. Bellonik, D. Bauchard and G. Vaill Ancourt, In *Proceeding of the 12th Canadian Symposium on Water Pollution Research*, 1977, Vol. 12, p. 91.
21. P.T.S. Wong, Y.K. Chau, L. Luton, G.A. Berguit and D.J. Swaine, Methylation of arsenic in the aquatic environment, in *Conference Proceedings on Trance Substances in Environmental Health XI*, University of Missouri, Columbia, 1977.
22. K. Reisinger and M. Stoeppler, *Nature*, 1981, 291, 228.
23. J.R. Grove, in D. Silverfobb, H. Branica and Z. Konrad (eds), *Lead in the Marine Environment*, Pergamon Press, Oxford, 1980, p. 45.
24. B.G. Murdock and D. Taylor, The acute toxicity and bioaccumulation of some alkyllead compounds in marine animals, in *Proceedings of the International Expert Discussions on Lead Occurrence Fate and Pollution in the Marine Environment, Rovini, Yugoslavia, 1977*, Pergamon Press, New York, 1980.
25. J.S. Thayer, *Occurrence – Biological Methylation of Elements in the Environment*, ACS Advances in Chemistry Series No. 1982, American Chemical Society, Washington, DC, 1978, p. 188.
26. P. Grandiean and T. Nielson, *Residue Review*, 1979, 72, 1979.
27. P.T.S. Wong, K. Chanug and P.L. Luxe, *Nature*, 1975, 253, 263.
28. C.R. Siroto and J.E. Uthe. *Analytical Chemistry*, 1977, 49, 823.
29. Y.K. Chau and P.T.S. Wong, Lead in the marine environments, in M. Branica and M. Konrad (eds), *Proceedings of the International Expert Discussions on Lead Occurrence Fate and Pollution in the Marine Environment, Rovini, Yugoslavia, 1977*, Pergamon Press, New York, 1980.
30. D.O. Reamer, W.M. Zoller and T.C. O'Haver, *Analytical Chemistry*, 1978, 50, 1449.
31. R.N. Ram and A.G. Sathyanesan, *Environmental Pollution*, 1987, 47, 135.
32. R. Kirubagarun and P. Joy, *Ecotoxicology and Environmental Safety*, 1988, 15, 171.
33. M. Czuba, R.W. Seagull, H. Tran and L. Cloutier, *Ecotoxicology and Environmental Safety*, 1987, 14, 64.
34. I.M. Davies, W.C. Graham and S.M. Pirie, *Marine Chemistry*, 1979, 7, 11.
35. A.M. Kienmeji and J.G. Kloosterboer, *Analytical Chemistry*, 1976, 48, 575.
36. K. Minagawa, A. Takizawa and I. Kifune, *Analytica Chimica Acta*, 1980, 115, 103.
37. S. Monoaro, *Water Research*, 1979, 13, 503.

38. J.F. Uthe, F.A.J. Armstrong and K.J. Tam, *Analytical Chemistry*, 1971, 54, 866.
39. K. Backman, *Talanta*, 1982, 29, 1.
40. L. Liang, J.-B. Shi, G.-L. Jiang and C.-G. Yuan, *Journal of Agriculture and Food Chemistry*, 2003, 51, 7373.
41. D. Cassa, M. Harmelin-Vivien, C. Mellon-Duval, V. Loizeans, B. Avertys and S. Crocket, *Environmental Science and Technology*, 2012, 46, 4885.
42. J.L. Kirk, V.L. St Louis, H. Hintelmann, I. Lehnhart, B. Else and L. Poissant, *Environmental Science and Technology*, 2008, 49, 8361.
43. D. Minchin, C.B. Duggan and W. King, *Marine Pollution Bulletin*, 1987, 18, 604.
44. W.J. Langston, G.R. Burt and Z. Mingjiang, *Marine Pollution Bulletin*, 1987, 18, 634.
45. G.V. Pickwell and S.A. Steinert, *Marine Environmental Research*, 1988, 24, 215.
46. M.H. Roberts, *Bulletin of Environmental Contamination and Toxicology*, 1987, 39, 1012.
47. J.S. Weis, J. Gottlieb and J. Kwintkowski, *Archives of Environmental Contamination and Toxicology*, 1987, 16, 321.
48. E. His and R. Robert, *Marine Biology*, 1987, 95, 83.
49. J.A. Zischka and J.W. Arthur, *Archives of Environmental Contamination and Toxicology*, 1987, 16, 225.
50. M. Guidici, S.M. De N Migliore, S.M. Guarino and C. Gambardella, *Marine Pollution Bulletin*, 1987, 18, 454.
51. J. Hu, H. Zhen, Y. Wan, J. Gad, W. An, L. An, F. Jin and Z. Jin, *Environmental Science and Technology*, 2006, 40, 3142.
52. H. Rudel, P. Lepper, J. Steinlanses and C. Schroter-Kermani, *Environmental Science and Technology*, 2003, 37, 1731.
53. K. Kannan, S. Tanabe, R. Tatsukawa and R.J. Williams, *International Analytical Chemistry*, 1995, 61, 263.
54. P. Tremolada, S. Bristean, D. Mozzi, M. Sugni, A. Barbaglio, T. Dagnac and M.D.C. Carnevali, *International Journal of Environmental Analytical Chemistry*, 2006, 86, 171.
55. J.L. Gomez-Aris, I. Giraldez, D. Schangez-Rodas, T. Acunta and E. Morales, *International Journal of Environmental Analytical Chemistry*, 199, 75, 107.
56. P. Navarro, S. Clemens, V. Perrot, V. Bolliet, H. Tabourel, T. Guirin, M. Monperreu and D. Amourioux, *International Journal of Environmental Analytical Chemistry*, 2013, 93, 166.
57. J.S. Han and J.H. Weber, *Analytical Chemistry*, 1988, 60, 316.
58. T. Suzuki, I. Yamamoto, H. Yamada, H. Kaniwa, K. Kondo and M. Murayama, *Journal of Agriculture and Food Chemistry*, 1998, 46, 304.
59. S. Tanache, M. Prudente, T. Mizuno, J. Hasegawa, H. Iwata and M. Migazaki, *Environmental Science and Technology*, 1988, 32, 193.
60. X. Kannam, K. Senthikumak, B.G. Hoganathan, S. Takahashi, D.K. Odell and S. Tanabe, *Environmental Science and Technology*, 1997, 31, 296.
61. K. Kannan, K.S. Guruge, N.J. Thomas, S. Tanabe and J.P. Giosy, *Environmental Science and Technology*, 1998, 32, 1169.

11 Toxicity of Pollutants in Photoplankton and Algae

11.1 TOXICITY OF ORGANIC POLLUTANTS

Di-*n*-butyl phthalate has a distinct adverse effect on the distribution of survival of marine phytoplankton. It also markedly affects the growth and aggregation behaviour of algae and diatoms [1].

Rhee et al. [2] studied the long-term response of phytoplankton (*Selanastrum capricornatum*) to 2,5,2′,5′-tetrachlorobiphenyl in water. This compound caused a reduction in the percentage of fixed carbon incorporated into the cells, and this carbon was probably excreted.

Concentrations of permethrin between 0.75 and 1.5 µg L^{-1} in water caused a decline in populations of *Daphnia rosea*, and at 10 µg L^{-1} a complete elimination of this species. *Acanthodiaptomus pacificus* behaved similarly. *Tropocyclops prasinus* was slightly more tolerant [3].

11.1.1 ZOOPLANKTON

Ali et al. [4] obtained no evidence that very low concentrations of diflubenzuron had any adverse effects on zooplankton and benthic invertebrates in ponds which had been contaminated by this insect growth regulator present as an air drift from a nearby citrus grove.

Day and Kaushik [5] studied the effect of short-term exposure to the synthetic pyrethroid fenvalerate in water on the rate of filtration and the rate of assimilation of *Chlamydomonas reinhardii* by three species of zooplankton, namely *Daphnia galeata mendotae*, *Ceriodaphnia cacustris* and *Diaptomus oregonensis*. Rates of filtration of *Chlamydomonas reinhardii* by all three species were decreased significantly at sublethal concentrations (≤0.05 µg L^{-1}) of fenvalerate in water after a 24 h exposure. Rates of assimilation of algae by the three species were decreased at lethal concentration of more than 0.05 µg fenvalerate. A change in rates of filtration and assimilation can be used to monitor the effects of sublethal levels of toxicants.

Applications of 1000 µg L^{-1} of carbaryl insecticide to pond water killed off all zooplankton but had no effect on phytoplankton, though changes in zooplankton densities affected phytoplankton community structures [6]. Lindane (gamma-benzene hexachloride [γ-BHC]) has no significant effect on natural zooplankton populations, but the population density of zooplankton was reduced even at concentrations of lindane as low as 20 µg L^{-1} [7]. Rotifers and nauplii were particularly adversely affected.

11.1.2 Algae

Exposure of periphyton communities from brackish water mesocosinus to $1-10\,\mu g\,L^{-1}$ 4,5,6-trichloroguiacol produced no evidence for adverse effects. Exposure of natural periphyton communities to atrazine, alachlor, metolchlor and metribuzin reduced the growth rate and rate of uptake of nutrients at least temporarily.

Minimum concentrations of terbutryn, diuron, monouron and atrazine for chlorobiphenyl resistance decreased the tolerance of the strain to lower salinities and nitrogen limitation, but increased its tolerance to lower temperatures.

Hamilton et al. [8] have studied the effect of up to 2 years of exposure of periphyton communities to concentrations of atrazine in the range of $80-1500\,\mu g\,L^{-1}$. Chlorophyll a, freshwater biomass, ash-free weight, cell numbers, species diversity, community carbon uptake and species-specific carbon uptake were measured. There was a shift from chlorophyte to a diatom-dominated community over the 2-year period, but *Cylindrospernum stagnate* and *Tetraspora cyclindrica* showed evidence of resistance to atrazine at $1560\,\mu g\,L^{-1}$.

Community productivity was reduced by 21% and 82% in the low and high exposures, respectively, returning to control levels in 21 days. The productivities of the larger algae were most affected. Reduced growth rates were obtained after exposure to the herbicide. Other workers have reported on a growth rate depression when green algae are exposed to atrazine [9].

Marine unicellular algae *Skeletonema costatum*, *Thalassiosira pseudonana* and *Chlorella* sp. have been exposed to water containing the brominated organic compounds decabromobiphenyl oxide and pentabromomethyl benzene. The corresponding LC_{50} values were greater than 1.1, 1.0 and $0.5\,mg\,L^{-1}$, respectively, which was the highest exposure concentrations tested [10].

The effect has been studied of atrazine ($50-30,000\,\mu g\,L^{-1}$) combined with either ethanol (0.1%–3% v/v) or acetone (0.1%–5%) on the growth of the green alga *Chlorella pyrenoidosa* [11]. Acetone and atrazine interacted antagonistically, but only at solvent concentrations exceeding 4%–5% with both solvents. Atrazine EC_{50} values (calculated using growth data in the additive solvent range) were between 50 and $80\,\mu g\,L^{-1}$.

The effect of $0-100\,mg\,L^{-1}$ concentrations of lindane (γ-BHC) in water on the alga *Scenedesmus obliquus* has been studied [12]. Daily samples were examined for algal growth, pigment content and accumulation and degradation. At concentrations above $50\,mg\,L^{-1}$ lindane in water, the algal pigment content was affected. Accumulation was enhanced by exposure time and vibration.

Walsh et al. [13] evaluated the effect of 21 pesticides in water on five different algal species by determining EC_{50} values.

The effects have been examined of the organophosphorus insecticide phosalone on the sexual life cycle of the algal *Chlamydomonas reinhardii* [14].

The formation of gametes, young zygotes and mature zygotes and the meiotic division of mature zygotes were examined following a 2 h exposure to $36.7\,mg\,L^{-1}$ phosalone. The formation of gametes and young zygotes was not affected by the treatment. Unlike control groups, the mature zygotes thus formed did not exhibit the ability of meiotic division in the first days of light exposure but remained in the same state for 5 days and then underwent meiotic division on the sixth day of exposure. Stratten [15] has studied the inhibitory effect of from 0.1% to 30% of six organic solvents (methanol, acetone, hexane, ethanol, dimethyl sulphoxide and N,N-dimethyl formamide) in water towards five species of blue-green algae (*Anabaena* sp., *Anabaena cylindrica*, *Anabaena variablilis*, *Nostoc* sp. and *Anabaena inaequalis*). Acetone and dimethyl sulphoxide were of intermediate toxicity as regards growth inhibition (EC_{50} values 0.36% and 4.4%, respectively). Dimethyl sulphoxide and ethanol were highly toxic.

In 10- to 14-day growth experiments, methyl formamide and ethanol have been confirmed as the most toxic organic solvents towards the green algae *Chlorella pyrenoidosa* [11] (EC_{50} of 0.84% and 1.18% v/v, respectively), followed by dimethyl sulphoxide, hexane, methanol and acetone (EC_{50} values of 2.01%, 2.26%, 3.02% and 3.60% v/v, respectively).

Chlorella vulgaris cultures exposed to *p*-nitrophenol or *m*-nitrophenol in water at concentrations between 5 and $20\,mg\,L^{-1}$ for 20–30 days exhibited inhibited growth in the case of *p*-nitrophenol at $10\,mg\,L^{-1}$ and stimulated growth in the case of *m*-nitrophenol at $5\,\mu g\,L^{-1}$ during 20 or 30 days of exposure but inhibited growth at $15\,mg\,L^{-1}$ during a 15-day exposure [16].

11.1.3   WEEDS

Thorhang and Marins [17] studied the effect of three oil dispersants (Corexit 9527, Arcochem D609 and Conco K(K)) on the subtropical and tropical seagrasses *Thalassia testudinum*, *Halodule wrightii* and *Syringodium filiforme*. At concentrations below 1 ml dispersant per 10 ml oil in 100 L of seawater, mortality rates were low even for long exposure times. At 10 ml dispersal per 100 ml oil in 100 L of seawater, *Syringodium filiforme* and *Halodule wrightii* died. Conco K(K) was far more toxic than either of the other two dispersants.

11.1.4   DIATOMS

Goutex et al. [18] studied the effects of $50\,mg\,L^{-1}$ 9,10-dihydroanthracene and its biodegradation products on the marine diatom *Phaeodactylum tricornatum*. Growth of the diatom was inhibited. Synergistic effects between 9,10-dihydroanthracene and its biodegradation products increased the toxicity of the hydrocarbons. Resistance to polychlorobiphenyls and cross-resistance to DDT were induced in a polychlorobiphenyl-resistant clone of *Ditylum brightwelli* by a 30-day exposure to $10\,\mu g$ of polychlorobiphenyl or polychlorobiphenyl concentrations, which increased progressively from 10 to $30\,\mu g\,L^{-1}$ over 30 days.

Polychlorobiphenyl resistance persisted for 2 years. The polychlorobiphenyl-resistant *Ditylum brightwelli* exhibited greater tolerance to polychlorobiphenyl than did the sensitive strain under all environmental conditions which permitted its growth, even when the conditions of salinity, temperature and nitrogen availability were very different from those maintained during induction.

Berrojalbiz [22] studied the accumulation of polycyclic aromatic hydrocarbons in 300 plankton.

Barga and Bidleman [19] examined the effects of enantiomer fractions of organic chlorinated pesticides in arctic ice fauna on 300 plankton and benthos.

Zang et al. [20] carried out an investigation of the effect of selected organic pollutants on marine plankton using multidimensional fluorescence measurements.

Multidimensional fluorescence is ideally suited to rapidly measure algal fluorescence generated by both chlorophyll a and accessory pigments, as well as any changes induced by pollutants. Laboratory cultured and natural algae samples from classes Chlorophyceae, Bacillariophyceae and Cyanophyceae were exposed to substituted nitroaromatics and fluorescence spectra of the algae recorded. Notable spectroscopic changes and fluorescence quenching were observed. In addition, a novel method for rapidly preconcentrating dilute natural marine sample is described.

Lee et al. [21] developed a sensitive method for monitoring the cell concentration of red tide phytoplankton *Chattonella marina*, which frequently caused fish death via an obscure mechanism. 2-Methyl-6-(p-methoxyphenyl)-3,7-dihydroimidazol[1,2-a] pyrazine-3-one (MCLA), a *Cypyridina* luciferin analogue which emits light strongly at 465 nm in the presence of superoxide, was applied as a chemiluminescence probe. The MCLA-dependent chemiluminescence of room-cultured axenic *Chattonella marina* was efficiently suppressed by adding superoxide dismutase (20 units ml^{-1}) to the phytoplankton suspension but not by filtering out the plankton cells, indicating the chemiluminescence was due to the superoxide released from *Chattonella marina*. Several species of other tested phytoplankton, such as *Heterosigma akashiwo*, *Skeletonema costatum*, *Chaetoceros sociale* and *Porphyridium cruentrum*, did not induce this chemiluminescence, whereas *Chattonella antiqua* showed characteristics similar to *Chattonella marina*, suggesting that this method has specificity for the *Chattonella* genus. As compared with the cell density of *Chattonella marina* during the red tide seasons, the detection limit of 200 cells ml^{-1} based on this luminescence method is considered to be applicable for *Chattonella marina* detection in the early stages of its red tides.

The Mediterranean and Black Seas are unique marine environments subject to important anthropogenic pressures due to riverine and atmospheric inputs of organic pollutants.

Berrojalbiz et al. [23] reported the results obtained during two east–west sampling cruises in June 2006 and May 2007 from Barcelona to Istanbul and Alexandria, respectively, where water and plankton samples were collected simultaneously. Both matrices were analysed for hexachlorocyclohexane, hexachlorobenzene and 41 polychlorinated biphenyl congeners. The comparison of the measured hexachlorobenzene and hexachloro- and cyclohexane concentrations suggests a temporal decline in their concentrations since the 1990s. On the contrary, polychlorobiphenyl seawater concentrations did not exhibit such a decline, but show a significant spatial variability in dissolved concentrations, with lower levels in the open western and southeastern Mediterranean, and higher concentrations in the Black, Marmara and Aegean Seas and Sicilian Strait. Polychlorobiphenyl and polychlorobiphenyl organochlorine pesticides in concentrations in plankton were higher at lower plankton biomass, but the intensity of this trend depended on the compound hydrophobicity (K_{ow}). For the more persistent polychlorobiphenyls, the observed dependence of organic pollutant concentrations in plankton versus biomass can be explained by interactions between air–water exchange, particle setting and bioaccumulation processes, whereas degradation processes occurring in the photic zone drive the trends shown by the more labile hexachlorocyclohexane. The results presented provide clear evidence of the important physical and biochemical controls on organic pollutant occurrence in the marine environment.

There is a need for an assay that can detect known and unanticipated neurotoxins associated with harmful algal blooms. Kulagina et al. [24] described an attempt to monitor the presence of brevetoxin-3 and saxitoxin in a seawater matrix using the neuronal network biosensor. The neuronal biosensor and biosensor saxitoxin rely on cultured mammalian neurons grown over microelectrode arrays, where the inherent bioelectrical activity of the network manifested as extracellular action potentials can be monitored noninvasively. Spinal neuronal networks were prepared for embryonic mice, and the mean spike rate across the network was analysed before and during exposure to the toxins. Extracellular action potentials from the network are highly sensitive not only to purified brevotoxin and brevotoxin-3, but also when in combination with matrices such as natural seawater and algal growth medium. Detection limits for brevotoxin and brevotoxin-3, respectively, are 0.031 and 0.33 nM in recording buffer and 0.076 and 0.048 nM in the presence of 25-fold diluted seawater. These results demonstrated that neuronal networks could be used for analysis of *Alexandrium fundyense* (saxotongur producer) and *Karenia brevis* (grevotoxin-3 producer) algal samples lysed directly in the seawater-based growth medium and appropriately diluted with buffered recording medium. The cultured network responded by changes in mean spike rate to the presence of producing algae, but not to the samples of two nonsaxitoxin and brevotoxin-3 isolates of the same algal genera. This work provides evidence that the neuronal network biosensor has the capacity to rapidly detect toxins associated with cells of toxic algal species or as dissolved forms present in seawater, and has the potential for monitoring toxin levels during harmful algal blooms.

Baffi [25] has described a procedure for the quantitative determination of 17 amino acids in a marine matrix using high-performance liquid chromatography. Precolumn derivatisation with *o*-phthalaldehyde, separation on C_{18}-bonded silica with phosphate-buffered (pH 7.2) acetonitrile as eluent and fluorescence detection have been used. The good variation coefficient (average 2% with working curves in a real matrix) and the low detection limit (1–5 fmole) make the procedure suitable for the determination of total or free amino acids in photoplankton marine cells.

Crystobal et al. [45] have determined the bioaccumulation of hexachlorobenzene polychlorinated biphenyls and hexachlorocyclohexane in phytoplankton from the Southern Ocean and South Scotia.

Aristide et al. [46] reported that weak organic ligands enhance zinc uptake in marine phytoplankton.

11.2 TOXICITY OF ORGANOMETALLIC POLLUTANTS

Birnie and Hodges [26] have given details of a procedure for the determination of down to 0.01 mg kg^{-1} of ionic species of alkyl lead in marine organisms by solvent extraction and differential pulse anodic stripping voltammetry. The sample is homogenised in the presence of a mixture of salts (lead nitrate, sodium benzoate, potassium iodide, sodium chloride and EDTA), which effectively releases the di- and trialkyl lead species present, and facilitates their transfer into toluene as a preliminary to back extraction into dilute nitric acid ready for differential pulse anodic stripping voltammetry. Recoveries were in the range of 70%–90% (Et_3Pb^+, Et_2Pb^{2+} and Me_3Pb^+) to 10%–40% (Me_2Pb^{2+}).

White and Englar [27] determined organoarsenic compounds in marine brown algae by a procedure based on generation of arsine and quantification by the silver diethyldithiocarbamate spectrophotometric method.

The sodium borohydride reduction–atomic absorption spectrophotometric method has been applied to the determination of monomethyl arsenic acid and dimethyl arsenic acid in macroalgae.

11.3 TOXICITY OF ORGANOMETALLIC AND METALLIC POLLUTANTS

High concentrations of inorganic metals and some organometallic compounds are found in algae and phytoplankton and fish where pollution is occurring. This is important in the sense that these are the foodstuffs of fish and crustacea, which by eating them increase their body burden of these contaminants. In Table 11.1 are compared typical values of inorganic copper and arsenic and organolead found in different samples.

Organolead compounds have been found in phytoplankton and in fish tissues (Table 11.2). In both cases, the organolead levels are considerably higher than those found in the water in which the fish or phytoplankton live.

The importance of complexing agents in the mineral nutrition of phytoplankton and other marine organisms has been recognised for many years. Complexing agents have been held responsible for the solubilisation of iron, and therefore its greater biological availability [31]. In contrast, complexing agents are assumed to reduce the biological availability of copper and minimise its toxic effect [31–45]. Experiments with pure cultures of phytoplankton in chemically defined media have demonstrated that copper toxicity is directly correlated with copper ion activity and independent of the total copper concentration. In these experiments, copper ion concentrations were varied in media containing a wide range of total concentrations through the use of artificial complexing agents. When the copper(II) concentration was calculated for earlier experiments with phytoplankton in defined media, it appeared that copper(II) was toxic to a number of phytoplankton species in concentrations as low as $10^{-6}\,\mu mole$ [41]. Since copper concentrations in the world's oceans typically range from 10^{-4} to $10^{-1}\,L^{-1}$, complexing agents and other materials affecting the solution chemistry of copper must maintain the copper(II) activity at sublethal levels.

Current water quality criteria regulations on copper toxicity to biota are still based on total dissolved ($<0.4\,\mu m$ membrane filter) copper concentrations with a hardness modification for freshwaters. There are, however, ongoing efforts to incorporate metal speciation in water quality criteria and toxicity regulations (such as the biotic ligand model) for copper and other metals. Etteberg et al. [47] showed that copper accumulation and growth inhibition of the Baltic macroalga *Ceranium tenuicorne* exposed to

TABLE 11.1

Comparison of Inorganic Metal and Organometallic Content ($\mu g\,kg^{-1}$) of Algae, Phytoplankton, Fish and Crustacea

Inorganic Compounds	Organometallic Compounds	Algae	Marine Crown Algae (*Laminaria*)	Organometallic Concentration in Metals		
				Phytoplankton up to $\mu g\,kg^{-1}$	Fish	Crustacea
Copper	—	660,000		40,000	3500	2600
Arsenic	—	56,000		2900	180	
		200–600 [28][a]				
		7600	40,300	—	—	—
		15,600 [28][b]	89,700 [28]	—		
		Organolead Compounds				
Me₃EtPb		—		38	<1 [27]	—
Me₂Et₂Pb		—		1500	57–1430 [28,29]	—
MeEt₃Pb		—		3610	140–290 [29]	—
Et₄Pb		—		16,500	780–7500 [29]	—
Et₃Pb⁺		—		650	53–3430 [29]	—
Et₂Pb²⁺		—		110	90–4300 [29]	—
Total		—		22,300	1400–21,100 [29]	—

Note: References are in brackets.

[a] As monomethyl arsenic acid.

[b] As dimethyl arsenic acid.

TABLE 11.2

Comparison of Concentrations of Organolead Compounds Found in Phytoplankton and Fish

Compound	In Fish ($\mu g\,kg^{-1}$) (From Table 11.1)	In Phytoplankton [30] ($\mu g\,kg^{-1}$)
Me$_3$EtPb	<1	38
Me$_2$Et$_2$Pb	57–1430	1500
MeEt$_3$Pb	140–290	3610
Et$_4$Pb	780–7500	16,500
EtPb$^+$	53–3430	560
EtPb^{2+}	90–4300	110
Total	1400–21,100	22,300

Note: References are in brackets.

copper in artificial seawater at typical coastal and estuarine dissolved organic carbon concentrations (similar to 2–4 mg LC^{-1} as fulvic acid) are better correlated to weakly complexed and total dissolved copper concentrations rather than the free copper concentration [Cu^{2+}]. Results using a combination of competitive ligand exchange–adsorptive cathodic stripping voltammetry measurements and model calculations (using visual MINTEQ incorporating the Stockholm Humic) show that copper accumulation in *Ceranium tenuicorne* correlates linearly well to [Cu^{2+}] (at relatively high [Cu^{2+}]) and in the absence of fulvic acid. Thus, this technique fails to describe copper accumulation in *Ceranium tenuicorne* at copper and dissolved organic carbon concentrations typical of most marine waters. These results seem to indicate that at ambient total dissolved copper concentrations in coastal and estuarine waters, *Ceranium tenuicorne* might be able to access a sizable fraction of organically complexed copper when free copper concentration to the cell membrane is diffusion limited.

Stretson and Nonova [48] determined heavy metal contents (Fe, Mn, Cu, Pb and Cd) in two brown macroalgal species (*Cystoseira crinite* and *Cystoseira barbata*). Samples were collected during six seasons from spring 1996 to summer 2000 from five different sites of the Bulgarian Black Sea coastal zones – Tuzlata, Ravada, Sozopol, Ahtopol and Sinemoretz. The local and seasonal metal distribution and their variations were examined.

The data obtained indicate that the two investigated species demonstrate various degrees of metal accumulation and can be used as indicators for the type and quantity of anthropogenic contamination in marine ecosystems.

REFERENCES

1. R. Acey, P. Healy, T.F. Unger, C.E. Ford and R.A. Hudson, *Bulletin of Environmental Contamination and Toxicology*, 1987, 39, 1.
2. G.Y. Rhee, L. Shane and A. Denucci, *Applied and Environmental Microbiology*, 1988, 54, 1394.

3. M. Yasuno, T. Hanazato, T. Iwakuma, K. Takamura, R. Ueno and T. Takamura, *Hydrobiologia*, 1988, 159, 247.
4. A. Ali, N.H. Nigg, J.H. Stamper, H.L. Kok-Yokami and M. Weaver, *Bulletin of Environmental Contamination and Toxicology*, 1988, 41, 781.
5. K. Day and N.K. Kaushik, *Archives of Environmental Contamination and Toxicology*, 1987, 16, 423.
6. T. Hanazato and M. Yasuno, *Environmental Pollution*, 1987, 48, 145.
7. J.P. Lay, A. Muller, L. Peichl, R. Lang and F. Korte, *Chemosphere*, 1987, 16, 1527.
8. P.B. Hamilton, G.S. Jackson, N.K. Kaushik and K.R. Solomon, *Environmental Protection*, 1987, 46, 83.
9. C.M. Hersh and W.G. Crumpton, *Bulletin of Environmental Contamination and Toxicology*, 1987, 39, 1041.
10. G.E. Walsh, M.J. Yoder, L.L. McLaughlin and L.M. Lopes, *Ecotoxicology and Environmental Safety*, 1987, 14, 215.
11. G.W. Stratton and T.M. Smith, *Bulletin of Environmental Contamination and Toxicology*, 1988, 40, 736.
12. L. Yi-xiong and S. Bo-zen, *Hydrobiologia*, 1987, 153, 249.
13. G.E Walsh, C.H. Deans and L.L. McLaughlin, *Environmental Contamination and Toxicology*, 1987, 6, 767.
14. M.D. Pednekar, S. Gandhi and M.S. Netrawate, *Environmental International*, 1987, 13, 219.
15. G.W. Stratten, *Bulletin of Environmental Contamination and Toxicology*, 1987, 38, 1012.
16. M. Megharaj, K. Venkateswarlu and A.S. Rao, *Ecotoxicology and Environmental Safety*, 1988, 15, 320.
17. A. Thorhang and J. Marins, *Marine Pollution Bulletin*, 1987, 18, 124.
18. M.M. Goutex, M. Al-Mallah and J.C. Bertrand, *Marine Biology*, 1987, 94, 111.
19. K. Barga and T.F. Bidleman, *Environmental Science and Technology*, 2005, 39, 3464.
20. J.B. Zang, R.L. Woodlee, M.R.S. Fub and I.M. Warner, *International Journal of Environmental Analytical Chemistry*, 1990, 41, 149.
21. T. Lee, N. Getob, E. Xiki, K. Yokoyama and M. Meuro-ku, *Analytical Chemistry*, 1995, 67, 225.
22. B. Berrojalbiz, *Environmental Science and Technology*, 2009, 43, 2295.
23. B. Berrojalbiz, J. Dachs, S.D. Vento, J. Ojeda, M.C. Valle, J. Castro-Jimininez, G. Mariani, J. Wollgast and G. Hanke, *Environmental Science and Technology*, 2011, 45, 4315.
24. N.V. Kulagina, C.N. Mikulski, S. Gray, W.M.A. Gregory, J. Daucette, J.S. Ramsdell and J.F. Panerazio, *Environmental Science and Technology*, 2006, 40, 578.
25. J. Baffi, *International Journal of Environmental Analytical Chemistry*, 1990, 41, 173.
26. S.E. Birnie and D.J. Hodges, *Environmental Technology Letters*, 1981, 2, 43.
27. J.N. White and J.R. Englar, *Botanica Marina*, 1983, 26, 159.
28. W.A. Maher, *Analytica Chimica Acta*, 1981, 126, 157.
29. J.K. Chan, P.T.S. Wong, G.A. Bengert and J.L. Dunn, *Analytical Chemistry*, 1984, 56, 271.
30. R. Capelli, C. Fezia and A. Franchi, *Analyst*, 1979, 104, 1197.
31. A.G. Lewis, R.H. Whitfield and A. Ramnarine, *Marion Biology*, 1972, 17, 215.
32. E.W. Davey, M.J. Morgan and S.J. Erickson, *Limnology and Oceanography*, 1973, 18, 993.
33. A. Prakash, M.A. Rashid, A. Jensen and D.V. Subha Rao, *Limnology and Oceanography*, 1973, 18, 516.
34. M. Gnassia-Barelli, M. Romero, F. Laumond and D. Pesando, *Marine Biology*, 1978, 47, 15.

35. D. Hongve, O.K. Skogheim, A. Hindar and H. Abrahamsen, *Bulletin of Environmental Contamination and Toxicology*, 1980, 25, 594.
36. N.S. Fisher and D. Frood, *Marine Biology*, 1980, 59, 85.
37. W. Sunda and R.R.L. Guillard, *Journal of Marine Research*, 1976, 34, 511.
38. D.M. Anderson and F.M. Morel, *Limnology and Oceanography*, 1978, 23, 283.
39. L.S. Murphy, R.R.L. Guillard and J. Gavis. Of resistant phytoplankton strains through exposure to marine pollutants, in G.F. Mayer (ed.), *Ecological Stress and the New York Blight: Science and Management*, Estuarine Research Federation, Columbia, SC, 1980, p. 401.
40. G.A. Jackson and J.J. Morgan, *Limnology and Oceanography*, 1978, 23, 268.
41. R. Chester and J.H. Stober, *Marine Chemistry*, 1974, 2, 17.
42. J.J. Alberts, D.E. Leyden and T.A. Patterson, *Marine Chemistry*, 1976, 4, 51.
43. F.A. Boyle, F.R. Sclater and J.M. Edmond, *Science Letters*, 1977, 37, 38.
44. H.L. Winlom and R.G. Smith, *Marine Chemistry*, 1979, 7, 157.
45. J. Crystobal, G. Malagon, S.D. Vento, N. Berrojalbiz, J. Ojeda and J. Dacks, *Environmental Science and Technology*, 2013, 47, 5578.
46. L. Aristide, Y. Xu and F.M.M. Morel, *Environmental Science and Technology*, 2012, 46, 5445.
47. Y.L. Etteberg, J. Karlson, S. Hoppe, B. Erlund and K. Nelugu, *Environmental Science and Technology*, 2011, 45, 3145.
48. A. Stretson and T. Nonova, *International Journal of Environmental Analytical Chemistry*, 2003, 83, 1045.

12 Bioaccumulation

12.1 METALS

12.1.1 Bioaccumulation of Metals in Sediments

Sediments in rivers and the oceans have the property of adsorbing some types of dissolved substances in the overlying water so that the concentration in the sediment ($mg\,kg^{-1}$) is appreciably greater than that in the water ($\mu g\,L^{-1}$) with which the solid is in contact. Similarly, the concentration of dissolved substances in creatures that live in the water can exceed the concentration in the water.

A convenient method of expressing this phenomenon is by calculating a concentration factor expressed by:

Concentration of substance in sediment or creatures ($\mu g\,kg^{-1}$)
Concentration of substance in water ($\mu g\,L^{-1}$)

Observed concentration factors for a range of organometallic compounds and metal ions in different types of water are tabulated in Table 12.1. Where the concentration factor is apparently greater than unity, the dissolved phase shows a tendency to be absorbed by the sediment. Examination of the data in Table 12.1 allows the following conclusions to be drawn (see also Figures 12.1 and 12.2):

1. All the inorganic metal ions listed are strongly adsorbed onto sediments (Figure 12.1).
2. Organotin compounds are strongly adsorbed onto river water sediments (Figure 12.2). The data suggest that organotin compounds may not be as strongly adsorbed onto sediment in saline water, that is, coastal and seawaters, as they are in non-saline waters.
3. Organo compounds of mercury and arsenic are not adsorbed onto sediments in either non-saline or saline waters.
4. The concentration factors for organolead compounds range from very low values, that is, no concentration in sediments, to 1000, that is, some adsorption sediments.

Both sources of pollution, that is, dissolved or sedimentary, are capable of entering living creatures with possible adverse effects. The concentration of toxicants present in sediments is a measure of their concentration in the water over a period of time and is therefore a measure of the risk to creatures. In the case of bottom-feeding creatures, there is the additional risk of direct ingestion of contaminated sediments in the gills and mouth, with consequent adverse effects.

When a creature is exposed to toxicants in the water or sediments in which it lives, then the concentration of those toxicants in its tissues gradually increases as

TABLE 12.1

Concentration Factors for Organometallics and Inorganic Ions between Sediments and Liquid Phases in Water

Compound	In Water ($\mu g\,L^{-1}$)	In Sediment ($\mu g\,kg^{-1}$)	Type of Water	Factor = Sediment ($\mu g\,kg^{-1}$)/Water ($\mu g\,L^{-1}$)		
				Minimum	Maximum	Mean
Alkylmercury	0.06	<0.01	Coastal river	—	—	<0.17
	0.0067–1.15 (mean 0.57)	<0.01		<0.0086	<1.66	<0.02
Inorganic mercury	0.009–13.0 (mean 6.5)	910–41,800 (mean 23,850)	River	3600	101,110	3669
Alkyl lead	<0.00001	<0.01	Coastal	—	—	1000
	50–530 (mean 290)	<0.01	River	0.000019	0.0002	0.00003
Inorganic lead	0.02–200 (mean 100)	23,000–38,200 (mean 30,600)	Coastal	191	1.5×10^6	306
	0.13–60 (mean 30)	110–506,000 (mean 253,000)	River	846	8430	8433
Monobutyltin	0.035–0.050 (mean 0.042)	280	River	5600	8000	6660
	<0.000–0.3 (mean 0.15)	1–8 (mean 4.5)	Sea	26.6	>10,000	30
Dibutyltin	0.010–0.040 (mean 0.025)	140	River	3500	14,000	5600
	<0.001–1.6 (mean 0.8)	0.1–1.0 (mean 0.55)	Sea	0.62	>100	0.69
Tributyltin	0.005–0.015 (mean 0.010)	55	River	3667	11,000	5500
	0.06–0.78 (mean 0.42)	0.01–0.3 (mean 0.15)	Sea	0.17	0.38	0.36
Inorganic tin	<0.0001	1000–20,000 (mean 10,500)	Coastal	$>10^7$	$>2 \times 10^8$	$>1.05 \times 10^8$
Alkyl arsenic	2.5–2.6 (mean 2.55)	<0.01	Coastal	<0.0038	0.004	<0.0039
Inorganic arsenic	1.00–1.04 (mean 1.02)	1600–117,000 (mean 59,300)	Coastal	1600	112,500	58,140
	0.42–490 (mean 245)	220–28,000 (mean 14,100)	River	57.1	523	57.6

(Continued)

TABLE 12.1 (*Continued*)

Concentration Factors for Organometallics and Inorganic Ions between Sediments and Liquid Phases in Water

Compound	In Water (µg L⁻¹)	In Sediment (µg kg⁻¹)	Type of Water	Factor = Sediment (µg kg⁻¹)/Water (µg L⁻¹)		
				Minimum	Maximum	Mean
Inorganic copper	0.069–9.7 (mean 4.85)	5400–84,800 (mean 45,100)	Coastal	8742	78,260	9298
	0.11–200 (mean 100)	70–244,000 (mean 122,000)	River	636	1220	1220
Inorganic nickel	0.2–15.0 (mean 7.6)	30,000–57,000 (mean 43,500)	Coastal	3800	1.5×10^5	5723
	1.5–4.5 (mean 3.0)	1000–238,000 (mean 119,500)	River	666	52,888	39,883
Inorganic selenium	<0.01–0.08 (mean 0.04)	1500–9000 (mean 5250)	Coastal	112,500	150,000	131,250
Inorganic antimony	0.30–0.82 (mean 9800)	6200–134,000 (mean 9800)	Coastal	20,666	163,414	17,500
Inorganic manganese	0.08–0.42 (mean 0.25)	10–2900 (mean 1455)	River	125	6904	5820
	0.35–250 (mean 125)	21,800–750,000 (mean 386,000)	Coastal	3000	62,285	3088

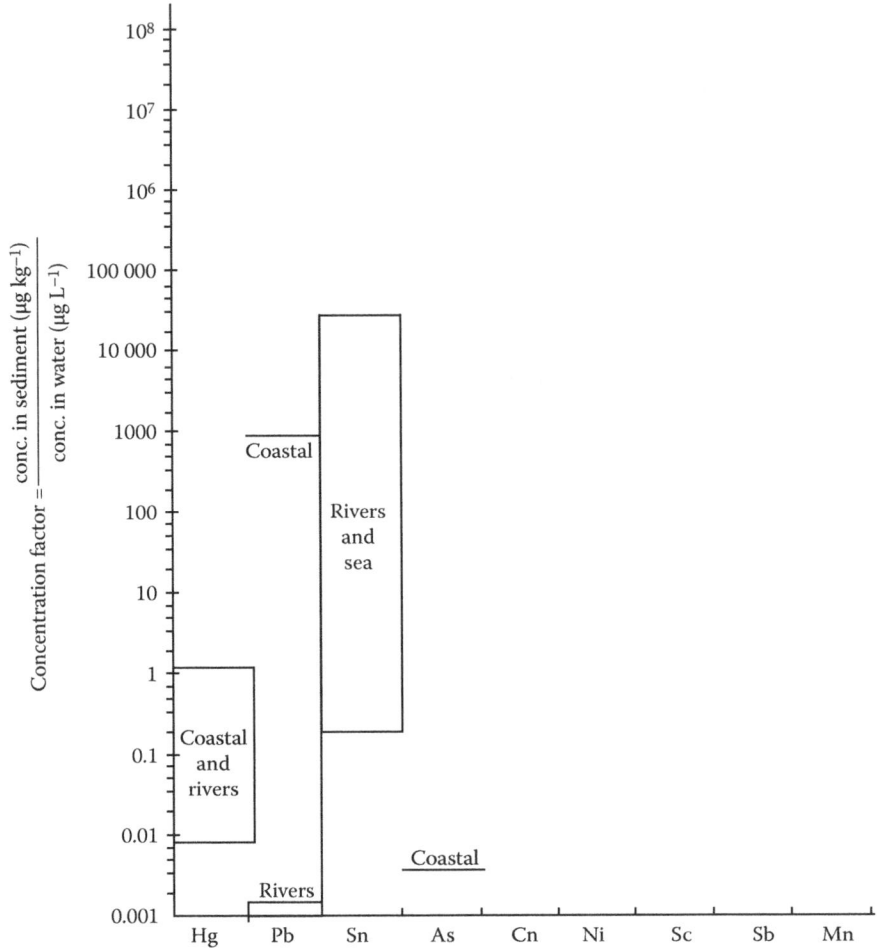

FIGURE 12.1 Concentration factors of organometallic compounds.

a function of exposure time and the concentration of toxicant in the water until the concentration in the tissues of the creature is many times that present in the water. This phenomenon is known as bioaccumulation, which is not to be confused with bioamplification, a process whereby an increase in toxicant levels occurs along a food chain, for example, plants–minute creatures–fish, as occurs for example in the case of chlorinated insecticides in animal tissues.

In bioaccumulation, it is found that the concentration of toxicant in the tissues, and particularly in some of its organs, such as kidney, liver and opercle, increases with both the concentration in the water and exposure time. Measurement of toxicant levels in tissues or organs provides an indication of the amount of exposure to toxicants that the creature has suffered over a period of time. Only limited information is available relating concentrations of toxicant in tissues and the onset of ill health or mortality. This is clearly an area where much further work remains to be done.

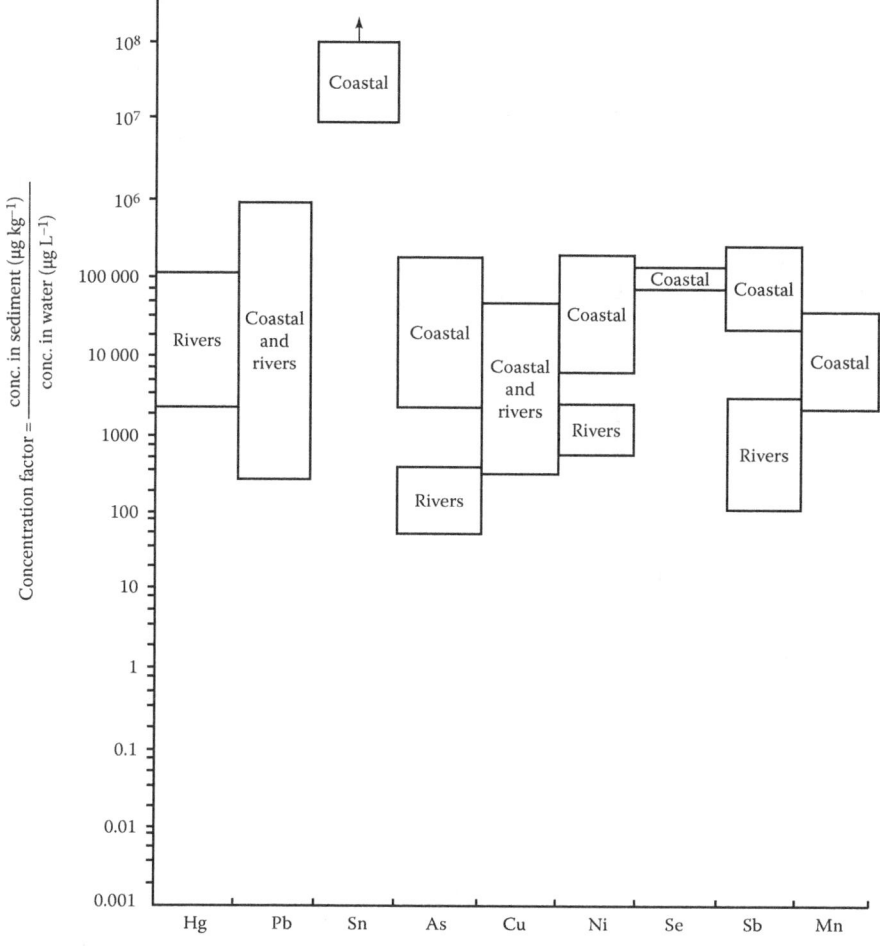

FIGURE 12.2   Concentration factors of metal in various matrices.

Monitoring of bioaccumulation of fresh and tidal waters as a trend in spacial monitoring has two purposes:

1. Macroscale, that is, the identification of potentially unknown areas of elevated contamination and assessment of the extents of the zone of contamination.
2. Monitoring of bioaccumulation in fresh and tidal waters as trends in time. These need to be monitored to identify trends in contamination, especially near effluent discharges, in order to identify stability, improvement or deterioration in contamination levels.

It has been previously stated that a bioaccumulation of inorganic metals by fish from water is greater in the case of fish opercle than in the case of fish muscle. This is

TABLE 12.2
Orders of Magnitude in the Bioaccumulation of Inorganic Metals and Organometallic Compounds by Fish, Invertebrates and Plants

Type	Metal	Inorganic Metals			Organometallic Compounds		
		Factor	Sediments (for Comparison)		Factor	Sediment (for Comparison)	
			River	Coastal		River	Coastal
Fish							
Trout	Pb	—	8433	306	934	<0.00003	100
Trout	Hg	8630	3669	—	—	<0.02	<0.17
Fish muscle	Cu	5.2 (0.5-day exposure)	1220	9298	—	—	—
Fish opercle	Cu	620 (70-day exposure)	—	—	—	—	—
Invertebrates							
Mussel	Hg	—	3669	—	1251	<0.02	<0.17
Barnacle	Cu	195,530	1220	9298	—	—	—
Barnacle	Zn	47,600	—	—	—	—	—
Clam	Cd	101 (29-day exposure)	—	—	—	—	—
Plants							
Phytoplankton	Hg	8630	3669	—	—	—	—
Algae	As	155	60	58,140	—	—	<0.0039
Algae	Cu	3500	1220	9298	—	—	—
Kelp	Fe	68,000	—	—	—	—	—
Kelp	Na	4	—	—	—	—	—

supported by the data in Table 12.2, which show a higher concentration factor of 620 after a 70-day exposure of fish opercle to inorganic copper compared with a value of 5.2 obtained for fish muscle. The mean concentration factor obtained in the case of the bioaccumulation of inorganic mercury into whole fish was considerably higher (i.e. mean factor 8630) than that obtained for copper and also about twice the value of 3669 obtained for the adsorption of inorganic mercury by river sediments. The bioaccumulation of inorganic mercury by whole fish (factor 8630) is considerably greater than the bioaccumulation of organomercury (factor <0.2–0.17).

The bioaccumulation of organolead by trout was relatively low (factor 934), and the adsorption by river and coastal sediments was also very low (factors <0.00003 and 100).

Spatial and time monitoring programmes of the type discussed above will also give the information needed to assess the risk to top predators and humans in a particular ecosystem.

A particularly useful way of expressing bioaccumulation is to calculate a concentration factor C which takes into account the concentrations of toxicants in water and the creatures:

$$C = \frac{\text{Concentration in creatures} \left(\mu g\, kg^{-1} \right)}{\text{Concentration in water} \left(\mu g\, L^{-1} \right)}$$

12.1.2 BIOACCUMULATION OF METALS IN INVERTEBRATES AND FISH

A very high degree of bioaccumulation occurs in the case of inorganic copper (mean factor: 195,530) and zinc (mean factor: 47,600) in barnacles. Both of these values are considerably higher than the concentration factors of 1120–9298 obtained, respectively, for the adsorption of inorganic copper by river water and coastal sediments. Bioaccumulation of cadmium by clams is quite low (factor 101 after a 2-day exposure).

12.1.3 BIOACCUMULATION OF METALS IN PHYTOPLANKTON AND ALGAE

The bioaccumulation of inorganic sodium by kelp was very low (factor: 4), while that of inorganic iron was very high (factor: 68,000) (see Table 12.2). The bioaccumulations of inorganic copper by algae and inorganic mercury by phytoplankton were intermediate (mean factors of 3500 and 8630, respectively). Bioaccumulation of inorganic arsenic was relatively low (factor: 155), a similar order of magnitude to the bioaccumulation of cadmium by clams (factor: 101).

The bioaccumulation of inorganic mercury by phytoplankton (mean factor: 8630) is considerably greater than the adsorption of inorganic mercury by river sediments (mean factor: 3669). The bioaccumulation of inorganic arsenic by algae by phytoplankton (mean factor: 155) is similar to the adsorption of inorganic arsenic by river sediments (mean factor: 60), but is considerably lower than the adsorption of inorganic arsenic by sediments in a marine environment (mean factor: 58,140) (see Table 12.1). The bioaccumulation of inorganic copper by algae (mean factor: 3500) is intermediate between the values obtained for the adsorption of inorganic copper by river and coastal sediments (mean factors of 1220 and 9298, respectively).

A summary of the available data in Table 12.3 shows that, in general, bioaccumulation of inorganic metals in water creatures such as fish and invertebrates and plant life in rivers and the oceans is higher than in the case of bioaccumulation of organometallic compounds, at least in the case of mercury and lead. As shown in Table 12.1, this is also the case in the adsorption of inorganic metals onto river and coastal sediments; that is adsorption factors in the case of mercury, lead and many other metals adsorbed onto river and ocean sediments are greater in the case of inorganic metals than in the case of corresponding organometallic compounds. In Table 12.3 is shown a comparison of the concentration factors of inorganic elements and organometallic compounds in water creatures and plants.

12.1.4   BIOACCUMULATION OF METALS IN FISH AND FISH ORGANS AND INVERTEBRATES AND ALGAE

The results shown in Table 12.4 demonstrate the increase in toxicant levels found in rudd fish organs compared with the levels found in the surrounding waters. After a 10-week exposure to water containing $11 \, \mu g \, L^{-1}$ inorganic copper, the opercle, liver and kidney contain, respectively, 12,000, 7000 and $60,000 \, \mu g \, kg^{-1}$ copper, that is, concentration factors of 1090, 636 and 545 in these three organs. Thus, analysis of the organs provides a very sensitive method of ascertaining the cause of death when fishkills occur in the environment.

Examination of rudd fish muscle after exposure to copper is not nearly as sensitive a method of indicating copper pickup by the fish. Thus, when fish are exposed for 12 h to water containing $250 \, \mu g \, L^{-1}$ copper, the concentration of copper found in the muscle is $2000 \, \mu g \, kg^{-1}$, that is, a concentration factor of only 8, whereas the concentrations found in the opercle, liver and kidney were 52,000, 22,000 and $30,000 \, \mu g \, kg^{-1}$, respectively, that is concentration factors of 208, 88 and 120.

Exposure of rudd to $250 \, \mu g \, kg^{-1}$ copper in the water for 12 h causes 100% mortality. The sensitivity of using the opercle to diagnose copper contamination ($52,000 \, \mu g \, kg^{-1}$ copper in organ, concentration factor: 208) as opposed to using fish muscle ($2000 \, \mu g \, kg^{-1}$ copper, concentration factor: 8) is very apparent. Fish gills and skin, similarly, are poor indicators of copper contamination, that is, low concentration factors.

The effect of copper content of the water is illustrated in Table 12.5 by data for fish exposed for 12 h to water containing 250 and $1600 \, \mu g \, L^{-1}$ copper, respectively. The copper contents of the opercle are 52,000 and $104,000 \, \mu g \, kg^{-1}$, respectively, that is, concentration factors of 208 and 65.

The copper contents of the muscle when exposed to 250 and $1600 \, \mu g \, L^{-1}$ copper in water, respectively, were 2000 and $4000 \, \mu g \, kg^{-1}$, that is, concentration factors of 8 and 2.5. Concentrations of copper of 250 and $1600 \, \mu g \, L^{-1}$ in water for 12 h cause 100% mortality in fish.

The results in Table 12.5 show the combined effect of exposure time to the contaminated water and the copper content of the water on the occurrence of bioaccumulation. This illustrates that an increase in both parameters increases the copper content of the organ.

TABLE 12.3

Concentration Factors of Inorganic Elements and Organometallic Compounds in Water Creatures and Plants

Element	Type of Sample	Inorganic Element		Organometallic Compounds	
		Range of Concentration Factors	Mean Concentration Factor	Range of Concentration Factors	Mean Concentration Factor
Lead	Sediment (coastal)	$191–1.15 \times 10^6$	306	—	100
	Sediments (river)	846–8430	8433	0.000019–0.0002	<0.00003
	Rainbow trout	—	—	124 (1-day exposure) 934 (7-day exposure) (as $PbMe_4$)	—
Mercury	Phytoplankton	$6231–3.46 \times 10^6$	8630	—	—
	Brook trout	10,000 (at $0.09\ \mu g\ L^{-1}$ Hg in water) 13,276 (at $0.93\ \mu g\ L^{-1}$ Hg in water)	—	—	—
	Mussels	—	—	1170–1333 (21-week exposure)	<160 (1-day exposure)
	Sediment (river)	3600–101,100	3669	0.008–1.66 (as alkylmercury)	<0.02
Arsenic	Sediment (coastal)	—	—	—	<0.17
	Algae	11.4–47.6	155	—	—
	Sediment (coastal)	1600	5140	50.0038	<0.004
	Sediment (river)	57.1–523	57.6	—	—
Copper	Phytoplankton	200–363,630	400	—	—

(Continued)

TABLE 12.3 (Continued)
Concentration Factors of Inorganic Elements and Organometallic Compounds in Water Creatures and Plants

Element	Type of Sample	Inorganic Element		Organometallic Compounds	
		Range of Concentration Factors	Mean Concentration Factor	Range of Concentration Factors	Mean Concentration Factor
	Algae	3300–4545	3500	—	—
	Opercle (fish)	65–620	65–208 (0.5-day exposure) 620 (70-day exposure)	—	—
	Muscle (fish)	2.5–8.0	5.2 (0.5-day exposure)	—	—
	Barnacle	7060–384,000	—	—	—
	Sediment (coastal)*	8742–78,260	9298	—	—
	Sediment (river)*	636–1220	1220	—	—
Zinc	Barnacle	10,660–84,600	—	—	—
Cadmium	Clam	—	101 (29-day exposure)	—	—
Iron	Kelp	—	68,000	—	—
Sodium	Kelp	—	4	—	—

TABLE 12.4

Effect of Copper Content of Rudd Fish Organs on Mortality

	Exposure Time				Maximum Concentration Found in Organs of a Wide Variety of Creatures		
	10 Weeks	3 Weeks	<12 h	>12 h	μg kg⁻¹ (dry weight)	Comments	
Concentration of metal in water, $\mu g\,L^{-1}$ (a)	11	50	250	1600	—	—	No mortalities
Rudd fish organs, $\mu g\,kg^{-1}$ (b) dry weight							
Opercle	12,000	31,000	52,000	104,000	124,000	100% mortalities	—
Liver	7000	20,000	22,000	40,000	48,000–62,000	During 12 h exposure	2000
Kidney	6000	28,000	30,000	100,000	6000	—	700
Muscle	—	—	2000	4000	—	—	—
Concentration factors: $\dfrac{\text{Concentration organ}}{\text{Concentration water}} = \dfrac{b}{a}$							
Opercle	1090	620	208	65	—	—	—
Liver	636	400	88	25	—	—	—
Kidney	545	560	120	62.5	—	—	—
Muscle	—	—	8	2.5	—	—	—
Condition of animal at end of test	Good	Good	100% mortality	100% mortality	—	—	—

TABLE 12.5

Effect of Exposure Time and Copper Content of Water on Copper Content of Fish Organs

	Copper Content of Opercle ($\mu g\,kg^{-1}$)				Copper Content of Muscle ($\mu g\,kg^{-1}$)			
Exposure (days)	0.5	0.5	21	70	0.5	0.5	21	70
Copper content of water, $\mu g\,L^{-1}$ (a)	1600[a]	250[a]	50[b]	11[b]	1600[a]	25	50[b]	11[b]
Copper content of organ, $\mu g\,kg^{-1}$ (b)	104,000	52,000	31,000	12,000	4000	2000	—	—
Concentration factor (b/a)	65	208	620	1090	2.5	8	—	—

[a] 100% fish mortalities.
[b] no fish mortalities.

Experiments with brook trout have been conducted in which the trout were exposed to waters containing 0.09, 0.29 and 0.93 $\mu g\,L^{-1}$ inorganic mercury for 9 months and the gonads of the fish then analysed for mercury. Concentration factors so obtained are shown in Table 12.6. Mortalities can occur of bivalve molluscs, crustacea and fish when they are exposed to 0.1–10 $\mu g\,L^{-1}$ inorganic mercury even during short-term (4–14 days) exposure.

Some fatalities, or at least signs of ill health, would be expected to occur under the conditions quoted in Table 12.6, that is, where mercury contents of gonads have increased to 900–12,300 $\mu g\,kg^{-1}$ and concentration factors are between 10,000 and 13,000. A comparison of these concentration factors attained during a 9-month continuous exposure to mercury with the values of 545 to 1090 was attained in the case of copper (Table 12.4).

The concentration of copper during a 10-week exposure is interesting in that it follows the expected trend of increase of concentration factor with exposure time.

Fish in contact with water containing 0.01 $\mu g\,L^{-1}$ inorganic mercury and sediments containing 30 $\mu g\,kg^{-1}$ of inorganic mercury have been found to contain 341 $\mu g\,kg^{-1}$ mercury in their flesh, that is, a 34,000 bioaccumulation factor. At Minimata Bay, Japan, mercury levels in some fish attained 50,000 $\mu g\,kg^{-1}$ wet weight, while levels around 20,000 $\mu g\,kg^{-1}$ inorganic mercury were common.

Just as fish organs demonstrate the phenomenon of bioaccumulation, so do phytoplankton and, indeed, freshwater algae. This makes these species very sensitive indicators of the occurrence of pollution in waters. It is seen in Table 12.7 that mean concentration factors range from 155 (arsenic in algae) to 8630 (mercury in phytoplankton).

The bioaccumulation factors obtained for copper (400 in freshwater phytoplankton and 3550 in freshwater algae) compare with the mean values of 545 to 1090 obtained for fish organs in long-term (10 weeks) exposure to copper (Table 12.4).

Renzi et al. [1] collected samples of surface sediments and tissues (liver and muscle) of commercially available European silver eels (*Anguilla anguilla*) from Varano lagoon (Italy). These were analysed for trace element contents. Univariate and multivariate analyses were performed to highlight both the differences between sampling sites and the influence of channel discharges. Atomic ratios indices for sediment data and biological enrichment factors for eel tissues were calculated in order to evaluate the enrichment factors due to human activities. The highest levels of arsenic ($11.9\,\mu g\,g^{-1}$) and zinc ($14.1\,\mu g\,g^{-1}$) were observed in the southeastern zone of the lagoon, which is influenced by urban and agricultural discharges. The low levels of mercury observed in this study ($0.04\,\mu g\,g^{-1}$) led Renzi et al. [1] to conclude that both natural and human sources of this element occur. Trace element concentrations of all elements were lower in muscle than in liver tissue. Significant enrichment of copper and zinc was found in livers.

TABLE 12.6
Bioaccumulation of Inorganic Mercury in Brook Trout, 9-month Exposure

	1 week	2 weeks	3 weeks
Mercury content of water, $\mu g\,L^{-1}$ (*b*)	0.09	0.29	0.93
Mercury content of gonads, $\mu g\,kg^{-1}$ (*a*)	900	2900	12,300
Concentration factor (*a/b*)	10,000	10,000	13,226

TABLE 12.7
Bioaccumulation of Inorganic Metal Ions in Freshwater Phytoplankton and Algae

	Compounds		
	Mercury	Arsenic	Copper
Concentration of metal in water, $\mu g\,L^{-1}$ (*a*)	0.009–13.0	0.42–490	0.11–200
Concentration of metal in freshwater algae, $\mu g\,kg^{-1}$ (*b*)	—	20,000–56,100	50,000–660,000
Concentration of metal in freshwater phytoplankton, $\mu g\,kg^{-1}$ (*c*)	31,200–81,000	—	40,000

Concentration Factor	Range	Mean	Range	Mean	Range	Mean
Freshwater algae (*b/a*)	—	—	114.5–47,619	155	3300–454,500	3550
Freshwater phytoplankton (*c/a*)	6231	8630	—	—	200–363,630	400
	3.4×10^6					

Rahman et al. [2] investigated heavy metal concentrations (Pb, Cd, Ni, Cr, Cu, Zn, Mn and As) in industrial effluents, water, sediment and fish samples collected around the Dhaka Export Processing Zone, Saver, Bangladesh, to evaluate the level of contamination. The metal concentrations in the industrial effluents and the water samples of Dhalaibeel (lowland cum lake) and Bangshi River were significantly higher than the guideline values for industrial effluents and drinking water (World Health Organization [WHO] and US Environmental Protection Agency [USEPA]), respectively. The sedimentary metal concentrations were found to be lower than the respective probable effect concentrations following the sediment quality guidelines. Furthermore, in comparison with the fish standards, fish species studies were not found to be contaminated by heavy metals. Principal component analysis and cluster analysis demonstrated that the wastewater from the numerous industries and the domestic sewage might have a possible impact on heavy metal contamination in the study area. Pearson correlation analysis showed significant correlation ($p < 0.01$ and $p < 0.05$) between most of the metals in the samples of effluents, water, sediments and fish muscles. The percentage enrichment factor and geoaccumulation index (^1geo) were followed to evaluate metal contamination in the sediment samples. Dhalaibeel sediment was maximally enriched for chromium (53.55%) and the Bangshi River sediment for zinc (54.37%). The geoaccumulation index values for the sediment samples were less than zero, indicating that the sediment samples were free from contamination. This study could be used as a model to assess the impact of anthropogenic activities on heavy metal contamination in aquatic ecosystems.

Galindo et al. [3] investigated trace metal accumulation tissues of sole (*Solea senegalensis*) and their relationship to the biotic environment.

The distribution patterns and the organ-specific accumulation trends of 10 trace metals (iron, manganese, zinc, copper, chromium, nickel, cobalt, lead, cadmium and silver) and 4 major elements (sodium, potassium, calcium and magnesium) in 10 different tissues (heart, muscle, kidney, stomach, intestine, liver, gill, gonads, white skin and dark skin) of a benthic fish (*Solea senegalensis*) from a densely populated coastal area affected by anthropogenic activities, the Bay of Cadiz (southwest Spain), have been investigated. High variability of metal concentrations among tissues was found for Ca, Fe, Zn, Cu, Pb and Ag. Factor analysis was applied to study this variability. Five principal components were found explaining 92.95% of the total variants, and similarities in behavioural patterns of bioaccumulation were described. They associated Mg, Cr, Ni and Mn to intestine and stomach tissues; Ag, Cu and Cd to liver; Zn, K and Co to gonads; Na, Fe and Pb to gill, heart and kidney tissues; and Ca, Pb and Mb to gill and dark skin. The metallic concentration in the sediment and water was also studied. The pollution in this area was found to be moderate with outstanding values of Zn, Cu and Pb (average values of 139, 50.4 and 75.6 $mg\,kg^{-1}$, respectively) in sediment and dissolved Cu (average value of $2.5\,\mu g\,L^{-1}$). Metal bioaccumulation trends followed the order Zn > Cu > Cd > Pb for dissolved metals in seawater, Cu > Zn > Cd > Pb ≈ Mn > Fe ≈ Ni ≈ Co for metals associated to particulate matter and Zn ≈ Cu > Cd > Mn > Co ≈ Fe > Ni ≈ Pb > Cr for metals in the sediment. Higher values were found for copper in liver, zinc and gonads and lead in gill, showing the relationship between the biotic and abiotic environment.

In addition, Cd bioconcentration factors were found to be high in liver and gill, showing the sensitivity of sole to this metal even at low concentrations.

Stary et al. [4] studied the accumulation of zinc and cadmium in fish (*Poecilia reticulala*). Zinc-65 and cadmium-115m were used for the investigation of the uptake and the release of zinc and cadmium in fish. It was found that the uptake of these elements directly from water is relatively low. Zinc and cadmium, consumed by fish in food, are released with the biological half-lives of 0.25 ± 0.05 days (94%) and 65 ± 9 days (6%) for zinc and 0.25 days (98.5%) and 36 ± 6 days (1.5%) for cadmium. The comparison with data obtained for mercury species shows that the accumulation of these elements in fish increases in the order cadmium, zinc, inorganic mercury(II), phenylmercury and methylmercury.

Ugart et al. [5] evaluated the bioaccumulation of trace elements in tuna species by correlation analysis between their concentrations in muscle and first dorsal fin.

Environmental pollution is a recognised problem worldwide. As a result of the exposure to this pollution, marine species may bioaccumulate metals in both muscle and fish bone, as has been demonstrated in some species of tuna. The object of this study was to develop and optimise an inductively coupled plasma–mass spectro-metric method which allows the quantification of 21 elements, including priority pollutants and biological essential elements (B, Mg, Al, V, Cr, Mn, Fe, Co, Ni, Cu, Zn, As, Se, Rb, Sr, Pd, Cd, Ba, La, Hg and Pb) in muscle and in the first spine of the first dorsal fin of albacore (*Thunnus alalunga*) and bluefin tuna (*Thunnus thynnus*). A microwave-assisted digestion was developed for sample treatment, which has been evaluated using isotope dilution analysis of Cr, Se, Cd, Ba and Pb. Evaluation of the analytical method in terms of sensitivity (sensitivity between 0.002 and $1 \, mg \, kg^{-1}$), accuracy and precision within and between days (coefficient of variation [CV] < 11.3%) was conducted. The development method has allowed information to be obtained on levels of these metals in both matrices. The correlation analyses per-formed for each of the metals in both matrices shows a positive linear relationship between the concentrations in muscle and fish bone for Zn, Se, Rb, Cd, As and Hg, which could be due to a higher bioaccumulation of these elements in muscle, as it is concluded from the low spine/muscle ratios observed for these elements. The 34 specimens of tuna analysed show that while the levels of Pb, Cd, Ni, Zn, Cu and Cr in muscle are below the limits set by the WHO/Food and Agriculture Organization (FAO) and USEPA, Hg shows a higher concentration than the limits set by the Economic Community (EC) in four samples, indicating a potential risk to human health.

Amato et al. [6] showed that diffusive gradients in thin-film techniques provided robust prediction of metal bioavailability and toxicity in estuarine sediments.

Bioavailability studies have also been done on copper and zinc in sediments [7], zinc and cadmium in scallop (*Chiamys nobilis*) [8] and a range of heavy metals in arctic marine biota [9] in fish and invertebrates.

12.2 ORGANIC METALLIC COMPOUNDS

Bioaccumulation has been observed in the case of tetramethyl lead in rainbow trout (Table 12.8). Analysing whole fish tissues 1 day and 1 week after exposure to water

TABLE 12.8
Accumulation of Tetramethyl Lead in Rainbow Trout

Exposure Day	Weight of Fish, g	Fish Alive or Dead	Water Averaged, $\mu g\,L^{-1}$	Fish, $\mu g/kg$ Wet Weight	Concentration Factors[a]
1	0.1211	Dead	3.46	430	124
2	0.3661	Dead		1080	312
	0.7982	Dead		2000	578
3	0.4116	Dead		1320	382
	0.6300	Dead		2090	604
7	1.3045	Alive		2940	850
	1.5466	Alive		3230	934
	0.8100	Alive		2250	650
	0.4926	Alive		1730	500

[a] Concentration factor = Concentration of Me_4Pb in fish, $\mu g\,kg^{-1}$/Concentration of Me_4Pb in water, $\mu g\,L^{-1}$.

containing tetramethyl lead gave concentration factors of 124 (1-day exposure) increasing to between 500 and 934 (1-week exposure); that is, the tetramethyl lead content of the tissues in micrograms per kilogram was between 124 and 934 times greater than that of the water in which the trout lived.

12.2.1 ORGANOMERCURY COMPOUNDS

The Mussel Watch Program is run by the US Environmental Protection Agency. This is a programme in which caged mussels are immersed in environment waters. A small number of mussels are removed periodically from the cage at known time intervals from the start of the experiment, homogenised and analysed for the toxicant of interest. A sample of the surrounding water is also taken so that concentration factors in the mussel (bioaccumulation) can be calculated. The experiment is statistically designed to detect 10% changes in toxicant concentrations in the mussels with a confidence of 90%.

In one such programme, *Mylitus* mussels were suspended in cages in the Firth of Forth. The methylmercury concentration of the mussels increased from less than $10\,\mu g\,kg^{-1}$ 1 week after the start of the experiment to $60-80\,\mu g\,kg^{-1}$ after 21 weeks of exposure to water with a mean methylmercury content of $0.06\,\mu g\,L^{-1}$. Calculated concentration factors were <160 at 1 week from the start of the experiment to 1170–1333 after 21 weeks of exposure. These values are similar to those quoted above for tetramethyl lead, that is, a concentration factor of 124 after a 1-day exposure and 500–850 after a 1-week exposure.

Richman et al. [10] have pointed out that the mechanisms responsible for higher concentrations of mercury in fish from acidic lakes were poorly understood. However, several hypotheses have been proposed: mercury might enter the catchment with acid deposition; acidification might mobilise mercury bound in lake sediment and catchment soils; lower pH could favour the production of the more bioavailable monomethylmercury species; pH could influence the rates of mercury methylation or demethylation by microorganisms; biotic characteristics of acid lakes could influence mercury transfer and biomagnification; and acidification might directly affect lake biota, altering the ability of organisms to bioaccumulate or excrete mercury. Evidence for and against these hypotheses is discussed, and it was concluded that mercury cycling and uptake in aquatic systems were governed by a variety of interconnecting and sometimes covarying factors, the relative importance of which could differ from lake to lake.

Tsuda et al. [11] have discussed results obtained in the determination of bioconcentration factors in carp (*Cyprinus carpio*) and the octanol/water partition coefficients for triphenyltin chloride, diphenyltin dichloride and monophenyltin trichloride. The further metabolism of triphenyltin chloride in the fish was also investigated, and the concentrations of phenyltin found in various tissues are discussed.

Anil and Wagh [12] collected samples of barnacles monthly (March 1983–May 1984) from two stations (the shipyard and the harbour) of the Zuan Estuary. Total copper and zinc concentrations in the water were 1–11 and 13–46 μg kg^{-1}, respectively. Copper concentrations in barnacles were 47,300–86,4800 μg kg^{-1} in the 0.1 cm size group and 39,700–625,700 μg kg^{-1} in the 1–2 cm size group. The corresponding zinc concentrations were 203,600–1,937,500 and 204,300–384,300,000 μg kg^{-1}. Concentration factors were 7060–384,300 for copper and 10,660–84,600 for zinc. In general, concentration factors in the 0.1 cm size group were higher than those in the 1–2 cm size group.

Langston and Zhou [13] collected samples of tellnid clams (*Macoma balthica*) from Whitehaven, Cumbria, in June 1984 and acclimated them for use in laboratory experiments on the bioaccumulation and bioelimination of cadmium by these species. At cadmium concentrations of 100 μg L^{-1}, accumulation by the soft tissues was linear (350 μg kg^{-1} per day, dry weight) throughout a 29-day exposure. Amounts of cadmium accumulated by the shell were low and elimination rates high (retention half-life: 7 days) compared with soft tissues (retention half-life: 70 days). Gelchromatographic profiles of cytosol extracts from control and experimental groups of clams provided no evidence for the involvement of either metallothionein or metallothionein-like proteins in cadmium accumulation. Most cadmium was bound to high-molecular-weight ligands, although the small amount (less than 15%) associated with low-molecular-weight ligands might be important in regulating cadmium uptake and elimination phases. The absence of a recognised detoxifying system in this species might be compensated for by the slow rate of cadmium accumulation.

Lyngby and Brix [14] investigated heavy metal contamination in Limfjord, Denmark, and eelgrass (*Zostera marina*) and marine mussels (*Mytilus edulis*) and compared these as indicators of heavy metals in shallow coastal areas. Background levels and threshold values were calculated for the organisms and sediments. Significant elevations of heavy metal concentrations were found in the Nissum Broad (mercury) in Veno Bay (cadmium) and Aalborg (mercury, zinc, lead and copper). Positive correlations between concentrations of mercury, lead, cadmium and zinc were found in eelgrass leaves and root rhizomes, mussels, and sediment, but the copper concentrations in mussels did not correlate.

Higgins and McKay [15] collected samples of *Ecklonia radiato* bimonthly, during July 1982–March 1984, from the shallow sublittoral zone of the kelp beds in Port Hacking Estuary, Australia. Tissue concentrations of iron and manganese were approximately 60% higher in late summer than during the rest of the year. Zinc, cadmium, copper, potassium, calcium, magnesium and sodium concentrations showed no seasonal variations. Concentration factors ranged from 4.0 (sodium) to 68,000 (iron). Seasonally averaged concentrations of sodium, magnesium, calcium and potassium were relatively uniform throughout the kelp tissues. Concentrations of iron, manganese and zinc were highest in the extremities (eroding tip and holdfast tissue) and lowest in the meristematic tissue. Cadmium concentrations were elevated in the extremities, but uniformly distributed in the other tissues. Copper concentrations were highest in holdfast tissue and lowest in the eroding tip. Experiments using ethylenediaminetetraacetic acid (EDTA) indicated that approximately 90% of total cadmium and zinc was associated with the apparent free space, whereas corresponding values for copper and iron were 25% and 7%, respectively. In view of the rapid exchange between seawater and the apparent free space, it was concluded that *Ecklonia radiata* would not be of general value in the assessment of long-term integrated changes of metals in the water column.

12.3 ORGANIC COMPOUNDS

12.3.1 BIOACCUMULATION OF METALS AND ORGANOMETALLIC COMPOUNDS IN FISH AND INVERTEBRATES

Musk xylene is used as a fragrance in commercial toiletries. Bioaccumulation of musk xylene into 2-amino musk xylene and 4-amino musk xylene metabolites in rainbow trout haemoglobin has been described by Mottola et al. [16].

The close response relationships of toxicokinetics of the metabolites as adducts on the haemoglobin were determined by gas chromatography–electron capture negative chemical ionisation mass spectrometry and gas chromatography–electron ionisation mass spectrometry using selected ion monitoring. The trout were subjected to a single exposure of 0.01, 0.03 or 3 ng musk xylene of fish. Haemoglobin samples were collected from exposed and control fish, and analysed subsequently to exposure at intervals of 24, 72 and 168 h. Alkaline hydrolysis released 4-amino musk xylene and 2-amino musk xylene metabolites from the haemoglobin, and the solutes were extracted into *n*-hexane. The extracts were preconcentrated and analysed. The presence of the metabolites in the haemoglobin extracts was confirmed based on agreement of similar mass spectral features from electron capture negative chemical ionisation and electron ionisation mass spectrometry spectra and retention times of the metabolites with standards. The electron capture negative chemical ionisation results were used for close response and toxicokinetics measurements. For close response, the concentrations of adducts of the metabolites increased with dosage, and a maximum adduct formation was observed at the 0.1 mg g^{-1} level, beyond which it decreased. The average concentrations of 4-amino musk xylene and 2-amino musk xylene at a dosage of 0.1 mg g^{-1} were 700 and 7.4 ng g^{-1}, respectively.

For toxicokinetics, the concentration of the metabolites in the haemoglobin reached a maximum in the 3-day sample after administration of musk xylene. Further elimination of the metabolites exhibited kinetics with a half-life estimated to be 1–2 days, assuming first-order kinetics. Quantitations were made based on an internal standard and a calibration plot. In control samples, non-hydrolysed haemoglobin and reagent blank extracts, the metabolites were not detected. The limits of detection for 2-amino musk xylene and 4-amino musk xylene and the haemoglobin were approximately 1.7 and 1.4 µg L^{-1}, respectively, based on a signal-to-noise ratio of 3.

Microcystins are widespread cyanobacteria toxins in freshwater systems and have been limited to both acute and chronic health effects. A growing number of studies suggest that microcystins can bioaccumulate in food webs, although several methods by enzyme-linked immunosorbent assay (ELISA) and LC-MS have been developed for analysis of microcystins in water. Extraction (for subsequent analysis of the toxin from biological matrices such as animal tissue) is impeded owing to the covalent bonding of toxins and the active sites of their cellular targets, that is, protein phosphatises. As an alternate approach, chromatographic methods for analysis of an unique marker, 2-methyl-3-methoxy-4-phenylbutonic acid, the product of the Lemieux oxidation of microcystins, have been developed and shown to measure total bound and unbound microcystins. Applications, however, has been limited by poor recovery of the analyte.

Suchy and Berry [17] proposed an improved recovery method – specifically the use of solid-phase microextractions. 2-methyl-3-methoxy-4-phenylbutanoic acid and oxidised microcystin were used to develop methods.

Specifically, a method employing postoxidation methyl esterification, followed by headspace recovery of 2-methyl-3-methoxy-4-phenylbutanoic acid, was developed and subsequently applied to analysis of environmental samples (i.e. fish tissues) previously shown to contain microcystins. The method shows high linearity for both water and tissues spiked with microcystins and an improved limit of quantitation of approximately 140 ng g^{-1}. Evaluation of field samples by gas chromatography–mass spectrometry detected considerably higher levels of microcystins than those detected by conventional methods (i.e. ELISA), and it is proposed that this technique reveals microcystins (particularly in the bound form) that are not detected by these methods. These results indicate that the method provides improved detection capability for microcystins in biological matrices, and will enhance our ability to understand bioaccumulation in freshwater.

Bioaccumulation studies have been conducted of polychlorobiphenyls and polycyclic aromatic compounds in tidal and marine biota [18], polychlorobiphenyls in marine calanoid copepods [19], hydrogenated bipyrroles and methoxylated tetrabromodiphenylethers in tiger shark (*Galevcerolo curvier*) [20], persistent organic pollutants in bottlenose dolphins (*Turslops truncates*) [21], perfluorooctane in marine tucuxi dolphins (*Sotalia guianensis*), hydrocarbons [22] in bivalves [23], d-hexachlorocyclohexane in arctic 300 plankton [24], organochlorine compounds in lower-tropic-level arctic marine biota, [25] perfluorochemicals in pacific oysters [26] and Dechlorane (mirex) in oysters [27].

REFERENCES

1. M. Renzi, A. Specchiuli, D. Barony, T. Scirocco, L. Cilenti, S. Focardi, P. Breter and S. Focardi, *International Journal of Environmental Analytical Chemistry*, 2012, 92, 676.
2. M.S. Rahman, N. Saha, M. Molla and S.L. Al Reza, *Soil and Sediment Contamination*, 2014, 23, 353.
3. M.D. Galindo, J.A. Jaradin, M. Garcia, M.L. Gonzales de Canale, Y. Oliva, F. Lopez, M.D. Gronallo and E. Espada, *International Journal of Environmental Analytical Chemistry*, 2012, 92, 1072.
4. J. Stary, K. Ratzer, B. Havlik, J. Prasilova and J. Hannsova, *International Journal of Environmental Analytical Chemistry*, 1982, 11, 117.
5. A. Ugart, Z. Abrego, N. Unelta, A. Giocolea and R.J. Barrio, *International Journal of Environmental Analytical Chemistry*, 2012, 92, 1791.
6. E.D. Amato, S.L. Simpson, C. Jarolimek and D.F. Jolley, *Environmental Science and Technology*, 2014, 8, 160.
7. A Turner, N. Singh and M. Millard, *Environmental Science and Technology*, 2008, 42, 8740.
8. K. Pen and W.X. Wang, *Environmental Science and Technology*, 2008, 42, 6285.
9. F. De Leander, D. Van Oevelen, S. Frantzen, J.I. Middeberg and K. Soeteert, *Environmental Science and Technology*, 2010, 44, 356.
10. L.A. Richman, C.D. Wren and P.M. Stokes, *Water, Air, and Soil Pollution*, 1988, 37, 465.
11. T. Tsuda, H. Nakonishi, S. Aoki and J. Takebayashi, *Water Research*, 1987, 21, 949.
12. A.C. Anil and A.B. Wagh, *Marine Pollution Bulletin*, 1988, 19, 177.
13. W.J. Langston and M. Zhou, *Marine Environmental Research*, 1987, 21, 225.
14. J.E. Lyngby and H. Brix, *Science of the Total Environment*, 1987, 64, 239.
15. H.W. Higgins and D.J. McKay, *Australian Journal of Marine and Freshwater Research*, 1987, 38, 307.
16. M.A. Mottola, T.W. May and J.H. Zimmerman, *International Journal of Environmental Analytical Chemistry*, 2006, 86, 743.
17. P. Suchy and J. Berry, *International Journal of Environmental Analytical Chemistry*, 2014, 92, 1443.
18. H. Nakata, Y. Sakai, T. Miyawaka and A. Takemura, *Environmental Science and Technology*, 2003, 37, 3513.
19. K. Borga, T. Aeron, B. Hargraue, P.F. Hockstra, D. Swackhamer and D.C.G. Muir, *Environmental Science and Technology*, 2005, 39, 4523.
20. K. Harrjuchi, Y. Hisamichi, Y. Kotaki, Y. Kato and T. Endo, *Environmental Science and Technology*, 2009, 43, 2288.
21. J.A. Lutz, L.P. Garrison, A. Lynne, A. Martinez, J.P. Contillo and J.R. Kucklick, *Environmental Science and Technology*, 2007, 41, 7222.
22. P.R. Dorneles, J. Lailson-Brito, A.F. Azevedo, J. Meyer, L.G. Vidal, A.B. Fragozo, J.P. Torres, O. Main, R. Blust and K. Das, *Environmental Science and Technology*, 2008, 42, 5368.
23. C. Porte, Y. Biosca, D. Pastor, M. Sole and J. Albaiges, *Environmental Science and Technology*, 2000, 34, 5067.
24. M. Pucko, W. Walkuzz, R.W. Macdonald, D.G. Barber, C. Fuchs and G.A. Stern *Environmental Science and Technology*, 2013, 47, 4155.
25. B.T. Hargreaves, G.A. Phillips, W.T. Vass, P. Brucker, H.E. Welch and T.D. Siferd, *Environmental Science and Technology*, 2000, 34, 980.
26. J. Jeaon, K. Kanna Hyhim, H.B. Moon, J.S. Ra and S.D. Kim, *Environmental Science and Technology*, 2010, 44, 2695.
27. H. Jia, Y. Sun, X. Lin, M. Yang, D. Wang, H. Qi, L. Shen, E.D. Sverko, E.J. Reiner and Y.F. Li, *Environmental Science and Technology*, 2011, 45, 2613.

Index